Hovels to high rise

'Many people are asking whether Britain took the wrong road to housing lower-income families. This well documented history of what a number of other European countries did enables this question to be answered.'
Brian Abel-Smith, Professor of Social Administration, London School of Economics.

Hovels to high rise traces how governments became involved in replacing industrial revolution urban slums with mass high-rise, high-density concrete estates; how huge inflows of rural and foreign migrants ended up reoccupying slums, then being housed in the poorest estates; how governments set up rescue programmes to reverse spiralling conditions on the worst estates; and how state-sponsored housing made a significant comeback in all countries at the outset of the 1990s.

Using detailed research on five countries – France, Germany, Britain, Denmark and Ireland – and carried out with close reference to the central housing bodies in each area, Anne Power highlights the convergence in housing experience between the countries and the lessons that can be learnt. By describing the historical foundations of the developments in housing structure, covering the different traditions of inner-city development, tenant-oriented housing, co-operatives, mass-rented building, rural and suburban owner-occupation, private landlords, and large local authority projects, she provides a comprehensive study of housing systems in each country showing that despite the differing backgrounds the problems are often the same. She looks at the attempts governments have made to fund state housing initiatives that try to avoid the development of ghettos found so often in America. As the similarities and differences throughout Europe are explored and the changing structure of social housing is charted, the author is able to draw a picture of European-wide political commitment to urban integration, while warning that there is still a trend towards polarization that will affect the multi-lingual, multi-ethnic urban societies of the future.

Anne Power is a Reader in Social Administration and Director of the Post-graduate MSc and Diploma in Housing at the London School of Economics. She is a founding Director of the Priority Estates Project and the National Tenant Resource Centre and advises the Department of the Environment on housing issues. She has travelled widely in Europe and has extensive links with European housing organizations. In 1985 she was awarded an MBE for her work in the Priority Estates Project in Brixton. Among other books, she has written *Property Before People: Management of Twentieth Century Council Housing* (Allen and Unwin, 1987) and *Housing Management: A Guide to Quality and Creativity* (Longman, 1991).

Hovels to high rise

State housing in Europe since 1850

Anne Power

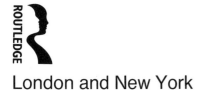

London and New York

First published 1993
by Routledge
11 New Fetter Lane, London EC4P 4EE

Simultaneously published in the USA and Canada
by Routledge
29 West 35th Street, New York, NY 10001

© 1993 Anne Power

Typeset in Times by Leaper & Gard Ltd, Bristol
Printed and bound in Great Britain by Biddles Ltd, Guildford and King's Lynn

British Library Cataloguing in Publication Data

A catalogue reference for this book is available from the British Library

ISBN 0-415-08935-2
0-415-08936-0 (pbk)

Library of Congress Cataloging in Publication Data
has been applied for

ISBN 0-415-08935-2
0-415-08936-0 (pbk)

Contents

Plates

Ireland

Figures

Tables

Acknowledgements

The main organisations that helped with the study in each country are listed in the Introduction (see pp. 16–17). Also, a number of people helped with the study in a major way and must be thanked individually:

Professor Brian Abel-Smith of the London School of Economics helped develop the idea for this book from the outset and guided the work throughout, making suggestions on drafts, probing the method and the findings, putting me in touch with valuable sources of information and questioning the different approaches of each country.

Monika Zulauf, research officer with LSE Housing, prepared background housing policy papers and researched international housing and social statistics; she also had a particularly important role in the German study, carrying out four visits, guiding the author through the upheavals of reunification, and explaining modern German social and political developments.

Peter Emms, Michael Burbidge, and Alan Holmans of the Department of the Environment were carrying out their own studies of European housing and were encouraging, helpfully critical, and painstaking in suggesting corrections and improvements.

John Hills, David Piachaud, Howard Glennerster, David Downes, Patrick Dunleavy, Bert Provan, Mark Kleinman, and Julian Le Grand of the London School of Economics forced me to rethink my approach to important international problems several times over, highlighted many of the difficulties of international research, and commented on all or parts of the book.

Over four years of visits, the following people were both inspiring and tireless in helping to locate information, in arranging visits to housing areas, and in correcting the text: Albert and Luce Martens, Soeren Villadsen and family, Jesper Nygard, Hedvig Vestergaard, Gert Nielsen, Anker Jensen, Henning Andersen, Soeren Thomasen, Ulli and Toni Pfeiffer, Rüdiger Rahs, Manfred Fuhrich, Iris Behr, Herr Hachman, Jean-Claude Toubon, Gilles Renaudin, Yves Burgeat, Jean-Michel Guenod, Jean-Pierre Schaefer, Marie-Christine Le Roy, Sylvie Harburger, John Blackwell, Peter

Whelan, Sean O'Cuinn, Larry Murphy, Flor and Judy O'Mahoney. My daughter Lucy, Colette Harrington, and Margaret Lonergan came with me on many of the visits.

Priority Estates Project workers debated the ideas presented in the book at several special residential meetings, helping to sharpen ideas and points of comparison.

Hilda Gage, with whom I have worked since 1974, was uniquely responsible for keeping documents and records in order, checking sources of information, and preparing the book for publication.

Wendy Lewsey of LSE Housing liaised with many European contacts in the course of the study and arranged several visits.

Without all these different forms of assistance, such a study would have been completely impossible.

The author takes full responsibility for errors and oversights and apologises for the inevitable limitations and inaccurate biases of a 'foreign' view of other countries' problems.

PERMISSIONS

The author and publishers gratefully acknowledge permission to reproduce the following illustrations.

France

CREPA H for *Les Minguettes, Venissieux, Lyon*; Photo Doisneau-Rapho for *A post-war shanty town at La Courneuve outside Paris*; Archiv OPHLM La Courneuve for *Crane on specially laid railway tracks to build La Courneuve*; Photothèque Union HLM for *The dinosaur launches the 1987 campaign against prejudice about HLMs.*

Germany

Archiv Kommunalverband Ruhrgebiet for *Early Ruhr housing*; Dr Ursula von Petz for *East German tenement housing.*

Britain

History Today Archives for *The cotton mills owned by Robert Owen at New Lanark, 1825*; Borthwick Institute, University of York (Rowntree's Survey of York) for *Working class by-law housing in 1900*; First Garden City Heritage Museum for *Typical cottage housing in Letchworth, the first garden city*; Popperfoto for *Ronan Point disaster marks the end of tower block building*; Al Cane for *Residents' committee meeting of the Charteris Neighbourhood Tenant Co-operative.*

Denmark

Niels Saliacath/Danish Ministry of Housing and Building for the following photographs: *Inner-city tenement blocks, Energy efficient cluster housing,* and *The big industrialised housing estates*; Boligselskabernes Kursusejendom – Haraldskær for *Haraldskær – Denmark's national tenant training centre.*

Ireland

Wolfhound Press/Michael Bannon for *The Dublin Artisan Dwelling Company Cottages, Rialto, Dublin*; Derek Speirs/Report for the following photographs: *The remains of the beautiful Georgian Mountjoy Square in Dublin, Some attractive local authority housing overlooked by St Patrick's Cathedral, Dublin,* and *Belcamp travellers' site, North Dublin*; Ballymun Task Force, Dublin for *The building of Ireland's only high-rise estate at Ballymum, North Dublin.*

Definitions

Europe includes the twelve member states of the European Community.

The Continent describes countries that are part of mainland Europe and does not include Britain or Ireland.

Ireland refers to the Republic of Ireland, excluding the six counties of Ulster.

Britain refers to England, Scotland, Wales, excluding Northern Ireland.

An estate refers to a group of dwellings built in a discrete area, normally by a single developer and/or owner.

High rise refers to multi-storey blocks of dwellings, normally above five storeys. Individual countries have varied definitions but in this study a broad general definition was used.

State-sponsored housing describes housing promoted or encouraged through state financial incentives, where the state plays a continuing role in regulation and use.

Housing company is a landlord organisation, where individuals and bodies, including public bodies, own shares. Shareholders control the board of the company. Social housing companies have shareholders whose object is not profit but the provision of low-cost housing – for example, local authorities, trade unions, employers, charities, etc.

Introduction

INTRODUCTION TO EUROPEAN HOUSING HISTORY

State housing for poorer groups

Housing has long dominated domestic politics in Europe. It is not only a fundamental human need, but also visibly reflects social progress and individual status within the society. It is strongly linked with urban tensions and disorder (Scarman 1986), and because it is so central to the life and well-being of any community, governments tend to intervene when for any reason housing systems break down.

The role of governments in helping to provide housing, particularly for poorer and more vulnerable groups, is the main theme of the study. Political tensions over the extent, style, organisation and funding of social housing are intense in all countries. National governments strongly defend their internal responsibility for housing, while the European Commission edges around housing problems *per se*, by stressing the social problems associated with it – segregation, race, migration, and poverty. The overlap between the two areas – housing and socio-economic problems – is almost total. This draws the European Community ever closer to housing issues, in spite of the principle of 'subsidiarity' which makes housing the responsibility of individual members rather than the EC as a whole (EC 1991a).

This study traces the history of urban housing patterns and state sponsorship of social housing up to the present day in five European countries, with a particular focus on how unpopular, 'mass' housing estates arose in all countries. The study underlines the re-emerging role of social housing, highlighting the concerted effort by governments to rescue mass housing estates. There is a growing international consensus that the role of social housing must be enhanced if poorer groups in Western cities are to be accommodated and the eruption of bitter urban conflict avoided. The experience of disorders in the United States in the 1960s has been periodically echoed in France and the United Kingdom in the 1980s and 1990s and, to a lesser extent, in Germany and Denmark, adding urgency to the problems of marginal communities.

Social housing has become a permanent form of provision; enthusiasm for owner-occupation has declined; and polarised urban conditions have brought social housing back onto the political agenda of the European Community and its member states. The role of governments in poor housing areas shows little sign of decreasing.

Overview of European social housing

Different approaches to housing, from free market to state control

There are different ways of approaching housing in industrialised countries. At one extreme – in the United States, for example, where the concept of the free market economy dominates – housing is considered a consumer good not a social right, something that the individual buys in accordance with income. If you 'make it' in America you will find housing; if you fail in America you may not (Mayer 1978). The state provides only a minute proportion of 'social housing' – 1 per cent of the total stock – very much as a complement to private urban renewal programmes, and its efforts are aimed at housing the 'permanent poor', one-parent, non-working families with several dependent children, or the low-income elderly. In America, homelessness and affordability are endemic problems, in spite of a large number of empty dwellings (Rossi 1989).

At the other extreme, in Eastern Europe, housing has been considered a universal right, not a consumer good, an essential part of the 'social capital' provided both for the workforce and for those unable to work. Because of the demands on state resources and because housing is linked to production rather than consumption, state housing has been provided to minimal standards for about 50 per cent of the population. Private initiative has made good some of the shortfall and the state has allowed, or even encouraged, private provision in the form of co-operatives and owner-occupation. Because a universal social right is not always easy to supply universally, the failure to deliver by the state on a number of fronts has propelled Eastern Europe into accelerated moves towards the West European market model over the last few years. Housing problems have been acute under direct state control.

Western Europe fits somewhere in between. Generally housing is considered a limited social right in all the countries in this study. British legislation on homelessness, giving local authorities statutory duties to help the unintentionally homeless in priority need (Housing Act 1985) is matched by similar responsibilities in other countries. In France, Germany and Denmark, governments use systems of nominations to social housing companies to ensure access for priority cases. France has recently passed a law enshrining the right to housing for all (Droit au Logement, May 1990). Ireland has similar homelessness legislation to Britain's.

Because housing is seen as a social right, the state has intervened on a broad basis and, since the war, a majority of households have been eligible for help of one kind or another. Between a fifth and a third of housing has been provided through social landlords in the five countries in this study, some of it for sale, and a majority of all housing has been subsidised in some way, often indirectly, through tax incentives.

Definition of social housing

In all the countries in this study, social housing can be defined as housing that is not provided for profit and is often let at below market rents; it is allocated to lower-income groups or to those whose incomes would not allow them to buy a home independently; the way it is produced – in quantity and quality – is laid down and regulated by the state (Ball et al 1988). The social landlords themselves are also regulated in the way they provide housing. Social landlords can include local authorities, housing associations, co-operatives, limited dividend companies and private land-lords.

Britain and Ireland: contrast with France, Germany and Denmark

Social housing in Britain and Ireland is further from market rents, more strictly allocated according to need, and more directly managed by govern-ment than social housing in Western Europe, where it is built, allocated and managed by independent landlords of many different kinds. This salient difference, based on large-scale, direct local authority production and ownership of social housing, may significantly affect its character. The contrast between government-provided housing (the British and Irish experience) and government-sponsored housing (the French, German and Danish experience) is one of the most interesting themes of this study.

Stages of housing development

We will begin our examination of European housing conditions with an outline of how we reached our present stage in housing development where the five countries in the study enjoy relatively high standards of housing.

All countries, as they industrialise and urbanise, go through the four main stages of housing development identified by McGuire in his study of international housing systems (McGuire 1981). Firstly, the rapid growth of cities creates a need for minimal housing units for each household. As this ambition is gradually realised, often under appalling stresses and strains, the aim becomes larger units with a single room per person. These first two stages are so painful and have so many casualties that there are intense pressures on the state to intervene.

The third stage shifts the emphasis from quantity to quality as the number of units roughly matches the number of households and the pace of urbanisation slows down. The shift to quality usually happens while there are still many dislocations and uneven conditions. Overall, however, crisis conditions no longer prevail while significant state subsidies still go to economically secure, well-housed citizens.

Finally, the state enters a period of attempted withdrawal, as the majority of the population is well housed and reasonably satisfied. At this point, help tends to gravitate towards two targets: special assistance for those on low incomes to help them pay for housing and special initiatives to encourage the growth of owner-occupation in order to involve individuals in saving and taking direct responsibility as a way out of state dependence.

Difficulties of state withdrawal

The four stages of development could be characterised as:

- intervention;
- provision;
- quality;
- withdrawal.

These changes in the role of the state, from intervention and direct provision to a narrowing of the state's role and attempted withdrawal from direct provision, do not seem to end public involvement in housing matters.

Firstly, government financial involvement continues in the shape of subsidies to individual households; tax concessions to encourage particular developments (for example, owner-occupation or private renting); loan guarantees; and other aids to saving and investment in housing.

Secondly, when once the state is involved in housing institutions (for example, council housing in Britain) it is very hard to get out. Therefore, a commitment to withdrawal may be hard to convert into practical reality. Even in Germany where radical legal changes aimed to turn social housing companies into private landlords, local authorities have been drawn into an active role as a result. Also, governments have found that disposing of social housing, or 'privatising', is not necessarily popular and that many vulnerable groups continue to need it.

Therefore the reduction in *direct* state provision does not lead clearly to state withdrawal. Beyond the fourth stage lie uncharted waters. The hope that each household will provide for itself in a near-autonomous way within a decreasingly regulated framework may prove illusory for many reasons. The interaction of urban change, demographic change, industrial change, and change in household composition creates new demands on social housing providers that make a simple withdrawal of government difficult.

Lasting government role

In addition to the issues of direct provision, access and affordability, the state has other roles in housing from which it may not be able to withdraw and for which its involvement may become more important. The legal framework within which housing, building and urban development take place, the financial and tax framework which shapes investment, and the planning framework which encourages or contains growth, are some examples. Even the freest economies, such as the United States, have a government framework covering these roles. The role of the state may therefore change radically, but it may not diminish significantly.

Europe since the Second World War

The main shifts in housing development in Europe since 1945 follow a fairly clear pattern, to bring us to our present conditions, though Ireland is different in many respects from the other countries in the study. It is much poorer, less industrialised and less economically developed than other parts of Northern Europe. For most of this century it has been a net exporter of people, with a large proportion of its people still living in small, rural communities. As a result, this section concentrates on developments in the four highly industrialised countries, although Ireland became involved in peripheral estate building and consequent problems a little later.

Emergency mass housing

At the end of the war the four countries were faced with a chronic housing shortage, although France and Germany were much worse than Britain or Denmark. The crises provoked strong state intervention with similar measures everywhere: rent freezes; generous subsidies to construction; and the commandeering of land, materials supplies and construction firms to the state programmes of reconstruction. Existing housing was also requisitioned for a time. Mass building to minimal standards was undertaken everywhere, often state-sponsored but often actually provided through non-state bodies. Careful planning and environmental issues were luxuries that simply could not be afforded. Much early post-war housing was monotonous, minimal, dense and 'public' in character. 'Mass housing' is an inexact term used to describe post-war, publicly sponsored, industrially built, large-scale housing estates for moderate- to low-income households. It gained credence in the early post-war years of housing shortage. Much European social housing was built in this form, as it was to a lesser but still significant extent in Britain and Ireland. 'High rise' came to symbolise the problems of mass housing, though only in France were a majority of social housing estates built in large multi-storey blocks.

Balance of supply

From the 1960s, the number of dwellings began to approach the number of households all over Northern Europe and the crudest shortages were over, in spite of a large population influx, rapid household formation, and rising standards. Progressively, rents were unfrozen, housing spending was reduced in public building programmes and directed towards special help for low-income households and towards encouraging private initiative. Cheap loans were made available to accelerate owner-occupation. The large estates of mass housing conceived in times of chronic shortage began to pose management, social and physical problems from the early 1970s, although it took at least a decade for these to be generally recognised. There appeared a crude surplus of housing, showing up most clearly on the least popular estates where there were increasing problems in keeping dwellings let.

By the mid-1970s, when the serious recession got under way following the oil crisis, the approach to housing had changed out of all recognition. There was growing disillusion with 'mass housing'. Sensitive urban renewal and renovation were coming into vogue in all the countries.

Owner-occupation

One trend was common to the five countries at the point when an acute and widespread housing shortage no longer existed – the growing importance of owner-occupation. As people became wealthier, they wanted to spend more of their income on housing in order to buy more of it at better quality. If they spent more, they wanted more security in their investment, more choice, and more control. Therefore support for owner-occupation grew. People were prepared to spend more of their income on their own home than on a rented home – this development was also witnessed in Eastern Europe.

Governments seem to find it easier to encourage individual owner-occupation through tax concessions and other indirect subsidies, such as reduced interest loans, than to provide housing for rent through direct subsidies for those who need it. Therefore, governments support owner-occupation and variants on it, as far down the income level as possible, to reduce demand on state provision as opposed to state support.

In spite of the popularity of owner-occupation, two factors limit its growth. Firstly, there is a growing minority of marginal households in societies moving into a post-industrial phase. Secondly, more and more of these fall to the state to be housed and social housing becomes increasingly targeted at them, as strong incentives to owner-occupation deplete the private-rented market and attract away economically active households. Social housing therefore serves an important new role as housing of last resort for marginal households. This role invites continuing government support, although it is only rarely the role that social landlords have aimed to play in the past. Private landlords continue to house some of the very poorest but they have been increasingly squeezed by the growth in owner-occupation.

Survival of non-profit housing

Throughout Europe, the non-profit housing tradition is strongly rooted both in private philanthropy and in social and political institutions such as trade unions, churches, local government. Non-profit housing has continued to flourish in market economies in much the same way as co-operatives and owner-occupation have flourished in Eastern Europe (Chamberlayne 1990).

The growing needs of marginal households have fitted closely with the social housing tradition, and there are signs everywhere of renewed interest in the role of housing associations and non-profit housing companies because they can provide for groups unable to become owner-occupiers where no longer housed by private landlords. The social housing programmes in the five countries were being reassessed at the end of the 1980s under political pressure to cope with increasing demand, which could not be met by increased owner-occupation.

Immigration and access to social housing

An element in renewed support for social housing is the volume of immigration. Throughout the 1960s immigration into Northern Europe and Britain had been strong, involving at least 10 million new workers from poorer countries. This began to place enormous strains on private housing provision, as well as raising the ugly spectre of racial violence.

Since the war, Britain, France and Germany each absorbed several million immigrants and their descendants from abroad, many of them of a distinct racial and cultural origin. Two million people of Algerian origin settled in France. A similar number from Turkey settled in Germany. Over 1 million from the Indian sub-continent settled in Britain and nearly as many of Afro-Caribbean origin. Other groups came too, but it appears that racial divisions were the most prominent, evoked the most prejudice, and caused the greatest hardship. In the 1980s, riots with strong racial under-currents occurred in Britain and France. Germany experienced growing tensions and attacks on foreigners increased sharply in the wake of reunification. Denmark experienced acute problems of integration, though on a smaller scale and more recently. Only Ireland was different because of the almost constant outflow of people and because of its ethnic homogeneity. Troubles in Northern Ireland, however, were a constant reminder of deep community and religious divisions.

A major cause of the vast scale of immigration was the growth of economic activity, the changing work patterns in Europe and the chronic labour shortages in 'dirty jobs' as a result of the post-war boom. Migration grew to such an extent that it became part of national life in all countries. It showed up most visibly in poor housing areas (McGuire 1981).

During the 1970s, these problems grew more intense as families joined the breadwinners. The anomaly persisted of temporary and *ad hoc* shelters but with vacancies in expensive, state-sponsored, high-rise social housing flats (Emms 1990).

Across Europe in the late 1970s and early 1980s, racial minorities and immigrant groups began to gain access to social housing on a large scale, invariably to areas being vacated by native-born households. A continued need for social housing was recognised. But conditions deteriorated in the unpopular areas that were used to house new groups; thus, inequality was highlighted and underlined through access to social housing rather than reduced. This remains an unresolved housing problem in *all* the countries of the study, which this book may serve only to underline. There are as yet no clear remedies or proven solutions, though many innovative measures are being tried to help stem the serious decline that is occurring.

Housing and social integration

Housing is the most conspicuous sign of inequality. Status, income, and racial and class identity are often tied in with the kind of housing that people occupy and with its location, condition and environment. Where societies are heterogeneous, housing segregation is often most intense. The United States is probably the most extreme example of deep inequality and intense housing segregation, along racial as well as income lines. Consumer sovereignty in the urban housing market has produced harsh outcomes (Lemann 1991).

As European countries increasingly house populations of diverse racial origins, the questions of segregation and inequality arise frequently. Segregation and inequality will grow in a situation where housing environments differ significantly. This issue affects mass housing estates very seriously.

Housing in the wider economy

Housing has an importance beyond its own boundaries. It is intrinsically linked to other aspects of social well-being, such as health and education. It has a major influence on economic development. Not only is housing a large part of national wealth, national production, and national income generation – it creates jobs, it has related supply industries, and involves maintenance – but it is also important to other kinds of investment. A good housing environment attracts investment. Conversely, a flourishing economy facilitates high housing production. In other words, whichever way we consider the role of housing, its importance is undisputed. Hence its growing fascination to social and economic investigators and its long-standing position in national politics. Whatever the housing system, its influence on other aspects of national life is bound to be significant.

Council housing or independent landlords?

British housing problems have been extremely controversial for a decade and have raised several questions: who should own social housing?; how much should people pay?; how far should owner-occupation go?; how should social housing be managed?; who is responsible for the homeless?; can housing estates be rescued? Continental countries argue over some of these issues, but less bitterly than the British and not over the ownership of social housing. Throughout the continent, single-purpose, independent landlords dominate state-sponsored housing. Therefore the moves in Britain to privatise council housing and to expand the role of housing associations as single-purpose, independent landlords has suddenly made European housing history more relevant to Britain (OECD 1988). At the same time, the strong welfare role of British social housing, the initiatives to break up its monolithic character through the right to buy for council tenants and other sales and transfers, and the strongly backed resident initiatives in some of the most run-down council areas, have attracted similar interest among continental social landlords and governments as they struggle with housing shortages for poorer groups and growing polarisation. Cross-fertilisation within the European community over issues affecting social housing, such as segregation and polarisation, management, resident involvement, racial tensions and rescue programmes, became intense in the late 1980s (Jacquier 1991, Burgeat 1989).

This study sets out to uncover both the historic divergence and the recent convergence in state-sponsored housing between Britain and her European partners.

Different patterns

The two main forms of social renting, through housing associations and through local authorities, are associated with differences in levels of private renting and owner-occupation. Therefore social renting – including a broad range of state-sponsored and funded landlords who are regulated by public bodies in how they operate, who they house and what units they provide – is inextricably linked to the two other main tenures: owner-occupation and private renting.

British housing is unique in Europe because over four-fifths of it comprises single-family houses, yet one-third of it was built directly by

Table I.1 Tenure in five countries in 1990 (per cent, rounded figures)

	Owner-occupation (%)	Private renting (%)	Housing associations (non-profit) (%)	Local authorities (%)
France	53	30	16	1
Germany[a] (United)	37	38	25	–
United Kingdom	66	7	3	24
Denmark	58	21[b]	18	3
Ireland	78	9	0.5	13

Sources: See country chapters.
Notes: [a] The figures for Germany are affected by reunification. See German section for breakdown for East and West.
[b] This figure includes private co-operatives. See Danish section.

local authorities. In Ireland, a similar though less pronounced pattern prevails. Continental housing is much more varied because governments rely on independent, non-profit and private landlords. The high proportion of flats, the late development of urban owner-occupation, and the relatively large, private-rented sector on the Continent, also raise puzzling questions about how such different urban housing patterns had grown up. Subsidies clearly played a determining role in the shifting housing patterns. Table I.1 illustrates the contrasts clearly.

A central difference

Continental landlords have a private tradition that makes them reluctant to take on a purely welfare role in modern society. They run their housing on business lines, creating enviable standards of supervision and repair. British council landlords, by contrast, accept a welfare role but strongly resist pressures to privatise or to adopt a business-oriented approach. For all the rhetoric of the Thatcher years, political and social change is slow. The welfare system is deeply embedded in the national culture and is proving hard to dismantle.

State ownership declines

The accelerating changes in Eastern Europe highlight housing problems from a different angle. A strong desire to escape from mass state provision and to encourage individual self-help and ownership challenges existing forms of state intervention. Britain's traditional council housing is in some ways closer to East European patterns than any of her European Community partners because of extensive direct state building and ownership and because of the political and administrative systems that evolved to cope with it. It was probably inevitable that direct state ownership of housing would be called into question, but disentangling a large, public-rented stock from state control was a complex and controversial task for which there was no model. The welfare role of British and Irish council housing proved more central than was originally conceived.

The past illuminates the path

Comparing the different patterns and discerning the emerging role of social housing involved tracing the history of each country. It was impossible to understand how different European countries arrived at present housing achievements and problems without examining the past. Therefore, we start by presenting a brief background history of each of five European countries in order to place in context the evolution of housing patterns. We then explore the emergence of social housing in each country, showing how state support for mass housing arose, how it gradually declined, and how eventually it re-emerged in the late 1980s with a crucial welfare role. The overlap between developments in state-supported rented housing, owner-occupation and private renting became clearer as the history unfolded.

The book discusses how the housing stock was built; what help governments gave; why government involvement remained important after crude shortages had disappeared, in spite of vigorous attempts to withdraw; why new problems emerged fast on the heel of solutions and how they were tackled; and why, increasingly, governments aimed to renew the existing environment rather than start again, even where mistakes were clear.

Deeper causes of polarisation

A fundamental question at the outset was whether the social welfare role of state-sponsored housing varied according to historic ownership and management patterns. The strong polarisation of social housing went beyond the narrow, party-political divides over council housing in Britain, for many social problems and divisions arose in their starkest form across Europe in state-funded housing.

Racial tensions and urban disorders in the 1980s were only two manifestations of a deep-seated malaise in modern urban society. The growing segregation of large state-sponsored housing estates was linked to many of the gravest problems, including Ireland where race and immigration were not issues of significance. Europe's different social housing systems had not prevented similar social problems from emerging.

Differences in wealth and condition

Any attempt at harmonising experience across countries is bound to be severely circumscribed.[1] Histories and cultures, however closely interwoven, diverge in their content, context and impact, sometimes in extreme ways, often even within countries. Legal, political and economic structures create different frameworks and throw up different patterns of leadership, policy and practice.

Economic conditions are usually a product and a cause of most other trends. Wealth and poverty may be the most visible and extreme manifestations of differences between, as well as within, countries. These factors all influence housing conditions and vary greatly (McGuire 1981). However, they are far from clear or strictly comparable, as many studies show (Holmans 1992). Crude comparisons suggest wide differences as Table I.2, which shows spending on housing and energy consumption as a percentage of total spending in the five countries, illustrates.

Table I.2 Expenditure per head of population on housing and energy as a percentage of the total, 1990

Country	France	W.Germany	UK	Denmark	Ireland
Per cent	17	20	15	27	11

Source: *The Economist Pocket Europe*, 1992.

It is far beyond the scope of this study to cover in detail all the interconnected areas that impinge on housing, or to make precise comparisons except in specific areas related to housing. Even this poses as many questions as it gives answers: rent levels, housing allowances, spending on management and maintenance, cost of housing, benefit systems, subsidy patterns and their relation to low-income earners, numbers of children, one-parent families, availability of work, and so on, are so complex and vary in so many different legal and financial details that valid comparisons are very difficult. For this reason, the five evolving histories are left to tell most of the story.

National stories

The evolution of social housing was bedded within the national framework of each country. The approach adopted here was direct and deliberately responsive – an exhaustive and in-depth quest, country by country, for live experiences and examples.

The approach was oriented towards 'discovery' rather than 'standardisation'. By following up on important issues as they arose, the author was able to uncover a pattern of urban development that was surprisingly uniform and that gave unexpected prominence to social housing in the current period.

Convergence of problems

The local approach makes it possible to go beyond the more standard international studies in pursuing a central theme to its roots – why are social housing areas both becoming more polarised and more popular across the European Community at the advent of the Single Market? But local developments in the ownership, management and social role of state-sponsored housing must be placed in the context of bigger issues. Outcomes, their causes and pre-conditions, have as many similarities as differences (Jacquier 1991). Urbanisation, state intervention, mass housing programmes, the stigma of welfare, social polarisation, were processes common to all countries. Yet each country's experience of these developments was unique. It was the common themes, in spite of different events, and different experiences, in spite of common patterns, that shed most light on the problems facing the European Community in its large urban centres in the 1990s.

In the fast-moving, internationally oriented world of the late twentieth century, nations are more, not less, conscious of their internal tensions as more of their national identity is shared with the international community. They are increasingly concerned about their internal stability and the integration of their weaker and poorer communities as they experience growing pressure and competition from the much wider international community and as they struggle to handle the increased social stresses of post-industrial urban settlements (Dauge 1991).

A broad canvas

To what extent can findings across such a broad canvas be considered valid? The main problems are the range of variables affecting developments; the complex relationship between variables that are often far removed from the direct issues being examined; the general unreliability of attempted scientific explanations for social phenomena; the possibly unmanageable detail involved in five country studies.

Theories in pure sciences, such as physics, are valuable because they can uncover simple, universal principles that explain a wide range of experiences across a huge range of events or manifestations: they have greatly advanced our understanding of life. Gravity and evolution are two examples.

Theories in economic life, though attractive because of the volume of measurable experiences or statistics, are affected by so many political and social developments that economic theories are bitterly contested and often proved wrong. Even the 'hard' facts are often disputed. The imperfect workings of markets amply illustrate this.

In areas of social policy, linking causes and effects is even more difficult. Many attempts at analysis and explanation have been proved wrong and many attempts are unintelligible to the lay person. The inability to find explanations simple enough to clarify human problems limits the relevance of much social theory. Understanding particular situations does not necessarily lead to widely applicable creative ideas. The multiple factors involved, and the unpredictable human events and reactions, often make attempts at 'universalising' circumscribed, arrogant, or simplistic. Yet the very coincidence of experiences in state-sponsored housing invited explanation. The areas selected for study were identified through the action of five governments, working independently of each other over a century. The role of residents and housing staff had evolved in thousands of separate communities with their separate languages and cultures. Therefore the findings should help our understanding of marginal communities in modern urban society.

AIMS AND APPROACH OF THE STUDY

Aims

The original aim of the study of state-sponsored housing in the five European countries was to establish whether ownership and management structures significantly affected the problems facing social landlords and to identify the extent to which common social, organisational and physical problems arose in spite of different structures. In the light of the experience from the five countries, the study attempts to outline common patterns, to identify underlying causes and to offer explanations for a common drift towards polarisation and segregation in all countries. The ultimate focus of the study is on problems. What were the origins of the current situation? Why and when did they emerge? What action was taken to tackle these problems?

The unpopular 'mass' housing estates serve to highlight wider intergroup tensions, often with a racial dimension, growing marginalisation, and renewed problems of crude shortage and access difficulties for vulnerable people within the European Community.

Approach

The investigation was focused on the European Community, because common interests between countries were being forged by the accelerated development towards a single European Market. This made the search for common experience relevant. Northern European countries were chosen because of their more comparable pattern of economic and urban development, thereby making differences in housing less dependent on factors that could not be compared. Southern European member states developed state-sponsored housing much later.

Selection of countries

Five countries were chosen. France, Germany, and Denmark share borders, are members of the European Community, have broadly comparable standards of living and operate their state housing largely through independent but publicly sponsored housing bodies. Britain and Ireland rely on local authorities directly for their state housing. Ireland's history is closely tied with Britain's and her forms of provision are broadly similar, yet in many ways Ireland is more European than Britain. In size, geography and rural base, Ireland is comparable to Denmark. As a poor country on the edge of a large and prosperous region of the world, her experience seems both relevant and unique.

Denmark is included for three reasons. Firstly, Denmark is the only Scandinavian member of the EC, with a very different approach to welfare and social housing. Secondly, Denmark is a small country, along with Ireland and at least five other EC members. Thirdly, Denmark is generally considered a success in economic and social terms.

Britain, France, and Germany have much more tense urban problems and are comparable in size and weight, although the reunification of Germany in 1990 clearly gave Germany a new and more dominant role. The study of these five countries would uncover a wide range of systems, affecting 200 million people.

Method of collecting information

The five countries were visited continuously between 1987 and 1991. Housing experts in government, housing organisations and research institutions were contacted; social housing areas that posed a range of problems in the eyes of policy-makers in each country were located and visited.

The author carried out visits to all five countries. Unifying the research across five countries was made easier by virtue of a single person carrying out most of the visits personally, although taking forward five national studies together over four years proved a daunting and often confusing undertaking.

Much information was only recorded in government documents or local records. It was not generally available in English on the Continent. The difficulties of penetrating and understanding foreign systems were partly overcome by the willingness of government officials, housing company representatives, national experts, local staff, and residents, to discuss problems, arrange visits, provide internal documents and check information. This story of five countries' housing relies heavily on those internal sources and could not have been written without that collaboration.

International sources

Three major constraints on the housing information collected in the five countries were: figures for production, tenure and housing stock were often incomplete, inconsistent and sometimes inaccurate within one country for certain periods (for example, pre-war Germany); some figures were out of date and the latest available figures were from different dates in different countries; different countries collected figures in different ways. Therefore, even national pictures were not completely accurate and comparisons were sometimes based on estimates derived directly from government sources. Where official statistics showed inconsistencies, the author attempted to reconcile conflicting figures with the help of national housing experts. Other studies were helpful in corroborating some of the findings (Emms 1990). All findings were checked with the national sources listed. Figures from international sources (EC, OECD) were invariably several years out of date and sometimes inconsistent with national sources. As far as possible, tables give the latest comparable information available in 1992.

Local documents, direct evidence, and internal information, were authenticated by a coherent sequence of developments. The very compact fit of the housing jigsaw that emerged was possibly the most convincing evidence to support the findings reported in this study. At the same time, local and internal information bore closer resemblance to actual conditions on the ground than information from existing international studies, which only rarely presented detailed national or local information.

The following organisations were the most important in providing information:

France. The National Union of HLMs; the Association of HLMs of Ile de France; the Délégation Interministérielle à la Ville; the Caisse des Dépôts et Consignations; SCIC; the former National Commission for the Social Development of Neighbourhoods; the Communes of Vénissieux, Val de Marne, Val Fourée and Marseille; the 'office public' of Romans and the OPACs of Rhône and Val de Marne.

Germany. Gesamtverband der Wohnungswirtschaft; LEG Wohnen, Institut Wohnung und Umwelt at Darmstadt; cities of Cologne and Bremen; Germany Ministry of Housing and Urban Development; Empirica and Büro Sachs-Pfeiffer.

Britain. Department of the Environment; Welsh Office; Scottish Homes; local authority housing departments; the Priority Estates Project; the National Federation of Housing Associations; Welfare State Programme

and Department of Social Administration (London School of Economics). *Denmark.* National Federation of non-profit housing companies; University of Roskilde; AKB housing company; Institute for housing and building research; Braband housing company; Haraldskaer national tenant training centre; the Danish Ministry of Housing.
Ireland. Dublin and Cork City Corporations; Irish Department of the Environment; Institute of Public Administration; Ballymun Task Force; Combat Poverty Agency; Focus Point; University College, Dublin.

Structure of the book

Each of the five studies opens with a brief introduction to the political and social development of the country, followed by the housing history. This division simplifies the story and allows the reader to make connections between major events and their impact on housing, or to skip the country's history and plunge straight into housing policy. The attempt to explain developments in chronological order was sometimes foiled by the complexities of the issues. To help readers, charts showing key dates have been used and summaries of conditions and progress are presented at the beginning and end of each country's history.

This study almost certainly raises more questions than answers. In order to stay within manageable limits, detailed information on difficult-to-manage estates and government rescue programmes in the five countries will form the basis of a follow-up study.

Terms

Each country has its own housing definitions. Where necessary, these are explained in notes at the end of each country study. The term *high rise* is used in the context of each country but generally refers to blocks of five or more storeys. All five countries in the study form part of *Europe.* Only France, Germany, and Denmark form part of the *Continent.*

The tables show variations on almost all measures between the five countries. These core conditions provide the framework for the different housing stories that follow. The snapshots conceal the common trends, such as towards smaller households. Those trends play a very big part in the development of housing, as the country studies show.

NOTE

1   The problems of international comparison have been extensively debated in a number of studies and that debate is not repeated here (Ball *et al.* 1988, Lundquist 1986, and Boelhouwer and van der Heijden 1992).

SUMMARY OF BASIC CONDITIONS IN FIVE EUROPEAN COUNTRIES

Tables I.3, I.4 and I.5 present in outline the composition of the population and conditions in the five countries.

Table I.3 Population in five European countries, 1990

	Population in millions	Projected population 2010	Population per sq. km	Population in urban areas (%)
France	56.4	59.4	102	74
Germany	78.8	75.1	221	84
United Kingdom	57.2	59.0	234	89
Denmark	5.1	5.1	119	87
Ireland	3.5	4.5	50	57

Source: *The Economist Pocket Europe*, 1992.

Table I.4 Composition of population in five European countries, 1990

	Crude birth-rate per 1,000 population	Crude death-rate per 1,000 population	Per cent under 15	Per cent over 65	Per cent belonging to racial and ethnic minorities
France	13.8	10.3	20	14	7
Germany	11.1	12.3	16	15	7
United Kingdom	13.6	11.8	19	15	6[a]
Denmark	10.8	11.3	17	15	2
Ireland	18.1	8.8	28	10	–

Sources: *The Economist Pocket Europe*, 1992; UN *Monthly Bulletin of Statistics*, 1992; *World Bank Development Report*, 1990.
Note: [a]This figure was obtained from *Population Trends 1992* from the Office of Population Censuses and Surveys.

Table I.5: Wealth and conditions in five European countries, 1990

	GDP per head ($)	GDP per head adjusted for purchasing-power parity	Unemployed of workforce (%)	Average no. per household
France	21,107	70.0	10.1	2.5
Germany	18,878	73.5 (West only)	6.4 (West only) 16.0* (East only)	2.2
United Kingdom	16,926	68.0	10.3	2.6
Denmark	25,458	69.3	8.9	2.2
Ireland	11,952	41.3	16.8	3.8

Sources: *The Economist Pocket Europe*, 1992; UN *Monthly Bulletin of Statistics*, 1992; *World Bank Development Report*, 1992; *Eurostat General Statistics*, 1992; * *Bundesanstalt für Arbeit Monatsbericht*, February, 1992.

Note: Income within each country cannot readily be compared with that of another country because of different costs. Therefore in addition to cash income, converted into dollars for ease of comparison, we show a measure of national income per head that is adjusted to reflect the typical cost of a common list of basic goods and services. In this measure, the United States, which has the greatest purchasing power, is represented by 100. This approach is referred to as purchasing power parity.

Part I
France

The state was king – popular French saying

France has a prominent place in the Europe that is unfolding. The French Revolution overthrew the old order of rigid privilege and exploitation, and generated a shift towards egalitarianism that created strong international waves. The rise and fall of the Emperor Napoleon built on the French sense of a grandiose international role and a preoccupation with state power and state-driven development. The rebuilding of Paris in the latter half of the nineteenth century gave France a uniquely planned and beautifully laid-out capital. France offered a model of civic development to the world.

The long-standing tensions with other European powers –Britain as well as the emergent Germany – made the French both proud, involved and aloof. The traditional neglect of housing for the growing urban populace in favour of civic and industrial works of great significance created a legacy in France which set her apart from the rest of Northern Europe.

France in the 1960s and 1970s built more high-rise, peripheral, mass housing estates than any other country in Western Europe. There have been more extensive violent disorders in French cities than elsewhere on the Continent. Racial tensions have often been intense. French housing experts are more outspoken about the problems of social segregation and polarisation that have resulted than any other national body.

The French have gone out of their way to foster partnership between central, local and regional government, stressing the vital role of the elected mayors of the local *communes* in which the difficult social housing estates are located. In spite of the still immense power of the state, they help create local civic pride and strongly defended local initiatives.

French housing policy since the war has advanced in three main directions: limited state support for private renting, as a result of which one-third of French households still rent from private landlords, although many of these are in furnished rooms and some in hotels; generous support to owner-occupation, offering grants as well as low-cost loans to households of modest means, leading to a dramatic increase in the construction of *pavillons* – single-family, detached surburban and semi-rural houses – and resulting in over half the population now owning their own homes; continuing ambitious social housing programmes which are still running at 50,000 units a year or more to help meet growing demand from more needy groups and indirectly to help the economy, as well as to create higher-quality, smaller-scale urban housing.

The major housing preoccupation of the 1980s was the *grands ensembles*, the social housing estates usually built as dense, high-rise blocks on the edge of cities. These troubled housing areas caught the headlines in 1981 as disorders broke out in a number of them, often between young people of North African origin and the police, generating support for renovation programmes. Renewed outbreaks of urban disorders in 1990 and 1991 led to a further expansion of government initiatives. The French experience of urban segregation and regeneration offers a wide canvas of very mixed experiences.

France today faces similar problems to other countries in this study of European housing: a high level of unemployment, particularly among young people of immigrant origin; strong social dislocation, stemming in part from very rapid urban expansion in the 1950s and 1960s, and continuing thereafter; increasing problems of cost, of loan default, and of urban sprawl in the owner-occupied sector; decline of the private-rented sector through inner-city renewal and renovation; intense polarisation in the society as a whole between the affluent majority and the large minorities of immigrant origin, of one-parent families, and of socially and economically marginal households; growing demand for social housing from lower-income groups; severe decline in large, peripheral estates where social and economic problems are most starkly and intensely exhibited.

Post-war France was the most serious protagonist in the evolution of the European Community, insisting on ever-closer relations with her most powerful ally and erstwhile enemy, Germany; engaging in active partnership with other members and potential members of the Community; and constantly pushing the pace on ventures like the Channel Tunnel and high-speed rail links across Europe. These international links belie her reputation for defending a strongly independent national identity and hark back to her historic role in forging the future of Europe. Her current pioneering approach to the wider social and economic problems surrounding mass housing is a microcosmic reflection of this ambitious role. The aim is to explain France's rapid post-war urbanisation and provide the history behind some of today's tensions.

Chapter 1

Background

BACKGROUND FACTS ABOUT FRANCE

While France is similar in size of population to Britain, with 56.4 million people, it has less than half the density, with only 102 people per square kilometre (*Economist Pocket Europe*, 1992). It has more young people and fewer old people than any other country in the study except Ireland, and a higher birth-rate than Germany or Denmark, though its birth-rate has fallen significantly. It has higher unemployment than Germany or Denmark, and about the same as Britain. France sits in the middle of the income range with higher income per head than Britain or Ireland (*World Bank Development Report* 1991).

France has experienced major waves of population immigration. French population statistics do not show French residents originating from French ex-colonies that became overseas provinces of the French Republic as they are considered indistinguishable from mainland French inhabitants and enjoy full French citizenship. Thus, many ethnic minorities from the French Caribbean and Indian Ocean islands are not counted and their numbers can only be guessed at. There are also growing numbers of second generation naturalised French citizens of ethnic minority origin. Therefore, official French immigration figures only reflect a part of the picture. None the less, 7 per cent of the French population is officially classed as of immigrant origin, with the largest group originating from North Africa. There are estimated to be a further 6 million French residents of recent foreign origin (Superior Council for Integration 1991). The French housing stock is shown in Table 1.2.

In the thirty-five years from 1955 to 1990 France moved from having some of the worst housing conditions, the grossest overcrowding, and an extremely low rate of building to having a massive level of production over the 1980s, totalling about 300,000 units a year. How this remarkable shift came about, and the unique forms that it took, constitutes the core of France's housing history.

Table 1.1 Basic demographic facts about France, 1985

Size of population	56,400,000
Density of population (inhabitants per km²)	102
Proportion of under-15-year-olds	20%
Crude birth-rate per 1,000 population	13.8
Unemployment	10.1%
People over 65	14%
Household size	2.5
Immigrants	7%
Population in urban areas	74%

Sources: Commission des Communautés Européennes 1987; *Economist Pocket Europe*, 1992; Eurostat General Statistics, 1992.

EARLY HISTORY

For many centuries France dominated Europe politically. French history was marked by Republican fervour, and international and sometimes imperial ambitions. The French Revolution of 1789 shook Europe and made France a pathfinder in a new, if sometimes cruel, social order where the populace was swift to exercise its power and where dictators like Napoleon could rise on the back of its acclaim. The storming of the Bastille goaded European governments, through fear of uprisings, to address urban conditions, if in a piecemeal way, well into the twentieth century.

France experienced three revolutions in less than a century but, ironically, French governments avoided significant housing responsibility for longer than any other country in the study. French social movements articulated demands for political rights and freedoms but appeared willing to tolerate urban conditions that were feared elsewhere.

Table 1.2 Distribution of French housing stock by tenure, 1991

	Millions	%
Total occupied units	20.83ª	100
Owner-occupiers	11.23	53
Private-renting (including furnished and tied accommodation)	6.00ᵇ	30
Social renting	3.54ᶜ	17
Total stock (including vacant and secondary residences)	25.00	

Sources: INSEE *Housing Survey*, 1988; *Tableaux de l'Economie Française*, 1991–92.
Notes: ªThere are 2 million empty dwellings and over 2 million second homes in France in addition to these figures (INSEE 1988).
     ᵇThirty-one per cent of private-rented units are either furnished accommodation, hotel rooms or tied flats; figures for this category of accommodation are not consistent and may be underestimates.
     ᶜEighty-seven per cent of these units are owned by HLMs.[1]

The Napoleonic legend, based on conquest abroad and a new civic order within France, helped bring Napoleon Bonaparte III, nephew of the great Emperor, to power in the mid-nineteenth century by popular plebiscite, making possible the modernisation of France. Rapid state-led industrialisation was facilitated through the building of an ambitious national railway network. Government was strengthened through the extension of the power of state-appointed prefects to control France on behalf of a highly centralist state. Napoleon III's ambition was to retain his country's precarious but dominant position in Europe by modernising France through the powerful alliance of the masses, the growing middle classes and the state. He also demonstrated concern for the impoverished urban dwellers.

Paris was at the hub of state-led change fuelled by colonial expansion. It grew rapidly from 1 million at the beginning of the nineteenth century to 2.5 million by 1870 and 3 million by 1910. Napoleon's plan for Paris – to make it the most beautiful city in the world – involved the building of large boulevards, public amenities and open spaces, the provision of a water supply and sewage system, the creation of elegant apartment blocks and an extension of the city boundaries to allow for growth. But it also involved the removal of the least-favoured classes to the outskirts, leaving the centre free to develop in a more modern and more ambitious way. The wide, new avenues were easier to keep clear of barricades and 'rebellious mobs'!

The rebuilding of Paris under Baron Haussmann made France's urban development appear more ordered than anywhere else in Europe. The rebuilding of Paris set in train a characteristic pattern of French urban development, a well-preserved, attractively planned centre with marginal housing around the edges in poorly planned developments, badly served by the urban infrastructure and lacking a longer-term vision.

While Haussmann's Paris stands as a landmark in urban planning, it tells us little about the bursting populace that actually carried out the plan while simultaneously being displaced by it.

THE PARIS COMMUNE 1871

The Paris Commune, a short burst of 'citizen control', shook the confidence of the state in locally based authority and coloured the special shape of state housing in France (see pp. 36–7). Crucial to the development of France in the last thirty years of the nineteenth century was the disastrous Franco-Prussian war, which led to the consolidation of modern Germany under Bismarck and the collapse of the Second Empire of Napoleon III. Paris was besieged for four months in 1870 and was forced to accept humiliating defeat at the hands of the German army by the provincial National Assembly at Bordeaux. The people of Paris rebelled against the French Bordeaux troops – the 'Versaillais' – who took over from the German occupiers, and the city was temporarily left to the

'Communards'. It was only won back after a week of bitter fighting, French soldier against French citizen, while German troops were still in the Paris suburbs. The Commune of Paris lasted for two months in the spring of 1871, but it proclaimed the rights of citizens in ringing phrases, echoing the earlier declarations of the French Revolution.

Paris was seriously damaged in the course of the upheavals; the troops massacred or deported many suspects; the Communards killed some hostages, including the Archbishop of Paris, and the Commune was brutally suppressed. France lost not only international standing but also, temporarily, control of her beautiful, proud, but overcrowded capital.

The new French government was deeply hostile to all movements from within Paris, mistrusted its people, and rejected any step that had a hint of socialism. Radical Parisians, on the other hand, fought bitterly over the political ideas that they had shed blood for. The main outcome was a bitter stalemate between Paris government and the government of France, which greatly impeded progress, particularly in tackling the chronic conditions of the 'peripheral' masses, and served to reinforce state power.

FRANCE AND GERMANY

The divisions within France compounded France's declining international position, not least her deep fear of Germany, by whom she was now displaced as European leader.

The First World War took 1.4 million French lives, wiping out maybe half a generation of young men, but France regained Alsace and Lorraine, a bitterly contested region between Germany and France.

After the First World War, France fought for the toughest possible of war 'reparations' by Germany. But France's invasion of the Ruhr in 1923 to exact dues from Germany further weakened France, militarily and politically, alienating her allies and helping Hitler's expansionist ambitions. The French economy grew very slowly between the wars and urban developments were very constrained. Paris stagnated.

The rise of Nazi Germany culminated during the Second World War in the German occupation of France. France paid heavily for the Second World War – a humiliating surrender; a puppet government; severe war damage; 400,000 houses destroyed and 1.5 million damaged; and a resultant block on economic development. But the heroism of the French Resistance during the war bequeathed on post-war France a sense of idealistic self-sacrifice and national purpose in spite of the huge problems. France entered the cold war era determined to see Germany's economic power constrained by peaceful European collaboration.

France gave political leadership to the creation of European economic institutions, such as the European Coal and Steel Community, as a way of developing political integration. In this way, France, more than any other major power, forged the emergent European Community. But France was involved in two crippling wars in the 1950s which seriously impeded progress; one in Indo-China, the other in Algeria. Both sapped her national strength, consumed vast industrial resources, and hindered economic development in the 1950s. They also divided France politically, particularly the Algerian War.

In the early post-war era, a powerful Communist Party, representing a large minority of the French population, often played an important brokering or blocking role, particularly representing the poorer, urban *communes*. A series of weak coalition governments paved the way for the return in 1958 of Charles de Gaulle, hero of the Resistance, fiercely nationalistic yet strongly European, and determined to restore the French state as a powerful, stable, modernising base, central to a new united Europe. He reinstated a presidential system and a preoccupation with French national interests. He established a masterplan for the Paris region and reorganised the *départements*.[2] From the days of Napoleon I, the French state had exercised enormous power in a highly centralist fashion – *L'état était roi* ('The state was king'). The *préfect*, the French central government representative in all the regions of France, played a crucial part. The reforms of de Gaulle reinforced centralist power.

Since 1982, there has been significant decentralisation of political power to the regions, the *départements* (sub-regions) and local authorities (*communes*), answerable to the local electorate, limiting the role of the *préfect*. But the state still exercises much greater direct power than in Britain or Germany.

Local authorities in France, by contrast, wield significantly less power. The very large number of often very small *communes* – 36,000 *communes* in all – limits the scope of local government, although mayors are often important local figures. The average *commune* has 1,500 inhabitants; many are in tiny rural hamlets of less than 500. The Commune of Paris, by contrast, has 3 million inhabitants. While the fragmentation and dispersal of power among the *communes* act as a counterweight to the centralising tendency of the state, some have argued that it has actually enhanced central power by creating a gravitational pull for wealth and ambition towards the centre, reinforcing the tendency to push poverty to the outer *communes* (HLM 1989a: 87).

The combination of a clear European commitment with an over-arching sense of national identity and strongly state-driven institutions made post-war France both powerful and somewhat turbulent. The impact of these important strands of history on her housing developments will become clearer in the following sections.

NOTES

1   HLM: *Habitations à loyer modéré.*
2   A *département* is a French sub-region, governed by a central government representative, the *préfect.* While decentralisation of French state power has led to the creation of directly elected regional government and limited the power of the *préfects*, the departmental structure is still in place and the *préfect* still wields significant authority on behalf of the state.

Chapter 2

Early housing developments

NINETEENTH-CENTURY UTOPIAN HOUSING DEVELOPMENTS – THE CITE NAPOLEON

In the year following Napoleon III's election in 1848, the very first workers' housing society was set up to build model dwellings, called 'La Société des cités ouvrières de Paris' (The Society of Parisian Workers' Housing Areas).

Napoleon gave direct aid to the earliest housing project, named Cité Napoléon after himself, built by the Société des cités ouvrières. It provided only 86 housing units, built around an enclosed courtyard in central Paris. The rents were about 15 per cent of wages, far below the average rents for very inferior workers' dwellings in the typical slum or '*taudi*'. The block was managed strictly through a concierge and incorporated not only solid rooms, a garden in the courtyard, heating and running water, but also a nursery for very young children, a school, and a medical service. Napoleon himself opened it.

The Cité Napoléon survives to this day at rue Rochechouart (Quilliot and Guerrand 1989). In spite of its minute scale, it was a pathfinder for later experiments and was the first of many ambitious schemes, inspired by the need to create better conditions for the workers. The aim was to build similar projects in each neighbourhood of Paris, inspired by the violent disturbances of the summer of 1848 and the fear of continuing unrest. In the event very little was actually done. The state was to hold back from intervening directly with money for another sixty years. And private initiative, while fertile in ideas, was slow to produce the money.

In spite of the historic importance of the state in France in sponsoring urban and industrial development, housing itself was strongly held to be a private matter for individuals or employers. This view was influenced by a deep-set fear of socialism, which revolutionary movements engendered. Therefore the Cité Napoléon was a rare experiment (Quilliot and Guerrand 1989).

A number of private housing associations and worker-led co-operatives did develop in this period – known as *sociétés anonymes* – but their contribution to housing development was extremely small and often in the form of single-family houses for individual workers – a few thousand homes at most. These societies gave a maximum 4 per cent return to investors and were funded entirely privately.

STATE-SPONSORED INVESTMENT

The self-appointed second Emperor continued to inspire social concern, conscious as he was of the political power of the masses and of their appalling conditions. Under his leadership, state sponsorship of housing investment, as well as infrastructure work, was made possible through the creation of the Crédit Mobilier and the Crédit Foncier. These central financial institutions were primarily responsible for funding the construction of modern Paris, as well as laying 16,000 kilometres of new rail track.

Individual savers at every level of French society paid into state-guaranteed savings banks, which then lent to state-sponsored bodies to carry out work on behalf of the government including, later, social housing. The habit of saving small sums regularly through the 'Caisse d'Epargne' is still a deeply ingrained habit of ordinary French citizens.

COLLECTIVE EXPERIMENTS

The ideas incorporated in the Cité Napoléon and other similar experiments were inspired by a handful of visionaries and industrialists who made a unique contribution to French housing history. They proposed a radical approach to the urban problems of cost, location, transport, hygiene and health. The rapidly expanding workforce was cramming into ever-worse conditions and the solution was sought in the collective ordering of housing, in parallel with the increasingly collective ordering of production and labour into growing factory units.

The most famous pioneers of social housing experiments were the industrialists Charles Fourier, Emile Meunier of chocolate fame, and Jean-Baptiste Godin, manufacturer of popular cast-iron stoves which are still in use today. Charles Fourier, in the wake of the 1848 rebellions, invented a totally new concept in urban housing called the Phalanstère, of which the Cité Napoléon was the earliest modest example. The aim was to unite capital and labour in a social *palais* that offered low-cost, modest but hygienic rooms for workers which opened onto balconies up to several storeys high, built around an enclosed courtyard. The open internal balconies were conceived as indoor streets (*rues galéries*). Fourier included revolutionary facilities, such as central heating for the dwellings as well as for the *galéries*, meeting rooms, library, church, mothers' and children's facilities, laundries, baths, and plans for a communications post, which at the time included a pigeon loft for carrier pigeons to fly from *phalanstère* to *phalanstère*. Each *palais* was to house 1,600 people. Although Fourier, like many utopians, was authoritarian in executing his projects, this dream inspired not only Napoleon but Godin, the builder of the most famous *phalanstère* at Guise and, to some extent, the precursor of modernist social housing concepts. Godin planned and executed ambitious communal housing projects with 1,800 inhabitants between 1858 and 1877 with many innovative features:

- The five-storey blocks were built around glass-covered courtyards.
- Each apartment had two rooms – a kitchen and living room – but the floors were built so that more rooms could be added to an individual dwelling if required (and if affordable).
- The housing developments were surrounded with gardens and recreation areas but were part of and adjacent to the factory and workshops to allow easy communication and access to work.
- Communal facilities, including nurseries and a swimming pool, were built in, with the impressively progressive motive of enabling a full social life for the masses, comparable with the upper classes and of 'freeing women from the bondage of constant domestic chores and responsibility' (Quilliot and Guerrand 1989).

Many enlightened details were thought of. The prevailing winds took the factory smoke away from the housing community (this of course supposed that there was no housing on the other side of the factory!). Hot water from the factory supplied the laundry, baths, and the heated swimming pool. Running water was piped to each balcony and a modern sewerage system was installed. Refuse collection was organised.

Le Familistère de Guise, as the model development was named (see Plate 2.1), was governed by the strong leadership and inspiration of its founder, helped by the exceptional social facilities and a belief in the possibility of co-operative relations between people at close quarters, aided by adequate conditions and communal structure. Unlike the Cité Napoléon, which was the precursor of the ubiquitous rule by *concierges* and *gardiens*, Le Familistère de Guise was organised through a 'self-policing' internal discipline that was powerfully inculcated into the workforce through the exceptional social leadership of Godin. This generous, if patriarchal, role was helped by his technical and industrial genius, making him a millionaire at a very young age (Quilliot and Guerrand 1989). To the modern eye, Le Familistère de Guise looks rather prison-like, but it continued to operate as communal housing on the lines inspired by Godin until the 1960s when the company hit serious financial trouble and was forced to sell the Familistère to a housing association.

Although Le Familistère and similar experiments in utopian workers'

Plate 2.1 Le familistère de Guise ou palais social (The social housing, community and factory development at Guise)

Source: Quilliot and Guerrand 1989: 34–5.

housing were privately funded, Godin himself advocated a form of state sponsorship combined with private backing and investment, called a *société d'économie mixte*. This was the precursor of a much more modern development. He also invented a social security system for his workers which foreshadowed state systems. Godin's ideas of collective provision, social responsibility and personal generosity, were unrealistically utopian. Godin enshrined the idea that renting was the most suitable housing tenure for an urban, industrial workforce, and that dense collective building, incorporating advanced social and educational facilities, would enable a modern and mobile workforce to meet their personal needs and realise their ambitions and potential. History was to prove his vision seriously limited in both concepts of co-operation and communal standards, if only because his principles were later only very partially applied.

Meunier, the rich chocolate manufacturer, built a model village of houses with gardens, incorporating similar facilities to Godin and introducing a form of rent allowances for the very poor.

GARDEN CITIES AND OWNER-OCCUPATION

French industrialists and developers went to Britain to learn about urban housing experiments there. An alternative model of housing for industrial workers was being developed. The aim was to diffuse the urban problem by locating factories in the country and surrounding them with single-family row houses with gardens, of which the workers would become the owners. In the past, each peasant family had owned and controlled through personal effort its own means of production and consumption. The sponsors believed that ownership was the only sure way to prevent the moral, economic and social decline of recent urban dwellers. This alternative utopian vision, backward looking as it was in the French cities of the mid- to late-nineteenth century, inspired a French garden city movement based on the construction of owner-occupied workers' *pavillons*.

The most renowned experiment was probably the pioneer workers' city built in the great industrial area of Mulhouse by the Société Mulhousienne des cités ouvrières between 1853 and 1897. It ended up with 10,000 inhabitants and 1,240 houses. Its construction was facilitated by state grants for the infrastructure development and launching funds, originating from Napoleon III's desire to see decent housing for the masses. After paying an initial deposit, workers paid off the rest in monthly sums over fifteen years, equivalent to under 17 per cent of an average worker's salary. This was a bargain, but lack of funds and absorption with other priorities prevented this ambitious model from being copied very widely. Similar developments in and around Paris quickly moved beyond the reach of manual workers. None the less, the concept of paying for ownership over time and generous state provision for infrastructure needs helped launch owner-occupied garden cities of worker housing.

Social housing ideas advanced slowly as *sociétés anonymes*, registered private companies, were set up to provide housing in Rouen, Lyon, Marseille, and other big cities. In Rouen in 1885, a combination of industrialists and traders invested in a large mansion block built near the cathedral, offering modern flats for rent, with previously unheard-of flush toilets. In Lyon around the same time, bankers and railway builders combined to form the 'Société anonyme des logements économiques', building the first concrete industrialised housing on a large scale. This society quickly became the second largest landlord in Lyon and by far the largest *société anonyme*, with 1,500 units by the end of the century.

The innovative breakthrough in Marseille was the involvement of the state-sponsored Caisse d'Epargne in better housing. The savings of workers in the Caisse were ploughed back into low-cost housing developments through the Société des habitations salubres et à Bon Marché de Marseille, starting from 1889. This last development was the direct precursor of the special funding mechanism for HLM organisations in the twentieth century.

Thus, by the last decade of the nineteenth century, some of the most exciting ideas for housing had been tried and tested, but only feeble efforts were made to extend these ideas beyond the occasional model.

At this stage, France was far behind Germany in the development of non-profit housing, with less than a quarter of the number of housing companies and very little housing built (Quilliot and Guerrand 1989). The total social housing effort over twenty-five years from 1870, the end of the reign of Napoleon III, to 1894, when the state first passed housing legislation, came to twenty-eight *sociétés anonymes*, each building a handful of *pavillons* for owner-occupation, and a total of less than 2,000 flats for rent. The revolt of the Paris Commune in 1871 had only made matters worse for the crowded masses by alienating the state from communal initiatives, enhancing the emphasis on almost unattainable owner-occupation.

BIRTH OF THE SOCIAL HOUSING MOVEMENT, 1889

The Universal Exhibition in Paris celebrated the centenary of the French Revolution in 1889 with the construction of the Eiffel Tower; there the French hosted the first international congress of working-class housing organisations. This congress, however, served to reaffirm the French commitment to no direct intervention by the state or the *communes* in housing provision!

Owner-occupation was roundly acclaimed as the best option, but good quality rented flats were needed too. The Société Française des Habitations à Bon Marché was born of this congress. It was to become the powerful National Union of HLMs thirty-six years later. The name given to social housing organisations was changed from '*cités ouvrières*' – (workers' housing developments) to '*habitations à bon marché*' (low-cost housing) to broaden its scope to include salaried and white-collar workers. The new organisation became the adviser and support to social housing companies, as well as the federal body representing them in their advance towards winning greater state involvement.

GOVERNMENT INTERVENTION

Five years later, in 1894, the first French social housing law, named the Loi Siegfried after its inspirational advocate, was passed amid opposition in the Senate, in spite of its modest aims. It set out to encourage building for owner-occupation by relieving both housing societies and purchasers of some tax burdens. Deep controversy surrounded such a law, which was seen as an infringement of the free rights of the individual, as it attempted to encourage investment by charitable bodies and by the Caisse des Dépôts et Consignations, the state-sponsored savings institution. The aim was to facilitate house-building for lower-income people.

The impact of the law certainly did not justify the fears of creeping socialism. A mere 1,400 houses were built with loans under the scheme by the turn of the century – a meagre 230 homes a year.

The disappointed French housing reformers constantly compared France's poor performance with success elsewhere. George Picot, Siegfried's socialist-leaning ally, berated French performance with the Belgian example – a sixth of the population and five times the investment in low-cost housing! Meanwhile, France's urban population was expanding rapidly, from 30 per cent of the total in 1870 to 44 per cent in 1914. In Paris at least, the only place to go was a crowded slum or the far edge of the city.

THE BEGINNINGS OF A MUNICIPAL ROLE

In 1906 – with conditions virtually untouched – the national census showed around half of all families in major towns occupying only one room each and the vast majority of French people occupying slums. This census marked the beginning of 'slum declarations' and municipal intervention (Quilliot and Guerrand 1989: 65, HLM 1989a: 33). Each *département*, by a new law passed in 1906, had to have a *comité de patronage des habitations à bon marché*, (a support committee to encourage low-cost housing). Local authorities began to help with land and loans by encouraging savings investment.

Loans on favourable terms were made available to low-income purchasers and to non-profit housing societies through the creation of *sociétés anonymes de crédit immobilier* (housing loan companies). A special purpose was to create allotments for working families – individuals or societies would borrow money to buy the land. The *crédit immobilier* finally unleashed 'popular savings for a seductive goal, buying houses and land' (Quilliot and Guerrand 1989: 67). Within twenty years, 137 lending societies had been created.

LOW-COST SOCIAL RENTED HOUSING

The mass of workers could not aspire to ownership. Finally, in 1912, the French legislators passed the Loi Bonnevay, named after the lawyer, *Député* and housing innovator who introduced the concept of public involvement in housing by providing money for the construction of low-cost flats for rent. The law was targeted at families with more than three children; it was to encourage new building and renovation, and to create garden cities and allotments as well as rented flats. It explicitly allowed local authorities and *départements* to create *sociétés d'habitations à bon marché*, low-cost housing organisations (HBMs[1]), known as *offices publics*, because of their public sponsorship. They were to supplement the work of the existing *sociétés anonymes* which were privately sponsored. But the new law did not allow government at any level to become the direct builders or managers of housing. Local authorities had to give this role to independent, legally autonomous HBMs. Bonnevay phrased this crucial division of responsibility in French social housing, between state sponsorship and autonomous legal status, in the strongest terms in advocating municipal intervention. It was a decisive point, both in allowing and encouraging at last municipal responsibility and in avoiding some of the political problems encountered under more direct public housing systems.

We quote Bonnevay in full because his concept of separate but sponsored organs for social housing provided a milestone in housing history:

> It would be possible for us to foresee the establishment of a public body for cheap housing that was autonomous and independent, permanent and disinterested, which would fulfil the role of developer and manager which many hesitate to give to political bodies.
>
> The direct management of working class housing by local authorities and '*départements*' would offer unquestionable threats in certain regions and at certain times to the proper regulation of public affairs and the healthy administration of money. Those who would with demagogy bid up on the conditions of tenure, the favours given to tenants based on the needs of local politics, the weakness of political managers facing the necessary eviction of certain tenants who are bad payers, but also voters, everything which risks being a corrupting element in the administrative or political processes, must be studiously avoided by the law-maker, anxious for the future of democratic government.

The public offices for low-cost housing will avoid this danger while allowing the necessary work of building healthy and economical houses for the less fortunate population of our major cities. But they will only achieve their objective with a double condition that *through real autonomy, they should be truly independent from the pressures of local politics, and that at the same time they remain in close collaboration over courses of action with the local authorities and départements* that have helped create them, so that by considering them their proper creation they will give them and assure them of indispensable resources. [Emphasis added]

(Bonnevay 1912, quoted in Cancellieri *et al.* 1986: 20)

This delicate balancing act was influenced by the level of mistrust of Paris, based on popular revolts against established order three times in a century and the deep implication of Paris authorities in some of these events. The desire to build with public help was firmly married with the need to separate and balance powers and responsibility between autonomous bodies and the state. Achieving this balance proved a constant problem, full of dilemmas and contradictions. But many argue that it eventually gave the organs of social housing greater power than they would have enjoyed as an integral part of the political framework.

Housing societies, whether public or private, were governed by an independent board representing the founding interests, employers, local *communes*, banks, charities and residents. The mayor of the *commune* often became president of the local *office public* and could, if he chose, wield considerable influence. However, the emphasis on legal and financial autonomy protected HBM organisations from serious interference – too much so was the later view (Dubedout 1983).

The first *office public* was created at La Rochelle a year after the 1912 law was passed, and the city of Paris founded its own HBM – 'L'Office Public des HBMs de Paris et de la Seine' – in 1914. But the First World War interrupted progress and devastated much of Europe, especially France.

BETWEEN THE WARS – INACTION

France was beset with economic problems in the period following the First World War and housing took a back seat yet again, while tight rent control made many pre-war owners give up renting property. France's economic problems were at the root of poor performance in housing. A very low birth-rate reduced some of the pressures, but also worried French governments.

The newly created *offices publics* developed slowly at first and, in Paris, the *office public* had only built 2,700 units by 1932. In 1925, based on the original Société Française des HBMs, the national union of Habitations à Loyers Bon Marchés was formed to promote low-cost housing and to help both private and public societies.

Suburban developments expanded more rapidly and fifteen garden suburbs were built around Paris, helped by the growth of railways as well as by the Office Public de La Seine and the co-operative housing societies. The co-operatives marshalled individual savings and investment, combined with special loans from the Crédits Fonciers. Ebenezer Howard's concept of self-contained, co-operatively built garden cities of individual houses and collective facilities beyond the existing city boundaries was the model, this being imported from Britain by French housing reformers (see Part III, 'Britain', pp. 163–238). The garden city movement in Paris was ironically another precursor of the high rise *grands ensembles* of the next epoch. Each *cité* had on average about 800 units, encompassing social facilities as well as housing. However, only 13,250 units were built by non-profit companies in these *cités*.

Alongside socially planned suburban developments, many Parisians, desperate for more space, simply occupied land on the outskirts, putting up huts in fields of mud, with no proper water supply. Tens of thousands settled 'on these ill-equipped plots where for many years most of the dwellings were nothing more than huts. Mud and the scarcity of water were nightmares for the inhabitants' (Lecoin, in Van der Cammen 1988: 70). There were 1.1 million rural immigrants into the Paris region between the wars, but new housing for only half of these was built – fewer than 150,000 dwellings. However, from 1930 onwards the city of Paris itself actually began to shrink in size and the Paris conurbation stopped growing between 1935 and 1950.

In 1928, the Loi Loucheur was passed with a specific housing target for the first time – 200,000 units to be built in five years, with special low-cost loans from the Caisse des Dépôts et Consignations. The aim was to help both owner-occupation, with the construction of houses on the outskirts, and renting, with the construction of flats around the old walls of the city. Five years later, the concessionary grants and heavy interest subsidies were reduced again, curtailing the impact of this first ambitious programme.

But HBM bodies had begun to grow rapidly and 300,000 units were built by social housing companies between 1918 and 1939, half of which ended up individually owned. The Office Public de Paris owned 24,899 rented units by 1936, a very big jump from just over 1,000 twenty years earlier. Only households in real difficulty were housed. An investigation in 1936 showed that the families being rehoused by the Paris HBM in 1936 had on average 3–4 children, that they were exclusively working class, and that 54 per cent came from a single room. Table 2.1 outlines progress in the formation of housing societies.

Table 2.1 Progress in the formation of housing societies in France, 1914–39

	Sociétés anonymes	Sociétés co-operativés	Caisses de crédit immobilier	Offices publics	Total
1914	100	170	3	2	275
1919	182	270	137	14	603
1925	285	429	157	294	1,165
1939	533	437	294	297	1,561

Sources: Quilliot and Guerrand 1989; HLM 1989a.

Twenty-five years of steady growth since the Loi Bonnevay had produced 150,000 rented flats owned by social housing companies, putting French social housing on a par with the British housing association movement, meanwhile, British local authorities had built over 1 million council homes.

France reached the brink of war with many economic problems, and with urban conditions certainly worse than those of her northern neighbours and possibly as bad as they had been at the outbreak of the First World War. But the social housing infrastructure, although relatively underdeveloped, was in place to allow for a post-war effort second to none.

France in 1939 had about 14 million housing units, over half of which were rented from private landlords and with less than 2 per cent from social housing societies; the rest were owner-occupied and mostly rural (estimate based on Ministère de l'Equipement et du Logement 1989).

NOTE

1   HBM: Habitations à Loyers Bon Marchés.

Chapter 3

After the Second World War

A WORSENING SHORTAGE

Housing was not a high priority after the war (Quilliot and Guerrand 1989). Building the economic infrastructure on a weak base came first and it was ten years before France began to start building housing again in earnest.

After the war, France entered a new period of rapid urbanisation. At the end of the war, less than half the French population lived in cities. About 1 million French peasants left the land for cities in the following ten years. By 1954, the proportion of urban dwellers had risen steeply to nearly 60 per cent. It went on rising into the 1980s, by which time 74 per cent of the population lived in urban areas (defined as towns over 5,000) – still a much lower proportion than other northern industrialised countries. It was the overcrowding and shanty building resulting from this rural exodus that finally made a mass building programme seem inevitable (see Plate 3.1).

At the same time, the previously sluggish birth-rate rose steeply and France's total population expanded rapidly. By the early 1960s,[1] it was predicted that the population would almost double by the year 2000 as a result of the post-war baby boom, producing a second large generation before then. The Paris region (Ile de France), which had been stagnant for nearly thirty years, was expected to double in size from around 8 million to 16 million (Van der Cammen 1988). in 30 yrs.

By 1954, France was experiencing a housing shortage of gigantic proportions: 14 million people lived in overcrowded accommodation; half a million families lived in hotels or furnished rooms; several hundred thousand lived in makeshift shanty settlements; and about 10,000 families were squatting. The majority of the French population was in grossly inadequate accommodation.

Investment in housing and house-building had been so slow for so long, reflecting the very weak economy and slow pre-war urbanisation, that the average French family spent less than 2 per cent of the family budget on rents in 1948 (HLM 1989a: 55). Tight rent controls in the face of serious

Plate 3.1 A post-war shanty town at La Courneuve outside Paris

Source: HLM 1989a

post-war shortage resulted directly in steeply deteriorating conditions. In 1948, rents were regulated upwards to encourage repair and investment and, for the first time, the *allocation de logement* (housing allowance) was introduced to make up to low-income families the difference between cost and affordability. This helped landlords to make good the deficits in superficial repairs, but it did not address structural problems or the serious lack of investment. Over time, many landlords of pre-1948 property sold up.

The most radical change in 1948 was the deregulation of rents for *new* properties. This helped create new units. By 1990, less than 10 per cent of private rents were regulated. Rented units were still disappearing very rapidly.

IMMIGRATION

Problems were greatly intensified by new immigration from overseas to swell the urban workforce, which accelerated from 1946 onwards. There were already nearly 2 million immigrants by 1954 (Emms 1990). The ending of the Algerian war in 1962 suddenly led to 1 million returning French settlers – *pieds noirs* – looking for jobs and housing, but long before then Algerians were erecting cardboard and corrugated iron shanties around Marseille and Paris. Ties between Algeria and France were strong and a cheap night boat-ride across the Mediterranean brought literally millions of workers. Growing demands for unskilled labour were ironically partly fuelled by the need to build! This was a self-generating circle which continued into the late 1970s as more people arrived to man the vast building sites that ended up housing much of the construction workforce.

The ambition to build a modern, technically advanced urban and industrial infrastructure, the huge national deficit in dwellings, and urbanisation which was more rapid and later than that of Germany or Britain, all led to an exaggerated emphasis on mass housing in France for thirty years after the war.

RECONSTRUCTION

By 1953, only 325,000 units had been built since the war. The Minister of Reconstruction, Eugène Claudius Petit, made an impassioned statement that the post-war aim of 'rebuilding France' was like 'a country facing the future with its eyes turned towards the past, from which it would not escape'. He said that rather than *rebuild*, France needed to *build* 14 million new units in the next twenty years (HLM 1989a: 158). This would roughly match the number of households in acute housing need.

An initial target of 250,000 new units a year was set. An important new tax exclusively for house-building – the 1 per cent employers' tax[2] was imposed on all firms with ten or more people on the payroll in order to generate a major fund for house-building. This employers' tax, the 1 per cent Patronal as it was called, became an essential source of money for the privately sponsored *sociétés anonymes* and for low-cost, subsidised owner-occupation. Employers who were liable for the '1 per cent Patronal' contributed 1 per cent of their payroll costs to social housing. This tax is still in place but now at only 0.5 per cent. CILs,[3] comprising employers and employees but operating independently of the state, invested the tax

Table 3.1 Expansion in construction of new housing units in France, 1948–55

Year	1948	1949	1950	1951	1952	1953	1954	1955
No. of units constructed	40,000	52,000	71,000	77,000	84,000	115,000	162,000	210,000

Source: Institut Français d'Architecture 1985.

revenues in HLMs, building and renovating housing directly or making individual loans to employees. Many private HLMs (*sociétés anonymes*) are controlled by CILs because they have received such significant investment money through the employers' tax.

Only half the building target was met in the first year but production was expanding as Table 3.1 shows. It leapt up after the ending of the Indo-China war in 1953.

The social housing organisations, renamed *habitations à loyer modéré*, dwellings for *moderate* rent to re-emphasise their wide role, were drawn into the new programme. The proportion of new housing that was built directly by HLMs was still small, but growing as Table 3.2 shows.

Table 3.2 New housing units constructed in France, 1950–5 (proportions built by HLMs and by other agencies)

	Total	HLMs	Reconstruction	Private-aided	Non-aided
1950	71,000	12,000	30,000	–	29,000
1953	115,000	20,000	40,000	35,000	20,000
1955	210,000	50,000	–	160,000	–

Source: Institut Français d'Architecture 1985.

ABBE PIERRE AND THE FOURTH WORLD

Just as the housing programme was getting under way, the housing crisis overtook the French government in the shape of a homeless child, a crusading priest, and a shack. Abbé Pierre, a Catholic monk and former elected politician (*député*) dedicated to the 'fourth world' as he called the marginal people living on the edge of rich societies, unleashed his full moral weight on an evasive, if not complacent, public conscience. According to the French census of 1954, in Paris alone about 160,000 families were living in 'unconventional' dwellings, including hotel rooms, maids' rooms, and temporary shelters (*abris de fortune*) (Ministère de l'Equipement et du Logement 1989). In January 1954, a child, staying in a makeshift shelter put up by Abbé Pierre, died of exposure.

The state launched an emergency programme – *cités d'urgence* (emergency prefabricated housing), *cités de transit* (temporary housing for newcomers), and *dortoirs* (dormitories) for migrant single workers, were all put up with state sponsorship. Around the large estates of outer Paris, Marseille, and Calais there are still occupied pockets of rudimentary barrack- and hut-like dwellings which were put up by the *offices publics* as *cités de transit, cités d'urgence* and *dortoirs* to cope with the acute post-war housing crisis. A special housing society – La SONACOTRAL[4] – was set up to build and manage accommodation for single immigrant workers who were being sucked in to help run the booming economy. By 1959, 320,000 units a year were being built, of which about a quarter were HLM dwellings.

THE MASS HOUSING ERA

France entered the mass housing era with an enormous deficit, huge political momentum and the pressure of shanty towns ringing her still preserved and much visited capital. Nanterre St Denis and Champigny were famous examples.

Big advances in building techniques, much smaller families, very rapid post-war urbanisation, and acute demand – not just for poor families but for households from a large cross-section of French society – influenced the type of building contemplated when France, relatively late, eventually began to develop a major housing programme. The pre-war pattern of house-building for owner-occupation and flat-building for rent prevailed. It seemed clear that flats would be cheaper and easier to produce in volume. Unquestioningly, the social housing programme became linked to industrialised flat-building.

In continental cities, urban living meant '*un appartement*' (a flat), not a house – unlike in Britain where even Londoners traditionally aspired to a house.

The need for 14 million new units involved the creation of large, dense, new housing areas in outer urban spaces around the edges of towns, between existing suburbs, or on industrial wastelands; technical ingenuity helped maximise size, speed, and density for minimum cost. French industrialised building became virtually synonymous with progress – the Camus system was vigorously marketed in eastern Europe while the Balancey system even showed up in rural Ireland! The vision of a modern France consisting of many millions of cellular housing units neatly packed into giant boxes had great allure, as the majority of French households were crowded into chaotic slums and many of them were forced into *ad hoc* solutions – hotel rooms or their equivalent for half a million, squatting and shanty towns for many others.

LE CORBUSIER

Le Corbusier has an over-sized place in French mass housing. The French did build a little mass housing designed by him, L'Unité in Marseille being the most famous of only four notable examples. But his schemes were greeted with vociferous opposition wherever he tried to plan them. They were rejected as inhuman, while he propounded them as a way of putting people back in contact with nature! However, many energetic young architects embraced and helped emplement his ideas on an overwhelming scale.

But he is wrongly famed for inventing streets in the sky. They originated with the *rues galéries* a century earlier. Nor was the idea of high-density building new; in fact, traditional Parisian streets often had higher densities than oppressive-looking high-rise. Urban patterns from medieval times had been extremely dense. Le Corbusier brought these ideas into play in dramatic contrast to the goal of single-family houses, which was undoubtedly the most popular housing form for most of the pre- and post-war period.

His ideas were promoted in an explosive way because the war had generated unprecedented support for state planning. His proposition of an 'urbanism free of the constraints of a paralysing past' could not have been more apposite to the situation facing France. His view echoed that of Eugène Claude-Petit, the post-war Minister of Reconstruction. His love of innovation rode roughshod over the consumer who 'in architecture objects to everything' (HLM 1989a: 58). This imposing view fitted precisely with the paternalistic French state, which by tradition imposed ambitious plans on the people in the name of progress and change.

The crucial modernist architectural concept, the 'machine for living' of Le Corbusier, encouraged the development of repetitive, monotonous, monochrome estates, although his ideas at the time hardly seemed monochrome. Many blocks were erected virtually without architects, so mechanised was their design and construction. The idea of mass-building made redundant the architectural input and facilitated the development of high-rise housing.

None the less, particularly in France, the legacy of high-rise housing cannot be separated from its most articulate and arrogant inspirer (Wolfe 1991).

ZONES A URBANISATION PRIORITAIRE

The construction of *grands ensembles*, outer estates of high-rise flats,[5] took off under intense pressure. The *cités de transit* were only a short-term stop-gap while ambitious modern developments were launched and completed. They were to provide housing for displaced residents until urban renewal bore fruit. In 1958 a new planning device was introduced – ZUPs[6] – which gave government the power to acquire land and establish large-scale housing developments through long-term favourable loans to HLM organisations, based on maximising the number of units on each site. The minimum ZUP was 500 units. The Caisse des Dépôts et Consignations lent money at very low interest. One hundred and forty ZUPs were declared by the government, many around Paris; all of them were around major urban areas. Special bodies, *sociétés d'économie mixte* with a blend of private and public investment, were established to develop giant estates of around 5,000 units each. Mantes-la-Jolie outside Paris had 8,500.

The scale of each ZUP led to evermore ambitious innovation. Railway lines – *chemins de grues* – were laid for construction cranes to move easily from block to block to facilitate scale, density and speed. When once this construction infrastructure was in place, vast scale was rationalised and the pace of building accelerated (see Plate 3.2).

Traditional Parisian blocks are six storeys high. In the new developments, at least eight storeys were commonplace and, from 1965, many were twice that. The blocks were massive: a single block could be 400 metres long and contain 900 units. The cost per unit fell sharply in direct relation to the number of units per site, and the density and the scale of the

Plate 3.2 Crane on specially laid railway tracks to build La Courneuve

Source: Quilliot and Guerrand 1989: 135

blocks. The average size of the French ZUPs was 5,300 dwellings. The 140 ZUPs contained three-quarters of a million dwellings.

Within ten years, the imposing scale and oppressive uniformity had lost political support. The strongly regimented imposition of a huge new industrialised housing development of 20,000 or more inhabitants was often opposed by the small semi-rural *communes* onto which they were grafted. The policy was modified to give more flexibility and adaptability but the wave of mass building had passed its zenith.

FRENCH SUCCESS IN MASS HOUSING

Three big attractions fuelled the mass approach to housing in France – speed, number of units, and cost. The French were highly successful on all three counts compared with other countries because of the way they applied the industrialised methods.

Speed

The time taken to produce a dwelling in France dropped from an average of 3,500 hours – nearly two man-years – in 1950 to 1,250 – under seven months – in 1960, reducing construction time per unit by two-thirds. On some sites, production time was reduced to 800 hours per unit (Institut Français d'Architecture 1985) – twenty weeks per unit. Three elements made such a huge gain possible: the use of large green-field sites meant that few impediments existed to the rapid mounting of a production factory on site; the size of developments meant that once the industrial machinery was in place, the multiplication of units became relatively easy, efficient and economic; and lack of developed urban planning prevented complex requirements being imposed on developers (Van der Cammen 1988).

Patrick Dunleavy's detailed analysis of British mass housing shows why the British failed to achieve similar success, most importantly because of the lack of suburban land, tight planning (including a strictly enforced green belt), and a highly complex slum clearance programme (Dunleavy 1981). For all that France suffered from an acute housing shortage, she did not have a serious land problem. Her most acute problems were to clear the shanty towns and replace them with modern housing, and to upgrade slums and relieve overcrowding. She did not enforce a green belt and allowed it to be largely eroded around Paris. She did not embark on a slum clearance programme because the shortages were too severe until the 1970s when the notion of mass clearance was losing favour. She only developed strict town planning regulations after the problems of the ZUPs had become all too apparent. By then the worst damage had been done (Van der Cammen 1988).

Two other elements favoured the speed of mass production. While there

Figure 3.1 Distribution of HLMs

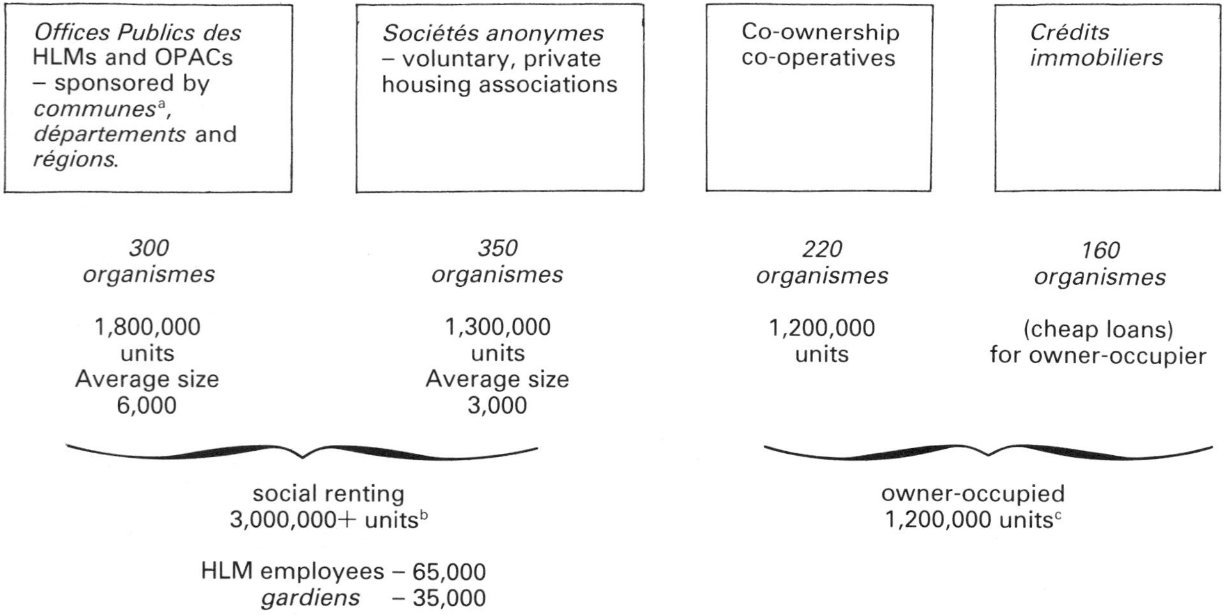

Offices Publics des HLMs and OPACs – sponsored by *communes*[a], *départements* and *régions*.	*Sociétés anonymes* – voluntary, private housing associations	Co-ownership co-operatives	*Crédits immobiliers*
300 organismes	*350 organismes*	*220 organismes*	*160 organismes*
1,800,000 units Average size 6,000	1,300,000 units Average size 3,000	1,200,000 units	(cheap loans) for owner-occupier

social renting 3,000,000+ units[b]

owner-occupied 1,200,000 units[c]

HLM employees – 65,000
gardiens – 35,000

Notes: [a] OPACs average 17,000 units each.
[b] There are fifteen regional federations of HLM associations, each with an office, giving individual HLMs financial advice and helping channel money for building and renovation.
[c] These organisations were historically important but today play a very limited role. Many of the early co-ownership societies have ceased to exist.

had been a long tradition of working class and rural owner-occupation through the construction of single-family houses, the vast mass of urban dwellers lived in tiny flats or single rooms. Even the majority of affluent middle-class urban households lived in flats. Over 90 per cent of mass housing was built in flats in high-rise tower blocks or in *barres* (medium-rise long blocks). This maximised the potential for industrialised techniques. Secondly, the HLM organisations proved to be ready and adaptable vehicles, and the direct financing structure through the national savings bank – the Caisse des Dépôts – simplified funding (see Figure 4.1). No complex local political processes were required. There was a rapid cascade effect. But it is important to remember that although ZUPs were built almost entirely in flats and predominantly by HLMs, 10 per cent or more of the units were sold to owner-occupiers, some with special subsidies. Many of the biggest ZUPs, like Mantes-la-Jolie and Les Minguettes, have large owner-occupied areas or blocks.

However, the scale of the ZUPs meant that it took fifteen or twenty years to build an estate of several thousand units. Speed per dwelling or per block did not necessarily mean that the planned area was built up and completed in a single phase. In fact, early unpopularity often stemmed from 'building site' conditions and incomplete facilities.

Number of units

The total numbers of housing units produced between 1960 and 1980 exceeded 9 million – nearly half a million units a year. HLMs, using almost exclusively industrialised techniques in that period, built over 2.5 million flats: a huge expansion on their performance in the first ten years after the war when they built only 110,000 social units in total. This represented a complete transformation of their minimal pre-war role. The fact that the construction industry was poorly developed in 1950 was possibly an advantage, facilitating the quick adoption of non-traditional techniques on a wide scale. There was little innate resistance over styles and methods since France faced an exploding urban population desperate for modern housing. Table 3.3 shows the rapid build-up of volume production by HLMs from 1956 onwards, parallelled by massive private building programmes.

French housing developments between 1960 and 1980 were visually and psychologically dominated by the pace, scale and volume of HLM developments. Ninety-five per cent of all HLMs were built in urban areas; 75 per cent of HLM estates had 800 dwellings or more. The outer estates normally contained at least 2,000 dwellings, a few reaching 10,000. Seventy-five per cent of all HLM units were in blocks of twenty or more units. Virtually the entire stock was made up of flats (Emms 1990).

Table 3.3 Volume production of modern housing units by HLM organisations in France, 1945–86

	Total	Numbers completed by HLMs	Numbers completed privately
Pre-1945	–	148,000	–
1945–50	199,100	12,000	187,000
1951–55	607,050	97,000	510,050
1956–60	1,435,780	324,000	1,111,780
1961–65	1,741,560	410,000	1,331,560
1966–70	2,131,267	625,000	1,506,267
1971–75	2,537,470	682,000	1,855,470
1976–80	2,126,700	394,000	1,732,700
1981–85	1,638,493	272,000	1,366,493
1986	300,000	66,000	234,000

Sources: Derived from Institut Français d'Architecture 1985; Ministère de l'Equipement et du Logement – Direction de la Construction 1989.

Cost

The cost of construction reflected the speed and volume. While the investment in massive cranes, site factories and metal tracks to move equipment round the sites was enormous, the pay-off as the number of units multiplied was equally significant. The initial outlay, spread over 4,000 units at La Courneuve in outer Paris or over 10,000 units at Les Minguettes in outer Lyon, reduced costs. Mass building techniques did for French housing what automated conveyor belts did for mass manufacturing everywhere.

A major problem with the favourable cost calculations for mass housing was that cost was only estimated in terms of immediate outlay. The scale and location of the *grands ensembles* dictated the need for comprehensive new urban facilities, including roads, transport services, schools, shops, health services and meeting places. The poorer outer *communes*, where the *grands ensembles* were concentrated, could not always provide for other needs in time. Commerce was often poorly developed in expensive premises and sometimes failed. As a result, the *grands ensembles* and ZUPs did not provide the comprehensive 'facilities for a full life' that had been the dream of the pioneers of social housing. This delayed the 'settling down' process, often disastrously, by causing an early population exodus in the 1970s and 1980s that generated a new instability. The long-term, hidden social costs did not form part of the original equation.

MANAGEMENT COSTS

The overwhelming monotony of the industrial housing production in France and the sheer scale of the blocks and estates had a quickly depressing effect on demand, leading by the early 1970s to slower take-up than had been foreseen. The drop in demand affected financial calculations adversely.

The biggest cost oversight was the cost of management, a problem still only partially overcome (Cancellieri *et al.* 1986). It was quickly discovered that recruiting *gardiens* or caretakers for large outer estates was more difficult than recruiting concierges for far more dilapidated property in inner areas. Repair requirements mounted quickly, contrary to expectations that the new materials would last, would remain attractive, and would function efficiently for a long time. Concrete had seemed indestructible.

As the programme advanced, the less visible costs rose too. The better sites were used up and transport links proved grossly inadequate for mobile workers. Increasing numbers of isolated, economically vulnerable households became the candidates for the least favoured *grands ensembles*. These problems were visible by the early 1970s. A conference was called in 1971 to discuss the problem. It preceded the clearing of the last *bidonville* (shanty town) in 1974. From 1972 onwards, no more giant *grands ensembles* were planned and new developments were limited to 2,000 units – still very large, concentrated estates. Anything bigger was banned by the Ministry of Construction. *Zones d'aménagement concerté* (areas of designated upgrading) were introduced to replace ZUPs with the aim of greater variety and flexibility. The tight grip of state-driven crane tracks was coming to an end.

THE NEW CRISIS

By 1975, France's supply of dwellings was roughly in balance with the number of households. France boasted her success 'without knowing that ten years later she would be confronted with ghettos, with financial problems, linkage problems and management problems' (HLM 1989a). For the next generation would want choice and quality and France's housing success in the 1960s and 1970s would cause great social and urban strains in the 1980s. The problem of 'exclusion' would dominate political debates (Zeldin 1991). ZUPs had broken the urban mould but their scale and oppressive architecture belied a curious lack of city life on the urban periphery, separating the often under-occupied estates from both town and countryside.

Thus, in the course of fifteen years of rapid progress France had managed to solve the crudest shortage in terms of numbers of dwellings, the inertia that had caused a build-up of slums and homelessness was gone,

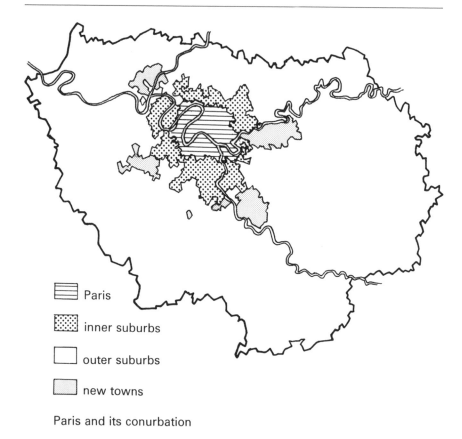

Paris

inner suburbs

outer suburbs

new towns

Paris and its conurbation

Figure 3.2 Map showing new towns around Paris, 1991

Source: Van der Cammen 1988

and the technical barriers to rapid, low-cost housing production had been overcome. The housing momentum was so great that France quickly responded to the new housing crisis in the troubled *grands ensembles*. In 1977, the French government launched the first of evermore comprehensive rescue programmes for the *grands ensembles* in difficulty under the name of 'Habitat et Vie Sociale'.

BACK TO HOUSES

The shift away from flat-building became significant from the 1970s onwards. Whereas in 1970 two-thirds of all building was in the form of flats, by 1980 two-thirds was in the form of houses. The fact that HLM landlords were building about a quarter of their new stock in the form of houses by the late 1980s reflected recognition of their popularity and an attempt to attract and hold customers (Quilliot and Guerrand 1989).

In 1965, the French government adopted a new towns policy, a move away from the monolithic estates, in an attempt to rationalise land use while building more socially mixed new communities. Eight new towns were planned between 1965 and 1972 – four within reach of Paris – and were executed in the 1970s and early 1980s to reduce regional pressures. These mixed developments took root more easily than the *grands ensembles* (see Figure 3.2). They were on a scale that encouraged employment, amenities, a mix of owning and renting, and they have generally been popular.

The new Socialist government elected in 1981 decided to promote social housing again, as well as to renovate the *grands ensembles*. The decline of social housing was too rapid, relative to private investment. As a result, 80,000 HLM units were built in 1983; by 1990 the number was running at about 50,000 a year. The emphasis had completely changed. Current developments are smaller, low-rise and often part of an urban renovation programme.

NOTES

1   Schema directeur d'aménagement et d'urbanisme de la Région Parisienne, 1963–5.
2   One per cent employers' tax is collected by a body called the Comité inter-professionnel pour le logement which represents all employers paying the 1 per cent tax. It can be used to reserve units in HLMs for employers; to build employers' housing directly; or it can be offered as loans directly to individual employees for home ownership. The tax is no longer 1 per cent, although it is often referred to as such, but 0.5 per cent. One-third is now used to help pay for APL.
3   CILs: Comités inter-professionnels pour le logement.
4   La SONACOTRAL stands for 'La Société nationale pour la construction de logements pour travailleurs algériens qui obvint ultèrement', which translated means 'the national society for the provision of housing for Algerian workers who arrived recently'.
5   In France, high rise is taken to mean ten or more storeys. However, cranes and other heavy mechanical equipment were required for industrialised building of blocks with six or more storeys (Dunleavy 1981). This would include almost all French mass housing. High-rise towers (*tours*) over ten storeys became fashionable from 1965.
6   ZUPs: *Zones à urbanisation prioritaire.*

Chapter 4

Social landlords

THE ROLE OF THE HOUSING COMPANIES

The social housing companies or associations were active partners in the mass housing boom but they were very much the tools of a bigger process. State-led development made most of the running. The instruments of national saving and of state-guaranteed finance, instituted under Napoleon III a century earlier, facilitated the funding of the vast building programme. The chronic shortage and the pace of change attracted powerful entrepreneurs and industrialists, often through the 1 per cent tax.

The creation of *offices publics* in all major French towns and cities had given the French government, local authorities and *départements* a vehicle for mass development. New powerful regional bodies – OPACs,[1] created in 1973 – accelerated further publicly sponsored social building. They had great operational scope; buying land, developing directly, and sometimes taking over local HLMs. Generally they were more financially secure, paid higher wages and were more versatile. Private *sociétés anonymes* were also used wherever possible. But 58 per cent of social housing was built by the local and regional *offices publics*; while 42 per cent was built by the *sociétés anonymes*. After the financial reforms of 1977, the private societies expanded more rapidly.

By 1990, there were 650 HLM landlords, over half of which were *sociétés anonymes*. These had on average a smaller stock with 3,000 units each, although the largest, la SCIC, had 150,000 rented properties. The publicly sponsored HLMs had on average 6,000 units each, but the regional OPACs were often very large, with an average of 17,000 units each. The large ZUPs were often owned and managed by several HLMs, public and private (see Figure 3.1).

THE FUNDING OF HLM ASSOCIATIONS

Two mechanisms made the funding of French social housing relatively straightforward. The first was the role of the Caisse des Dépôts et Consignations, founded in 1816, in making low-interest loans for building and

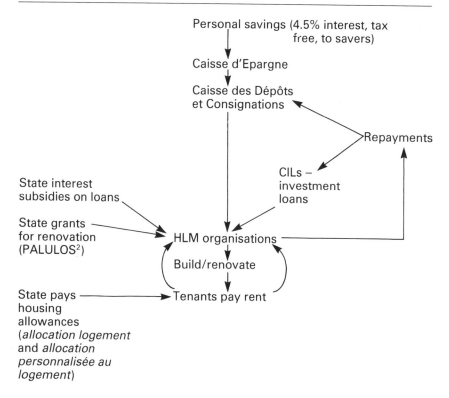

Personal savings (4.5% interest, tax free, to savers)

Caisse d'Epargne

Caisse des Dépôts et Consignations

Repayments

CILs – investment loans

State interest subsidies on loans

State grants for renovation (PALULOS²)

HLM organisations

Build/renovate

State pays housing allowances (*allocation logement* and *allocation personnalisée au logement*)

Tenants pay rent

The balanced revenue budget of an HLM organisation		
Income		*Expenditure*
Rents, including: Housing allowances Local taxes Grants Concessionary loans		– Management – Maintenance – Planned maintenance – Loan repayment – Contribution to new construction (if funds available)

Figure 4.1 Funding of HLM construction and improvement

renovating social housing. The bulk of building money came from this *independent* but state-guaranteed source. The second was the balanced revenue budget where rent income had to equal all expenditure, including loan repayments, and all management and maintenance costs. About 40 per cent of rent income now comes through housing allowances. However, rent increases were not sufficient to fund the renovation of the high-rise stock and special funds were made available for this (see Figure 4.1). Improvements were always reflected in some level of rent increase.

Public HLMs (OPHLMs[3] and OPACs) depended on the state for initial subsidies and grants for building; they borrowed money through the HLM Bank for the difference. Many voluntary HLMs (*sociétés anonymes*) received help through the employers' tax of '1 per cent' (now only 0.5 per cent of the wage roll), charity, and banks, as well as state-supported loans and grants.

The principle of independence, involving a self-contained rent account, an internally generated management and maintenance budget, and balanced annual accounts, was central to the viability of the HLM organisations. It conferred on HLM organisations a level of decision-making authority and incentive to performance that disciplined their organisation. For example, disrepair made higher rents for better-off households unacceptable; exclusive concentrations of very disadvantaged households in a block or estate made that area unlettable to better-off households; low standards of caretaking made high-rise flats unmanageable; rent arrears bankrupted the housing company; empty units did the same. HLMs had to manage within the limits of financial solvency year by year. This salutary discipline made for commitment, even in very difficult areas.

Rents rose steeply, with higher costs and greater management problems. In 1991, new HLM rents were in the region of £45 a week in Paris. Overall rents represented 16 per cent of average income, but there was a variation of 1:8 between the cheap rural and expensive Parisian rents (Schaefer 1992). Management and maintenance allowances – a target spending level set through negotiation – were £750 a year in 1989. The level of rent arrears nationally was 5 per cent of the rent due, and 7 per cent of tenants were in arrears at any one time (1989). Rent rises had a serious impact on tenants, as the most generous housing allowance only covered two-thirds of the cost; the tenant had to cover the other third. So although the allowances helped access for the poor, steep increases could drive them out.

ACCESS

Each publicly sponsored HLM organisation had a *commission d'attribution* (allocations committee), responsible for ensuring that bodies with nomination rights took them up and used them as intended, determining local priorities, ensuring fairness, preventing stigmatisation, segregation and 'dumping', supervising the decisions of the managers and considering their recommendations. In practice the whole process of allocations in HLMs was the most controversial aspect of their work, leading to certain estates becoming stigmatised, hard to let, and marginal. The allocations committees reflected the composition of the board of the HLM and represented the local authority, the *préfect*, local employers and other bodies involved in social housing, such as the Caisse d'Allocations Familiales, the French state agency that supports families with children. The allocations committee was a crucial instrument. Where allocations were a sensitive or difficult area, where there were racial tensions, where the HLM and the *commune* did not see eye to eye, they could be very controversial. But they provided a vital forum for ensuring that the nomination system worked (see pp. 57–9). The local *maire* was most likely to play an active role in the allocations committee, as it could affect the local constituency (for example, as at Mantes-la-Jolie). The bulk of lettings in high demand areas, particularly the Paris region, went to households nominated by employers, by the *préfect* or by the local authority. Factors taken into account in the nomination system were overcrowding, dilapidation, numbers of children, income, local links, and other kinds of need.

Private societies allocated more freely though often their funders could nominate; this applied particularly in the case of employers paying the 1 per cent tax. Since 1991, *sociétés anonymes* must play a bigger social role (Loi d'orientation pour la Ville, 1991).

Employers' nominations

Employers had the right to nominate their workforce to flats – up to 30 per cent of lettings depending on the employer's contribution which came from the 1 per cent employers' tax. This was a great asset to employers, giving them control over an important resource that influenced their development. In many *grands ensembles*, segregation was more pronounced because companies used nomination rights to channel particular groups of workers to particular estates.

It could complicate management if used rigidly. At Mantes-la-Jolie, the car factories, such as Renault, had nomination rights to a large number of flats and often wanted to keep their full quota, even if they could not use them all at a particular time. This led to empty flats and disputes with the *maire* and the HLM companies. It also led to heavy concentrations of foreign workers. Economic retrenchment affected immigrants more quickly and this exacerbated social tensions on some large estates where redundant workers were concentrated.

Préfecture nominations

The *préfect* had the duty to nominate homeless, seriously overcrowded, and very needy households when they applied to his office for help with rehousing. The *préfect* could nominate up to three applicants for each flat. Only if very strong reasons existed would an HLM be able to turn down all three. The *préfect* could nominate up to 30 per cent of HLM dwellings in areas of high demand; 5 per cent of these were for state employees.

Commune nominations

Where the local authority provided land, funding, or guarantees for loans, it could nominate up to 20 per cent as well. The local authority, since the 1980s, has developed measures to encourage priority for local residents or workers – a practice that is not strictly legal. This resulted from some more-working-class local authorities which had large HLM developments being used as 'dumping grounds' for the adjoining middle-class areas where HLMs had tended not to be built. The mayors of these working-class *communes*, often Communists or Socialists, fought very hard to retain their better-paid workers within the HLM areas because this made them more viable and more attractive. The crude political reality was that skilled French workers were more reliable Socialist or Communist voters than incoming Algerians. Obvious, though often unspoken, it was also an attempt to help existing residents, for whom the *commune* already had responsibility, and to force better-off *communes* to play a more significant social role. In 1991, a new law was passed to force more affluent *communes* to build a proportion of social housing if less than 15 per cent of their housing belonged to HLMs. The aim was to redistribute the burden of low-income households – 'La Loi d'orientation pour la Ville'.

Social mix

HLMs aimed for a balanced social mix – an aim reflected in their title and constantly articulated. The income limit set by government allowed over half of the French population to qualify for HLM property. Applicants had to demonstrate the ability to pay their share of the rent, as well as demonstrating need. The policy of achieving a social mix often ran counter to the growing pressure to house the most needy. There were conflicting aims in French housing policy – to encourage HLMs to cater for stable, middle-income households and to house the most needy. In practice, over time, HLMs housed ever-greater proportions of lower-paid and more disadvantaged households. Between 1978 and 1984, the proportion of tenants on incomes below the median increased from 48 per cent to 59 per cent. The proportion of unskilled workers was double that in the rest of the stock (Curci 1988).

Waiting list

Each HLM organisation kept its own waiting list. The HLM manager vetted all applicants and nominees. Nominees took priority up to the limit set. In the Paris region and other high demand areas, this might be virtually 100 per cent of lettings. But any individual could be turned down if there were reasonable grounds; the HLM retained the final control over each individual letting. Because of the large private-rented sector, with two private flats for each HLM flat, there was no question of the HLM being the only option. This made the screening and selection of tenants common. In spite of this, HLM bodies accepted growing numbers of more and more needy people, as more middle-income households moved into owner-occupation (Délégation Interministérielle à la Ville 1991).

On estates where there were lettings difficulties, the HLM manager might try even harder to select tenants or to screen out people who might cause problems, in an attempt to prevent spiralling conditions or to upgrade the area in order to attract more applicants. This approach, common for example in Marseille, did not work, as empty dwellings proved a great deterrent to lettings and caused the greatest blight.

Droit au Logement

The recently passed 'Droit au Logement' (May 1990) aims to ensure access to accommodation for all households resident in France, particularly previously excluded households. While not imposing direct obligations on HLM companies, it gives applicants a right to housing, a right which the allocations committees of HLM companies will be under increasing pressure to respond to, as far as they are able. The new law has been made possible by the more adequate supply of housing and made necessary by the loss of private-rented housing. It is an attempt to ensure access at the bottom, recognising the need to force French social housing bodies, not just HLMs but also local authorities, to help minority households and other marginal households.

The Right to Housing Law offers incentives to private landlords to house needy families, rather than forcing HLMs to take exclusive responsibility. Each *département* throughout France is now obliged to draw up a housing plan for disadvantaged people to ensure that they are catered for. Because the private-rented sector, traditionally very important, is now shrinking so fast, with the disappearance of 100,000 units in 1991 alone, this new law will keep social housing in the political arena for the foreseeable future, although it will not necessarily guarantee access for the poorest.

REHOUSING PROBLEMS IN HLM ORGANISATIONS

HLM bodies were set up to help people of limited means (*peu fortunés*). Ideal vehicles that they proved to be for attacking the post-war housing crisis, the companies were not ready for the huge task they faced in managing the development of such a complex stock or in running the huge estates after construction was completed. They were unprepared for the social responsibilities they encountered.

Table 4.1 Proportions of disadvantaged households in the population of HLMs in France, 1989

5-person households	22%
6-person households	26%
1-parent families	27%
Immigrants	25%
French from overseas provinces	41%
French households overall	17%

Source: INSEE 1989.

Because of the deficit in dwellings up till the early 1970s, affecting a huge number of French urban dwellers, better-off households in severe housing need tended to get housed. But as the supply of dwellings expanded, access became easier for poorer groups, greatly facilitated by the introduction of more generous housing allowances in the late 1970s. Even so, only families who could pay their share of the rent could get rehoused.

At the same time, the emergency housing programme – *cités de transit* – and the clearing of *bidonvilles* led to HLMs playing a growing role in housing the most marginal groups. The new estates were often built on the same or adjacent sites. Over time, rehousing people from temporary shelters became inevitable, although this process is still incomplete.

As the popularity of certain *grands ensembles* declined, an exodus got under way, making accelerated access for previously excluded groups possible. Over the 1970s, the social and economic profile of HLMs changed radically. There was a rapid increase in the numbers eligible for housing allowances, rising to 50 per cent by the late 1980s. HLMs came to house disproportionate numbers of large households, of one-parent families, of immigrants, and of French citizens from overseas provinces (see Table 4.1).

There is almost double the concentration of one-parent families and of large families in HLMs compared with the French housing stock as a whole. Also, by 1989, HLMs housed double the proportion of North Africans compared with the private rented sector, as Table 4.2 shows.

For much of the post-war period, more articulate, more steadily

Table 4.2 Proportion of North Africans in rented housing in France, 1989

	HLM dwellings	Other renting
French	88%	90%
North African	6%	2.5%

Source: Ministère de l'Equipement et du Logement – direction de la construction 1989.

employed, more qualified and therefore more economically secure house-holds gained access to social housing ahead of the most needy because they combined very poor housing conditions with moderate incomes and some leverage in the system. But as volume of production continued, so the better-off moved out. The shift was greatly influenced by developments in the private-rented sector and owner-occupation in the post-war period.

Before going on to look at the social housing crisis of the 1980s, we will briefly outline the post-war role of private landlords and the spread of owner-occupation.

NOTES

1   OPACs: Offices Publics d'Aménagement et de Construction.
2   PALULOS: Prime à l'Amélioration de Logements à Usage Locatif et à Occupation Sociale.
3   OPHLM: Office Public d'HLM.

Chapter 5

Private housing

PRIVATE-RENTED SECTOR

At the end of the Second World War, a majority of French households rented from private landlords, a very high proportion of their properties dating from before the First World War. Because of the acute housing shortage, rent controls were regarded as essential. But an active black market flourished and as people poured into the cities from the land, landlords subdivided units, responding to demand and doubling income, while official rents remained low. In spite of the black market, rent payments represented under one-fiftieth of family budgets and the French were renowned for not caring about their housing conditions (Quilliot and Guerrand 1989).

Conditions in private renting

Disrepair was very serious in the older stock. Parts of the private-rented sector earned ill-repute in the years of most acute shortage, when *marchands de sommeil* – literally 'traders in sleep' – sold bed spaces in eight-hour stretches to unfortunate workers unable to afford, or find more than, a bed for more than eight hours at a time. Small hotels in Paris, a form of private renting, are still crowded with single African men, working very early hours on the street hoses, cleaning the cities. Over half a million units in the private sector were furnished rooms or hotel lets; 1.2 million were occupied rent free, about a third of these being tied to jobs. (Tied accommodation is normally provided for the police, the *préfect* and other state functionaries, as well as concierges, *gardiens*, and others.)

Private flats at the cheaper end of the market – the majority – tended to be very small, particularly in the Paris region, with fewer rooms than HLM flats. In the higher-income, older rented areas of Paris, corridors of maids' rooms on the seventh floor above the elegant apartments were let out, tiny room by tiny room, thirty-five feet square, offering rudimentary shared kitchens and shared washing facilities. These rooms were originally for the

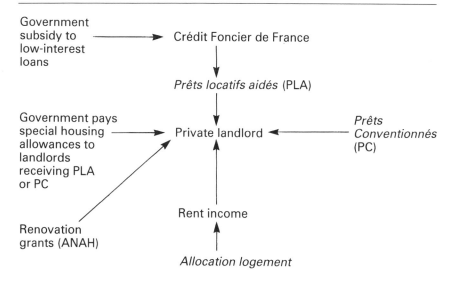

Figure 5.1 State and government funding support for private landlords in France

maids of the six floors below but were taken by students, immigrant single workers and sometimes childless couples, though a double bed would barely fit!

Support for private renting

The French government had little ambiguity from the outset about the role of private landlords; they were essential to the housing market, were encouraged, protected and supported, although not to a level that prevented their eventual decline. Housing allowances, introduced in 1948, were a major means of indirect support. The avoidance of slum clearance policies implied support for the survival of private landlords of even very low-quality housing. Improvement grants from the 'Fonds National d'Amélioration de l'Habitat'[1] were available from 1948. Reformed in 1972, it was replaced in 1977 by *prêts locatifs aidés* (assisted loans for renting), which private landlords could obtain on favourable terms, with government backing, from the Crédit Foncier. There were also *prêts conventionnés* from private banks, which made private tenants eligible for special housing allowances. The result was a significant building and renovation programme for private renting, providing roughly 2.5 million units since the war. Renovation of old rented units was also funded from special grants (ANAH[2]). Figure 5.1 shows how private landlords received support from the government and from state funding institutions.

Decline

At the same time, however, older rented units disappeared through three main routes:

1   Renovation of dilapidated property in and around town centres led to amalgamating units to produce higher standards, and often movement away from private renting altogether.
2   Demolition took place on a significant scale where property was obsolete or, more often, where the land was required for alternative uses such as roadbuilding, other public uses, or for commercial development. This removed possibly half a million units in the post-war era in Paris alone (Van der Cammen 1988: 77).
3   Many properties in declining rural and urban areas were abandoned, leaving nearly 2 million empty properties by 1991.

Private renting began to decline fairly rapidly in the late 1970s – leading to a loss of 95,000 units a year during the 1980s (Dauge 1990). The new emphasis on renovation targeted many of the poorest and most overcrowded private dwellings. As slums were upgraded, the number of occupants fell drastically. Existing inhabitants were frequently replaced by newer and better-off households.

The pressures on the private-rented sector were intense. Private rents for post-war flats were on average 14 per cent higher than social housing rents by the late 1980s (Emms 1990). In major urban centres like Paris, they were 50 per cent higher than in other towns. But in spite of decline, it still constituted about one-third of all dwellings, housing well over the official 6 million households in 1990, twice the number housed by HLM landlords. Both the government and social landlords were anxious that it should continue to play a major role, allowing flexibility, a variety of access routes, and a very large contribution to the overall supply. Therefore, the government talked of further increasing the level of support for private renting in the 1990s.

OWNER-OCCUPATION

Owner-occupation in single-family houses was supported in France for a century and a half by social reformers and the HLM housing associations, as well as by governments of all persuasions and the public at large. From the mid-nineteenth century, building for owner-occupation, for low-income as well as more affluent households, was encouraged by the state as an early ideal of the social housing reformers. In the inter-war years, suburban owner-occupation expanded in spite of very low house-building. When asked for their views on tenure in the 1960s, 80 per cent of French households said they preferred to buy a traditional house (HLM 1989a: 73).

Owner-occupation since the war grew more rapidly than any other tenure, with 6 million new owner-occupied units. Loose planning regulations facilitated a ready supply of suburban and rural land which kept down the cost of housing construction. The low population density of France assisted in this trend.

Government support for low-income owner-occupation

After the war, the government assisted owner-occupation on more and more generous terms. Low-income buyers were eligible for housing allowances, as well as favourable loans and grants, but they had to provide at least 5 per cent and, more often, 10 per cent of the cost directly. A majority of houses built for owner-occupation were constructed with direct government subsidy. HLM associations were encouraged to build for owner-occupation and from 1945–85 they produced 1 million owner-occupied units, about one-quarter of their post-war building total. Within the social housing movement, there were powerful advocates of owner-occupation. The *sociétés de crédits immobiliers*, which helped finance lower-income purchasers, were an integral part of the HLM movement itself.

Low-cost owner-occupation was also encouraged by the 1 per cent employers' tax, about two-thirds of which went to support funding for low-cost owner-occupation for lower-income employees, through the *crédits immobiliers*. The money was channelled both through HLMs and through loans to eligible house purchasers.

A higher proportion of working-class households than white-collar workers became owner-occupiers – 41 per cent compared with 34 per cent (Emms 1990). This may be because white-collar workers gained access to HLMs on a wide scale at an earlier date, thereby reducing their incentive to buy in the early 1960s. The very generous help to low-income buyers favoured blue-collar workers. More manual workers were willing or able to help build their own houses. This was certainly true in the inter-war and early post-war period (Van der Cammen 1988). It significantly reduced the cost of owner-occupation. Not more than one-third of income was expected to go on monthly repayments.

Funding

Figure 5.2 illustrates state support for low-income owner-occupation.

Universal owner-occupation

Behind all this support was the guiding principle of housing policy to 'make home ownership a possibility for every Frenchman'. In the words of an HLM Director:

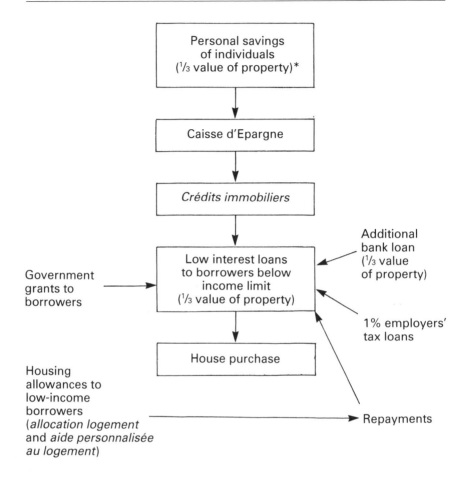

Figure 5.2 State funding in France for low-income owner-occupiers

Note: * It was possible to qualify with less (10 per cent), but one-third was the target.

Owning will lead to people spending more of their own money on housing. It will encourage people to respect property and therefore the wider environment, and it will, through the greater spending, enhance economic activity.

Three factors favoured the growth in owner-occupation in France:

1   High-rise flats on large estates did not appeal to families with children if they could afford to buy. Yet virtually the whole post-war social rented stock was built this way.
2   The private-rented sector, with its large supply of old, unmodernised

property, no longer satisfied people with secure and rising incomes. People who could, moved out into new homes.

3   Conversion of private-rented flats into owner-occupation accelerated as renovation programmes extended in the older urban areas in an attempt to improve conditions. This pushed more people into becoming owners.

Owner-occupation in France formed an essential part of social housing provision in three ways: homes could be built with subsidies by social housing organisations; buyers received government subsidies for the purchase costs; and eligible, low-income households could be helped through housing allowances on the same basis as tenants.

In all, two owner-occupied houses were built for each social rented flat after 1945.

HOUSING REFORM AND GOVERNMENT WITHDRAWAL?

The developments outlined in the last three sections were driven by government support, which grew steadily up to 1977.

The structures for funding and for building state-supported housing in France dated from before the First World War. The same systems applied till the 1970s. When the oil price crisis and the economic recession hit Europe in the mid-1970s, the high costs resulting from rapid inflation created slack demand and a sharp drop in house production. That coincided in France with a rough match of dwellings with people (Quilliot and Guerrand 1989). Simultaneously, vacant units were beginning to emerge in the largest peripheral estates. Although there were still significant pockets of acute need, particularly in Paris, the crudest shortages were over.

The government instituted major housing finance changes under the impetus of Raymond Barre, then Minister of Finance. His important reform proposals in 1977 attempted to address a number of problems: over-complex layers of financial support; over-centralisation of state intervention; help distributed too unevenly, with too little help to the very poor; the steeply rising cost of subsidy to buildings (*aide à la pierre*); a need – or desire – for a freer housing market. Barre advocated:

1   A reduction in state support for building with an emphasis on market forces.
2   Greatly increased government support for owner-occupation which therefore in a strict sense interfered with market forces, but which facilitated individual control, saving and investment through the increased financial costs to the individual household.
3   A targeting of aid on the households with lowest incomes who were in greatest need (*aide à la personne*). This was applied to all three main tenures: social renting, regulated private renting and low-income

owner-occupation. It opened the way to social housing bodies accepting many more of the lowest paid and most marginal households.

4   A simplification of housing finance and the administration of it.
5   A shift in emphasis from new industrialised building to rehabilitation of the existing stock.

A main vehicle in these reforms was the introduction of the *Aide Personnalisée au Logement,* an expensive personal housing allowance commonly known as APL, which would replace the earlier housing allowances where a property met certain standards of improvement or construction. It would be available to eligible owner-occupiers as well as to tenants. A condition of APL was that rents should rise to reflect the increased cost of an improved dwelling. This was to ensure more market-oriented rents and a higher level of repair. APL would be paid directly to landlords.

Grants also became available to meet around 25 per cent of the cost of renovating the HLM stock. These government grants, called PALULOS, could be paid as long as the total spending per unit was limited to £7,000, although the grant system allowed some discretion for additional exceptional costs. After renovation, the improved dwelling would be subsidised through APL.

The Barre reform also introduced specially advantageous loans to support low-income owner-occupiers called PAPs, short for *prêts d'accession à la propriété.* In addition, special renovation programmes for older housing areas were launched.

In 1982, under the new Mitterand government, the reforms were extended and targeted more directly at the improvement of the *grands ensembles.* Through the reform of housing finance, the government was able to agree with the HLMs a renovation contract, some of the costs of which would be reflected in the rent levels (*conventionnement*) but, in turn, the rent levels would be subsidised for people on low incomes through the *Aide personnalisée au logement.*

One of the main results of the Barre reform was that rent levels soared and the cost to the government of the new housing allowance system rose dramatically. Between 1978 and 1984, rents rose by 127 per cent. More importantly, households had to pay around 17 per cent of their income in rent, compared with 12 per cent before the reform.

The new financial system, directing help towards low-income households, gave a powerful incentive to the rapid spread of low-income owner-occupation through PAPs – about 300,000 new units a year were built throughout the 1980s; it also encouraged HLM organisations to speed up spending on their increasingly problematic stock, with nearly 200,000 PALULOS grants being paid in 1985 alone (Emms 1990).

The subsidised owner-occupier loans (PAPs) and the HLM renovation grants (PALULOS) undermined the government's desire to withdraw. The bill for housing allowances to cover part of the costs to tenants and owner-occupiers rose tenfold in the 1980s and the number of households becoming eligible for assistance rose to 3.5 million. Within the APL system, households qualifying for an allowance had to pay about one-fifth of their housing costs, no matter how low their incomes.

COMPETITION AND QUALITY

But as a result of the narrowed emphasis on personal support and targeted help, the numbers and proportion of new dwellings funded without state support rose significantly. The Barre reform achieved the aim of creating a more competitive housing climate. By pushing up rents to reflect real costs, it made HLM landlords more quality-conscious. The numbers game had run its course.

The government had intended to finance new housing investment and more generous allowances through higher rents, a bigger contribution from better-off tenants and from the HLM organisations. These measures were not only extremely unpopular with the *communes* and HLMs as well as tenants, they were also unrealistic because of growing unemployment and the growth in low-income households generally – particularly within social housing. By 1991, the government had managed to divert a proportion of the employers' tax to help meet the growing costs of APL.

Therefore the reforms did not lead to the hoped-for withdrawal of government. Instead they extended heavy government involvement in low-income owner-occupation through the more generous housing allowances, grants and subsidies. They also kept alive the government's commitment to support for the private-rented sector and extended it to inner-city renewal. At the same time, they led to major new initiatives to renovate the *grands ensembles.* The increased investment in all tenures, reflected in higher rents, was increasingly indirectly funded through the higher housing allowances.

Two million additional households bought their homes in the following decade. But the cost of the huge expansion in low-income owner-occupation resulting from the Barre reform proved prohibitive, while the impact of renovation and expanding owner-occupation on private renting was to intensify its erosion.

DECLINE IN SUPPORT FOR OWNER-OCCUPATION

The quality of owner-occupied, single-family houses, by comparison with rented flats, increased demand. But while its popularity pointed to continued growth, other factors limited its expansion. The high cost of direct government support, particularly the housing allowances, led to curtailment.

The growth in lower-income groups through rising unemployment, the growth in one-parent families and in second generation immigrants, all increased demand for cheap rented housing. By the late 1980s, social housing organisations virtually stopped building for low-income owner-occupation because it was too expensive. In addition, concern about urban sprawl due to lack of planning controls, transport problems due to ever-more distant suburbs, and evidence of extreme poverty among large, low-income families who over-extended their resources in the purchase of cheap *pavillons*, all curtailed the building of single-family houses (Ville et Banlieue de France Conference September 1990).

Owner-occupation was expensive for the purchasers as well as the government. The cash outgoings for an owner-occupied household were about three times those of a tenant in spite of generous allowances. By 1990, new low-cost, PAP-subsidised, owner-occupied units had tailed off to almost nothing, although conversions were still going on.

COMPETITION FOR SOCIAL HOUSING

The Barre reform of 1977 opened up HLMs to competition, aiming to equalise state support between sectors, forcing up HLM rents and targeting subsidies on private building. Cheaper owner-occupation attracted away more economically active households. HLMs could only prevent polarisation and decline by improving the quality of their services. This they set about doing.

NOTES

1   In 1948, Le Fond National d'Amélioration de l'Habitat – National Fund for Housing Improvement – was set up. It was replaced by l'Agence Nationale pour l'Amélioration de l'Habitat in 1972.
2   ANAH: Agence Nationale pour l'Amélioration de l'Habitat.

Chapter 6

Difficult-to-manage estates

ESTATE-LEVEL SERVICES AND *GARDIENNAGE*

The nature and scale of French housing estates made estate-level services central to the viability of the HLMs. French private renting developed the renowned 'concierge' system, the goal of which was to ensure private enjoyment of peaceful conditions to each tenant, while enforcing communal rules to maintain collective areas on behalf of the landlord. This sometimes oppressive control system was transferred to HLM estates through a system of *gardiennage.* By far the biggest group of HLM employees was the caretaking service, with 35,000 out of a total of 65,000 employees. This front-line, intensive management, vested with authority akin to estate superintendents in the old-fashioned British housing trusts, like Peabody and Guinness, kept most estates working with a reasonable level of basic services. Without it, they would have almost certainly lapsed into chaos.

One *gardien* or caretaker took responsibility for around 100 flats, making sure that day-to-day services on estates worked properly, liaising with other estate staff, providing a link to tenants, making sure that repairs, cleaning and ground maintenance were carried out, overseeing contract work, liaising with the landlord, and, in many cases, collecting rents.

The caretakers were paid higher wages than the traditional concierges in the private sector, reflecting their more complex task and also difficulties in recruiting. The title '*gardien*' was intended to convey higher status. Not only did they have direct responsibility for cleaning, repairs and open space maintenance; they were also the first line of contact for tenants and were expected to supervise estate conditions. They were supposed to report on noise and other forms of nuisance. They normally had a small office on the ground floor of the block they were responsible for.

MANAGEMENT DINOSAURS

In spite of intensive, front-line caretaking services, estate-level problems grew faster than the capacity of *gardiens* to contain or control them. There were many reports of caretakers feeling unsupported by their headquarters and suggestions that caretakers were not qualified to handle the difficult social task that had been added to their service role. The HLMs faced many problems of bureaucracy, over-rapid growth and remote administration. Even though the average size of an HLM was under 6,000 units, large estates could not be successfully run from central headquarters, and tenants needed a much more direct link with the landlord.

HLM companies and the National Union of HLMs embarked on management reform. They characterised the change of emphasis as abandoning the 'construction dinosaur' in favour of a 'customer service' (see Plate 6.1).

From the mid-1980s, the HLM movement became far more promotional, both internally and externally; HLM organisations set about reforming their management style in relation to tenants; the caretaking and estate-level services were upgraded and supported more intensively. The aim of recruiting younger, more flexible and more tenant-oriented staff became widely embraced.

The *grands ensembles*, overwhelming in their scale, demanded a revitalisation of management, particularly at the local level. As a result, managers were increasingly placed in local offices, close to tenants and caretakers, and in daily contact with problems. Often this happened as part of the government rescue programme for the *grands ensembles*.

Plate 6.1 'Certain prejudices against HLMs are a little out of date'

Source: Quillot and Guerrand 1989: 98

POLARISATION

The *grands ensembles* were built for large populations but in the late 1970s the number of residents began to dwindle as their popularity declined. This was often accompanied by a large increase in the proportion of families with more than two children and of one-parent families (Curci 1988). Table 6.1 illustrates the trend on a number of estates.

The population loss led to strong polarisation. HLMs became desperate to let property and reduced their 'entrance requirements'. As housing allowances targeted at poorer groups became more generous, so the poverty of residents actually increased. It was the poor who could at last 'afford' a social housing rent. Many of the better-off chose to buy. A very rapid acceleration of physical decay was an inevitable consequence of empty dwellings and income problems for the HLMs. The contrast between home ownership in single houses and HLM renting became acute.

The changing social composition of estates highlighted the dislocation between estate communities and the urban centres to which they were so loosely linked. Because of their size, often equivalent to small towns, the lack of social and commercial facilities and the alienation of residents became serious issues.

Table 6.1 Population loss on unpopular *grands ensembles* in France, 1975–82

Estate	Loss (%)	Estate	Loss (%)
Sarcelles	2.1	Orly	9.0
Massy	2.7	Meudon	8.4
Antony	5.1	Les Minguettes	25.0
La Courneuve	11.6		

Source: Tuppen and Mingret 1986.

CARETAKING PROBLEMS

Caretakers faced growing social pressures. The increasing poverty of households was compounded by the social and cultural problems associated with households from different backgrounds. Large numbers of children and young people placed a heavy strain on buildings, environment and facilities. But their job was regarded as central to the management of estates. They were the only group of housing staff for whom there was specialist training, run by the National Union of HLMs. Problems in recruitment threatened both the status and the survival of the local system of *gardiennage*, but its centrality to the survival of the estates themselves was universally accepted.

The most critical issues were:

- the ability of caretaking staff to cope with new problems;
- adequate training and status;
- sufficient 'clout' to play a supervisory role;
- sufficient social skills to build good relations with tenants;
- fear of rapid decline in caretaking morale due to the social pressures and increasing social problems within *grands ensembles*

Caretakers were in an increasingly exposed position, often far removed from their administrative headquarters, confronting problems that defied the wider society, as HLMs became more involved in housing more marginal groups.

The emphasis shifted to decentralising other management functions to estate level in order to support caretaking staff. The estate- and block-level supervision through the *gardiens* was more pivotal than ever to holding conditions on estates.

THE RESCUE PROGRAMMES – FROM HABITAT ET VIE SOCIALE 1976 TO DEVELOPPEMENT SOCIALE DES QUARTIERS 1984

Habitat et Vie Sociale had modest aims. It was set up in 1977 to try and help the most difficult *grands ensembles*. Its impact was limited, but it served two purposes. It targeted some of the most racially explosive and hard to let areas. It focused government attention on a problem it had played a large part in creating.

In 1981, a sequence of events, starting with disorders at Les Minguettes on the outskirts of Lyon, goaded the Union des HLMs into pressing for drastic government action. A major factor behind the disorders was the growing tension between North African youth and the police. The gap between first and second generation immigrants often weakened parental authority. Loss of manual jobs led to fewer opportunities for undertrained and disadvantaged youth. Their disaffection fed on problems in the wider society. But the explosion was a symptom of a many-sided problem – poor facilities, isolation, over-large-scale building, poverty, unemployment, unwieldy and insensitive institutions (Dauge 1991).

As a result, an inter-Ministerial Commission was set up under the active leadership of the Mayor of Grenoble, charged with reporting direct to the Prime Minister on the problem. The report, *Ensembles refaire la ville* ('Together, rebuild the town'), was presented to the Prime Minister and led in 1983 to the creation of the National Commission for the Social Development of Neighbourhoods (CNDSQ). Regional Commissions were established in 1984.

The National Commission was to provide the framework for a major renovation programme, environmental improvements, and social and economic initiatives. The Commission was to work on behalf of the government in partnership with regional and local government and the HLM organisations. At first it worked on twenty-two troubled ZUPs. The programme was then expanded to 120 areas, including some older inner areas (Dubedout 1983). The target was 350,000 dwellings in the first five years, though only 160,000 were actually renovated.

From the outset, the French programme recognised and reflected the wider social, economic and urban aspects of the problem. None the less, the bulk of the money was spent on physical renovation, much of it aimed at brightening the external appearance and changing the outside image of the estates, so that visitors and potential residents could recognise the investment and judge the estates by the visible attention they were receiving.

The money came from three main sources in fairly even proportions – the state, regional government, the HLM movement and other support bodies. Seventy per cent of it was spent on renovation and 30 per cent on support programmes. Social support covered a very wide range, including estate facilities, employment and training, help for foreigners, participation, and cultural and craft activities.

THE NATIONAL COMMISSION FOR THE SOCIAL DEVELOPMENT OF NEIGHBOURHOODS

There were four founding principles of the National Commission under-pinning the work they launched on the estates:

1 Rescuing the *grands ensembles* from further disorders depended on *central government/local government partnership.*
2 *The physical and the social conditions* had to be improved together.
3 In the course of renovating the buildings, the *management of the HLMs* had to be tackled.
4 The major actors in the *grands ensembles* were the local authorities or *communes*, the tenants themselves and the HLM organisations. All three had to agree to be *partners* in the programme.
    (Commission nationale de développement social des quartiers 1987)

The main characteristics of the original twenty-two estates and the later 148 estates identified by the National Commission as being in difficulty were pinpointed as:

* intense social *segregation*;
* *misrepresentation* in the Press and in the surrounding community;
* a high proportion of *children and teenagers*, often reflected in a high rate of delinquency and problems with schooling and youth unemployment;

- inadequate *management and maintenance* of the blocks.

    (Commission nationale de développement social des quartiers 1987)

The Commission adopted a central aim – to attack the *deeper causes of urban decay* through broad policy initiatives over:

- education;
- employment;
- health;
- the integration of cultural minorities and marginal groups;
- the development of cultural and sporting activities.

A ROLE FOR MAYORS

HLM societies were only responsible for direct housing provision, not for social amenities or wider needs. Local *communes* often had neither the cash nor the political will to make good this deficiency. The political separation of the local mayor from the large constituencies of the ZUPs, while avoiding political strife, also removed political responsibility. Given the huge problems of the *grands ensembles*, the French government and the Union of HLMs almost beseeched local mayors to become involved in estate problems. Only where this has happened did facilities, integration and upgrading become high priority (Ville et Banlieue de France September 1990).

IMPACT

The renovation work was strikingly visible and was dominated almost entirely by external work on the blocks and entrances or the environment.

The 'social' money stimulated employment, youth training programmes and community initiatives within the *grands ensembles*.

Special educational priority areas were declared, enabling schools to recruit additional teachers. Holiday programmes were funded, using local young people as helpers. Homework programmes were set up. These additions had some impact in some areas – for example in Vénissieux and Mantes-la-Jolie – but often progress was fragile and the impact was diluted by wider problems.

Some HLM organisations responded to the National Commission by becoming more thrusting and up-beat in their approach to management. They already had a built-in financial incentive that was tightened by the changes and pressures they were now under. But the grant money and the National Commission programme were often insufficient to tackle the roots of the problem.

There was huge variation in the response of the HLM organisations and the impact of the improvements, depending on a complex range of factors,

including housing demand, the HLM's style of management, the engage-
ment of the local mayor, the degree of stigma, and the history of the area.
The more long-standing and deep-set the problems, the more difficult they
were to reverse. Overall, the programme was claimed to have cut empty
units, upgraded conditions, enhanced the image of estates and helped
integration (Délégation Interministérielle à la Ville et au Développement
Social Urbain 1989; Délégation Interministérielle à la Ville et Délégation à
l'Aménagement du Territoire et à l'Action Régionale 1990).

Plate 6.2 Les Minguettes, Venissieux, Lyon

   But the urgency of carrying out further improvements accelerated. Some
HLM organisations appeared remarkably resistant to change, and some
estates did not look or work significantly better after upgrading in spite of
money for renovation, local project initiatives and some fanfare. Often in
these cases the vital partners to progress were missing: residents were not
always fully involved or consulted; the HLM organisation was sometimes
unresponsive and demoralised; the local mayor was not always supportive
or committed. Sometimes the estate itself was too decayed, or demand for
housing had evaporated with unremitting economic decline. Northern
France seemed particularly affected by the lack of demand and little could
be done in the short-term to reverse that (Provan 1991).
   There were a number of model estates demonstrating what could be
achieved. The impact in those areas was clear to see – façades restored and
repainted; entrances secured and upgraded; gardens planted; smart local
offices opened; tenant training and employment projects set up; youth
organisations supported; new and improved facilities installed; development and management services effectively integrated.
   One of the most successful initiatives to engage residents has been the

formation of contract organisations among the residents of the estates (*régies de quartiers*) to carry out work on the estates. Over fifty such companies are now operating. The French believe that this is a possible way forward, combining an economic initiative with improving the condition of the estate and local training and leadership (Behar 1987).

But there remained nearly 1 million similar dwellings on similar estates, possibly with less extreme conditions but sharing many of the same problems yet to be tackled.

The government restricted the amount that could be spent per dwelling to ensure that the large stock of unpopular flats could be covered. The rent rises would be limited by the amount that could be spared through housing allowances. The physical impact of the programme might be reduced and visible changes would be slower. These limitations followed from the sheer scale of the new programme.

INTERMINISTERIAL DELEGATION TO CITIES

In 1988 the National Commission was amalgamated, along with other government-sponsored bodies attempting to tackle the problems of social segregation and urban decline, into the Délégation Interministérielle à la Ville. In the new phase, it targeted 360 estates.

The drive of the Interministerial Delegation to Cities was to integrate the funding of programmes of the eighteen ministries involved in the peripheral estates. This administrative effort aimed to overcome the heavy state bureaucracy at the highest level; it revealed all the problems of inflexibility and top-heavy imposition that made it so difficult for mayors, *communes*, or residents to seize the initiative.

1990 – A YEAR OF DISORDERS AND POLARISATION

Just at the point where co-ordination was becoming possible and some gains were being made from channelling multiple funding sources into single sites, the programme was shaken in 1990 by widespread eruptions of violence to a level that shook the French establishment. The most serious riots lasted for five days at Vaulx-en-Velin outside Lyon, and at Mantes-la-Jolie in the Paris region. Both estates were very large, with over 5,000 units, and far from the main urban centres.

In early 1991, the programme was extended even further to 400 'sensitive' estates and a new Minister for Cities was appointed to drive the effort against 'exclusion' of all kinds:

- physical in the shape of decaying peripheral estates;
- social and economic in the shape of poverty and unemployment;
- educational in lack of opportunity;
- racial in the visible hostility and divisions within French society.

Table 6.2 Key dates in the development in France of the National Commission
for the Social Development of Neighbourhoods

1971	Early government meeting to discuss problems in large estates.
c. 1975	*Grands ensembles* start to depopulate.
1970s	Families joined immigrant workers on larger scale.
1976	'Habitat et Vie Sociale' – first rescue programme for *grands ensembles*.
1977	Barre reform – market orientation; rent rises; targeted benefits; more generous APL introduced.
1981	Disorders – clashes between immigrant youth and police in two areas. Serious problems on a number of ZUPs.
1981	Union des HLMs presses for Government action.
1981	Mitterand elected President – launches social housing spending programme.
1981	Banlieue 1989 set up to experiment and innovate in underprivileged suburbs to mark the bicentenary of the French revolution.
1982	Interministerial Commission to examine causes of decline.
1982	Founding of National Commission for the Social Development of Neighbourhoods – 22 target sites.
1983	Report to Prime Minister – *Ensemble refaire la ville* – identifies multiple pressures leading to intense social and economic segregation and alienation on 140 estates.
1984	Programme extended to 120 sites – both programmes focus mainly on *Zones à urbanisation prioritaire*.
1987	French government takes lead in developing a European Community-wide experiment to target run-down estates in member countries.
1988	Extension of the programme to 360 sites.
1989	The National Commission combines with other initiatives, such as the government programme to combat delinquency and becomes the Délégation Interministérielle des Villes, with more powers and more comprehensive programmes, still under the Prime Minister's Office.
1989	Villes et Banlieues de France is formed under the Delegation to involve the state and mayors of outer *communes* where problematic *grands ensembles* are located.
1990	Further major disturbances on the outskirts of Lyon and around Paris after over a year of racial attacks and growing tensions, mainly focused on the large estates.
1991	Minister for Cities appointed – programme extended to 400 sites.

It remained unclear whether the new spending would slow the greater polarisation already under way. The improvements made the areas more acceptable to residents and easier to manage. They did not remove the distinctive image of state-provided mass housing. Nor did they simplify internal relations between several thousand households in such close proximity. The need to police conditions spilled over into violent or potentially violent confrontations between authority and alienation, between state institutions and dispossessed youth (Dauge 1991).

Some HLMs were reacting by restricting rehousing for the new second generation of immigrant origin now needing housing. They were the group – poorly trained, often unemployed, lacking in a sense of belonging to either community, unable to leave home through discrimination against them in rehousing – who were taking to the streets in open rebellion against French authority.

Under intense pressure, the government was increasingly drawn directly into tackling racial problems and issues of integration. The serious disorders in 1990 and 1991 in France emanated largely from the *grands ensembles*, many of which now housed large concentrations of minorities. Therefore the Interministerial Delegation was almost swamped by wider political tensions, the rise of the extreme right, the policing problems, the high levels of unemployment, and the weaknesses of the education system in some of those areas.

Since the disorders were largely concentrated within the least popular estates, the government was constantly brought back to the need to tackle these areas. Housing decline, social segregation, and racial tension coincided in a series of easily identified estates that were 'on the boil, with the lid off'. The government had become more and more heavily engaged in these areas.

Table 6.2 shows the development of French government involvement in the outer estates.

Chapter 7

Current issues and conclusions

THE POST-WAR TRANSFORMATION

We have outlined the framework for the great changes in French housing conditions since 1945 and we have touched on some of the upheavals surrounding those changes. Owner-occupiers, private landlords and social housing companies together added 14 million new units, doubling the stock in the forty-five years from 1945 to 1990. All three tenures added significantly to their stock in the post-war boom, although many older, private rental units were lost, particularly in the 1980s and, overall, more private rented units were lost than gained.

Since the war, over 4 million homes were built by HLM organisations, all with state support; 1 million of these were for owner-occupiers. Altogether over 7 million new dwellings were built for owner-occupiers and 2.5 million for private tenants. Of these, less than half were built independently of state support.

Table 7.1 shows the way the housing programme accelerated up to

Table 7.1 Increases of housing stock in France, 1945–90

	Stock		Additions
1945	13,500,000	1945–54	651,100
1955	14,100,000	1955–64	2,978,900
1965	17,000,000	1965–74	4,926,100
1975	21,500,000	1975–84	3,949,200
1985	25,000,000	1985–90	1,500,000
		Total additions	14,005,300

Source: Estimates based on figures from Ministère de l'Aménagement du Territoire, de l'Equipement, du Logement et du Tourisme 1991.
Note: These estimates assume the loss of approximately 1 million dwellings due to demolition, conversions and amalgamation of dwellings between 1965 and 1985. The numbers include second homes and vacant units.

Table 7.2 Changes in housing tenure in France, 1945–91

Total no. of units (millions)	Owner-occupation (%)	Private renting[a] (%)	HLM renting and other social renting[b]
1945: 13.5	46	52	2 (estimate)
1991: 25.0	53	30	17

Sources: These figures have been calculated based on information from the Ministère du Logement 1989, from INSEE 1988, and from Tableaux de l'économie française 1991/2.

Notes: [a] The numbers of private rental units possibly conceal a larger number of households sub-letting and occupying rooms on an *ad hoc* basis, often in overcrowded conditions. There is also a large number of households occupying 'free' accommodation, about 1.2 million in all.
[b] HLMs own the vast majority of social rented units.
[c] This total includes over 4 million empty units or second homes originating from both owner-occupied and rented sectors.

1975, continuing to produce 400,000 units a year to 1980, and still building intensively up to 1990.

Owner-occupation and social housing expanded significantly in the forty six years from 1945–91, as Table 7.2 shows.

CURRENT SOCIAL HOUSING ISSUES

The transformation of French housing conditions has thrown up some major questions.

Monopoly landlords

The French stuck vigorously to the letter of the Loi Bonnevay, separating the role of the state and the direct provision of housing.

None the less, some outer *communes* in the bigger cities are totally dominated by social housing, with HLM tenants comprising 80 per cent of inhabitants. Others have very little. This makes the role of the local mayor crucial. The new 'Loi d'Orientation pour la Ville' reinforces this, as each *commune* now has to draw up a Plan for Housing, and there are attempts to ensure that future social housing is distributed in *communes* with very little at the moment.

The political links between HLMs and the Town Hall are sometimes unhealthily close and it is possible for one to dominate the other (Avery 1987). The fact that an HLM may provide virtually all housing within a *commune* makes for a near monopoly. This can enhance management problems, access problems and segregating tendencies. There is something of a balancing act between state, region, the HLMs, and the *communes* over the direction of social housing.

Tenant selection

The preoccupation with how to assure access for the poor *and* to keep better-off households within the social rented areas is a live issue. Because HLMs have to retain their viability as independent, financially stable organisations they constantly attempt to balance these two pulls. They stress that the creation of ghettos of minorities has an adverse effect on the social conditions of an area. They also stress the importance of local links so that community stability is generated after a period of rapid transition, population loss and difficulty in letting. Access for the poor is assisted through the generous housing allowance system (APL), although its future is far from certain because of its high cost. A limit has already been put on the level of APL (Ghekière 1988).

The French government and social housing companies alike openly debate the conflict between ensuring access for the most vulnerable and deprived members of society and attracting more secure residents in areas of declining popularity, such as the high-rise peripheral estates. Most organisations support a limit on the concentration of minorities and other disadvantaged groups.

Targeted help

Housing subsidy is oriented towards support for the poor and for large families, whether in the social rented sector, private renting or in the low-income band of owner-occupation through direct allowances. In addition, there are big incentives to save for owner-occupation with subsidised interest payments, as well as bonuses linked to savings. This encourages many lower-paid workers to move to owner-occupation. The result is many more very poor households becoming HLM tenants as many more lower-income households become owners. This has an impact on costs, standards, and household budgets. Everyone ends up paying more.

The standards of conversion are being cut in order to allow rents to be fixed at a lower level, so that the housing allowance bill will also be lower. This will clearly become a major problem if it is extended as it will reinforce the two-tier nature of housing and will dilute the impact of the Delegation to Cities and the impetus to rescue the distressed HLMs.

Peripheral estates

The large *grands ensembles*, located on the outskirts of towns and cities, built in a brutalist architectural form, are on the one hand difficult to change and on the other hand difficult to integrate. The strategies to modify the estates are threefold:

1   to change the appearance of the estate by external adornment;

2   to link the *grands ensembles* into the cities through transport, through urban development and through cultural and political links.

3   to give the people who live there a more central role.

Participation

The French have concluded that unless residents are directly involved in improvement and change, real progress is unlikely. Previous policies of imposing solutions are seen as unsuccessful. Particularly acute is the awareness in France that young people form a major force within the *grands ensembles* and their interest must be engaged as part of any change.

Consultation over improvement is now routine. Tenant representation on the boards of HLM organisations is an accepted feature, although in practice it often leads to minorities being largely excluded and the occasional political activist dominating. There is some renewed interest in the co-operative housing movement. New experiments in resident involvement are constantly debated, as the gap between marginal communities and mainstream society causes serious strains.

Racial problems

The most run-down and problem-prone housing areas in France are the ZUPs and other *grands ensembles*. These areas house disproportionate numbers of North and West Africans, as well as other minority groups. The migrants from Africa are often said to be the most difficult to integrate. Their women are often not allowed to participate. Second-generation young men have found particular difficulties in schooling, in employment, and in assimilating to either their own or to French culture (Dubedout 1983). The French government is troubled by the serious racial incidents, by growing problems of segregation and signs of alienation, by the threat of right-wing movements, and by the political organisation, still embryonic, of young North Africans.

The HLM companies are just beginning to address these problems with specific programmes (Renaudin 1991), but at every level of French housing the issue is discussed.

The thrust of government policy is to 'include' foreign groups, *to make them feel French*, without, in official language at least, obliterating their own sense of identity. The French government has recently reaffirmed the aim of 'integration' as against a separate cultural identity for distinct ethnic and racial groups. Because of prejudice, fear and discrimination, some minorities cannot put their faith in official institutions (Burgeat 1991, Dauge 1991).

Segregation

The most serious unresolved policy issue is the issue of segregation. While the generous income limit for access to French social housing appears to ensure the broad acceptability of social housing, there are problems as income rises. Many move away because of the desire for owner-occupation or better housing. Therefore, instability is generated.

Because a proportion of the HLM stock is intrinsically unpopular, it is hard to recruit tenants with a broad band of income and, disproportionately, new tenants are poor. As the poor can only afford to move in with the support of *aide personnalisée au logement*, the whole HLM system becomes extremely expensive for the state if the buildings are renovated. It also becomes increasingly dependent on state support.

When rents rise after renovation, better-off tenants are less attracted to social housing because of the high cost, relative to the quality of dwelling. This is a vicious circle. If *l'aide personnalisée au logement* became uncertain and the standards of renovation were reduced because of the costs, the whole integration programme would be threatened because many estates would become even more unacceptable. It is not clear how the French will find a way out of this dilemma.

CONCLUSIONS

French housing experience is unique in a number of ways:

Utopia

France experimented in the nineteenth century with communitarian, utopian housing developments, aiming to embrace every aspect of social and economic life. These heavily influenced modernist architects, such as Le Corbusier, who helped inspire the creation of maybe 10 million mass housing units in post-war northern Europe. France built more high-rise, large-scale estates than any other country in western Europe.

State systems

France's state-driven civic and infrastructure building led to the early 'peripheralisation of poverty' that concentrated low-income marginal communities around the outskirts of major cities, leaving the historic centres largely to fine churches and town halls, to beautifully preserved and restored buildings, to washed streets and manicured public gardens. A large stock of private-rented housing survived in the inner areas around the centre.

France rejected explicitly at an early stage the idea that the state should provide, own, or, in particular, manage housing directly, creating powerful independent entities to develop social housing.

Late urbanisation

France embarked on social house-building on a significant scale only in the mid-1950s, at least a generation after the other countries in this study in spite of intense urban problems. France was unique in northern Europe in experiencing in the post-war period a proliferation of self-help shanty towns around the beautiful cities of Paris and Marseille, as well as other industrial areas, to house the swelling urban workforce she desperately needed to fuel her booming economy. Her rehousing programme therefore accelerated and became more uniform than any other country in this study.

Powerful independent funders

France relied heavily on industrialists and employers of all kinds to help fund state-supported housing programmes, thus ensuring partnership between public and private initiative in urban developments. Social housing depended on French national savings banks and the small savings of virtually every French household. The independent sources of finance underpinned the state housing programme and the legal autonomy of the HLM housing movement. HLM organisations have shown remarkable resilience and political weight in the current climate of social tensions and rapid change.

Segregation and insertion

Immigration generated intense racial strife. At the most extreme, Algerians have been killed in Marseille and in Paris (Avery 1987). Because of the acute labour shortages, immigration was tolerated, but immigrants, almost exclusively, performed dirty or unskilled jobs and were excluded from popular housing areas. As the process of decline in the *grands ensembles* got under way in the 1970s and as families joined immigrant workers in the same period, so the racial tensions surrounding immigration became concentrated in the *grands ensembles.*

Thus race and racial tensions have been at the core or at the most visible end of the growing divide between an affluent and well-housed French majority and a significant minority of people who do not have access to prosperity and better conditions and who therefore occupy the housing relinquished by the economically successful.

Social housing, by its nature, will be more segregated, less popular, poorer. The French, with their strong national pride and desire to make their national institutions work, will continue to invest in its rescue. But complex social and racial problems will undermine progress and threaten stability.

The French are sanguine about the long haul they face in creating a new urban order that is more integrated, more pacific, more balanced than today's tense, over-exploited and sometimes chaotic urban landscape of conflicting needs. In 1983, the head of the Prime Minister's National Commission for the Social Development of Neighbourhoods – the influential mayor of Grenoble who died in 1987 – predicted ten years at least to restore reasonable conditions. Now the prediction is open-ended.

> Big public investment [in distressed areas] does not necessarily provide a mechanism for communicating with residents. If there is no real dialogue, then investment will not work. Neighbourhoods that are in difficulty can be antagonised by intervention unless some new way of bridging the social cleavage can be found … we face many years of hard work to achieve it.
>
> (Dauge 1991: 7)

Part II

Germany

The Founding Fathers of the Modern German State fought against the spectre of Weimar instability ... thereby blocking the road for rapid innovation.
Wollmann, in Von Beyme and Schmidt, *Policy and Politics in the FRG*

A sense of history is divisive in Germany. Sitting in the heart of a far more diverse and yet more unified Europe than was conceivable even five years ago, Germany is possibly the most misunderstood, most fascinating and most important of European states. Between 1989 and 1990, Germany became again the most politically and economically strategic nation in Europe. Arguably it has never ceased to be in this position since the time of Bismarck. And the efforts of other European states to the east and to the west of Germany to curb German might have only underlined its unique position in European cross-currents. The rapid maturing of the European Community, with Germany in centre stage, belies Ralf Dahrendorf's view that, unlike the United States, Europe is endemically slow to seize opportunities (Dahrendorf 1985). Recent events in Germany also belie the complacent notion that German politics are inherently dull (Von Beyme and Schmidt 1985).

Germany's housing system is of singular importance because it has grown up through greater civil and international conflict, through more extreme demands and crises and through more ambitious production programmes than any other European country. It has managed to retain diversity while overcoming appalling shortages.

German history did not lend itself to this success. The penal settlement terms of the First World War aimed at no less than paralysing and dismembering the German economy. The humiliations and collapse of German national confidence after the horrors of the Second World War, left a divided Germany with responsibility for 12 million refugees and half the housing stock of its major cities in literal ruins.

Germany built more houses, more quickly, than any other country in this study in the generation following the war. The German 'economic miracle' now faces the new challenge of over 1 million new immigrants from Eastern Europe between 1989 and 1991, and the mounting cost of German reunification.

The collapse of authoritarian government in Eastern Europe led to 700,000 ethnic Germans crossing the West German border in 1989. In early 1990, about 2,000 East Germans a day moved into the Federal Republic.

It is important for those who have grown up with the idea that East Germany was a separate country to remember that to Germans, while reunification may have long seemed an elusive if not unrealisable goal, East and West Germany belonged together, geographically, linguistically, culturally, and historically. The border crossings had a certain flavour of 'coming home' or visiting relatives and friends.

While the rest of Europe watches with amazement mingled with anxiety, Germany has entered a period of dramatic change, in total population, in economic power, in strategic weight and in internal and external tensions. It is hard to ignore these radical shifts in a review of Germany's housing history and current housing problems.

Even before reunification and the changes throughout Eastern Europe, Germany was reaching a crisis of housing supply and of housing costs coupled with a chronic shortage of skilled young workers (Pfeiffer 1989, The Economist Intelligence Unit 1989). Reunification both heightens the problems and gives Germany a unique opportunity to take a great leap forward. Germany's overall economic strength is underlined by reunification, but the direction of change remains unclear and much is still to be learnt and documented about post-war events and conditions in the former eastern states.

Therefore this study is written very much from the West German perspective. Almost the entire post-war sections cover only West Germany, and only the West German housing systems and conditions are considered in any detail. A brief resumé of the East German housing situation is presented, in full recognition that it certainly does not do justice to such a large and important area.

The German housing story is a complex mixture of problems and successes, political and social pressures. The role of private landlords, owning and controlling over 40 per cent of all housing in Germany; the prevalence of flats, coupled with a long-standing preference for single houses and enthusiasm for owner-occupation; the collapse of Neue Heimat, at its zenith the largest social landlord in Europe with over 400,000 housing units under its control; all offer important lessons for other countries, as they attempt to mix variety and market pressures with social goals. This section attempts to consider the origins of these diverse strands, leading to an examination of how and why Germany ended up with nearly 250 large post-war peripheral estates and a declared shortage of maybe 2 million dwellings.

Part II.1 Early Ruhr housing

Part II.2 East German tenement housing

Chapter 8

Background

BASIC FACTS

Germany has the largest population of all members of the European Community. It has almost as high density of population as Britain. It has a low birth-rate and a rapidly ageing population. This gives it a lower proportion of young people than any other EC member. Between 1970 and 1985, the proportion of young people in former West Germany dropped by 40 per cent and the number of elderly people over 65 rose by 1.4 million. In the 1980s, a rapid fall in Germany's population was predicted but is unlikely to materialise. There has been a steady and gradual rise in the population of foreign origin, which has partly made up for the low indigenous birth-rate. Although previous immigration from the former East Germany has become internal to the new Germany, it is likely that immigrants from Eastern Europe and elsewhere will continue to settle within Germany's new borders.

The former West Germany is richer than most other countries, with lower unemployment and a larger housing stock in proportion to the population. Even after the vast building programmes slowed down, production remained far higher than in Britain, but in 1987 a new shortage of nearly 1 million dwellings was revealed. This came fast on the heels of widely publicised lettings problems in unpopular modern housing estates, which were taken to signify a serious over-supply of housing.

The reunification of Germany brought very big housing and economic problems, with steep rises in unemployment and homelessness. The former East Germany has over 16 million people and 7 million dwellings, many of which are either old and unmodernised or extremely stark and minimal in style.

Throughout almost all areas of the united Germany there are steeply rising housing costs and rents. The pace of change, the degree of need, and the confusion of events, make it impossible to give a fully accurate and up-to-date picture of Germany.

Table 8.1 Basic demographic facts about Germany: former FRG,[a] former DDR,[b] and United Germany, 1990[c]

	Former FRG	Former DDR	United Germany
Size of population	62,640,000	16,160,000	78,800,000
Density of population (inhabitants per square km)	245	150	221
Proportion of under-15-year-olds	–	–	16%
Crude birth rate (per 1,000 pop.)	–	–	11.1
Unemployment (%) (1992)	6.1%	16%	7%
People over 65	–	–	15%
Household size (1990)	–	–	2.2
Population in urban areas	–	–	84%
Immigrant population (non-German)	–	–	7%
Stock of dwellings	27,000,000	7,000,000	34,000,000
Dwellings per 1,000	–	–	430
Private consumption on housing	20%	4%	–
Tenure (in rounded figures):			
Owner-occupation	40%	25%	37%
Housing associations[d]	18%	15%	25%[e]
Private renting	42%	25%	38%
Direct state ownership	–	35%	–

Sources: Commission des Communautés Européenes 1987, Gesamtverband der Wohnungswirtschaft 1991, The Economist Intelligence Unit 1989, *The Economist Pocket Europe* 1992, Greiff and Ulbrich 1992.
Notes: [a] Former West Germany – Federal Republic of Germany.
     [b] Former East Germany – Deutsche Demokratische Republik.
     [c] All figures relate to latest available information.
     [d] Housing associations in Germany are often partly state-owned with local and regional governments holding shares in the companies.
     [e] Includes former state-owned housing in former DDR.

A BRIEF LOOK AT HISTORY:

Developments before the First World War

Germany did not exist as a nation until 1871. It consisted of a loose confederation of small states and principalities, with a range of cultural and political ties and traditions from the eighteenth century, dominated by the Prussian Empire to the north and east. Wars with Denmark, the Austrian

Empire and France, forged an uneasy German unity under Bismarck, the first German Chancellor, built on an intense nationalism surrounding the German language and ethnic identity, on Prussian military might, and rapidly growing German economic power.

Based on the early unification of many states into one, Germany has tended to have a more federal government structure than most European states, with the exception of the twelve years of Nazi government when German state power became all-pervasive under 'the reign of terror'.

Germany's uneasy borders to the west (France and Denmark) and constantly changing borders to the east (Poland, Czechoslovakia and Hungary) led to great tensions and instability. Wherever the boundaries were drawn, ethnic Germans would be beyond them, or large minorities within them. Germany's own nationalist fervour was matched by waves of extreme nationalism among subject people within Russian, Austrian and German empires.

Insecure German boundaries had a significant impact on the fragile and inexperienced German state and on her imperial neighbours – Britain, France, Austria, Russia. Two world wars, whose impact is indelibly inscribed on the Europe of the 1990s, were in part an outcome of this. The intense nervousness that a newly united, large and powerful Germany at the hub of Europe generated in her ambitious neighbours was based on the continent's history of almost constant warfare, changing boundaries, and imperial domination from Roman times.

Germany arrived long after France, Russia and Britain as an imperial nation but grew extremely rapidly in population and in industrial might and then accrued power and international importance beyond the ability of a still rapidly developing Europe to accommodate. The total German population grew by nearly 50 per cent between the founding of the united German state in 1871 and the outbreak of the First World War.

While Germany was developing politically, she was simultaneously urbanising extremely rapidly. The Ruhr area, which formed the heart of modern Germany, illustrates the transformation of Germany from 1860 when industrialisation got under way, before the foundation of the German state, to 1913 when the Ruhr was at its peak prior to the outbreak of the war. The number of industrial workers (mainly in mining and steel) in the Ruhr grew elevenfold in this period; the output of coal grew thirtyfold; and the output of steel over a hundredfold.

In the incredibly fast-moving fifty years from 1865–1914, Germany both became a unitary state and established her modern economic and urban infrastructure.

Between the wars

The First World War was a catastrophe. At the end of it, Germany had lost large amounts of territory, her empire had collapsed, her industry was as far as possible paralysed by her vanquishers, and a new but insecure and unpopular coalition was in place under the Weimar Republic, founded as the second attempt to create a German state in her brief history of nationhood. The weak coalition was the only German political grouping prepared to accept the harsh 'peace' terms of 1918 imposed by France, Britain and other allies, based on war guilt, reparations, and territorial losses. The far-seeing John Maynard Keynes predicted as early as 1919, from the British Foreign Office where he worked, that major future upheavals in Europe were inevitable in and around the German 'peace' terms.

The economic recovery of Germany after 1920 was seen as vital by Britain, as the threat of world recession grew. But the French invasion of the Ruhr in 1923 to claim war reparations and the total collapse of the German economy between 1923 and 1929 meant that by 1933 6 million Germans were unemployed, leading to major food shortages in many parts of Germany. The Weimar coalition collapsed, the 'peace terms' were modified, and Hitler led the emergence of the Third Reich. This was the third attempt in living memory to establish a strong and united Germany. National Socialism took unemployment down to half a million by 1938 through a huge building programme, as well as mass military build-up.

Nazism led to an unheard-of escalation in nationalism, persecution, terror and conquest, throwing every corner of Europe into chaos, and eventually embroiling every major world state in a devastating conflict.

Germany between the wars swung from a political limbo to international tyrant. Her economic and urban infrastructure continued to grow and her population to expand in the face of huge problems of external hostility and internal extremism. But the entire political and urban infrastructure was shattered when Germany was again defeated and stripped at the borders of many territories whose status had never been internationally accepted although many of the inhabitants were of German origin. This precarious sense of national identity was to haunt Germany through the post-war decades, right up to the present.

Post-war Germany

At the end of the Second World War Germany was a devastated and partitioned nation. The immediate effect of the war on German morale was described by Dieter Hötker of the Ruhr Municipal Association in these words:

> the collapse of all material and idealistic values was so absolute that at first people had little hope in the future. Nor did the economy give any impulses. Production totally collapsed at the end of the Second World War.

> (Van der Cammen 1988)

In Berlin and Munich, a half of all property had been destroyed. In Hamburg, Hanover, Bremen, Dortmund, Essen, Cologne, Aachen, Dresden, Nuremberg, Frankfurt and many other smaller towns, between two-thirds and three-quarters of all property was destroyed (see Figure 8.1). Post-war Germany comprised only two-thirds of what had been Germany prior to Hitler's first invasions in 1937 (see Figure 8.2). There were huge movements of population across Europe. The area of German-speaking territory that made up post-war West Germany had about

Figure 8.1 Map showing war destruction to German towns in 1945

Source: Gesamtverband der Wohnungswirtschaft (German National Federation of Housing Associations) 1989

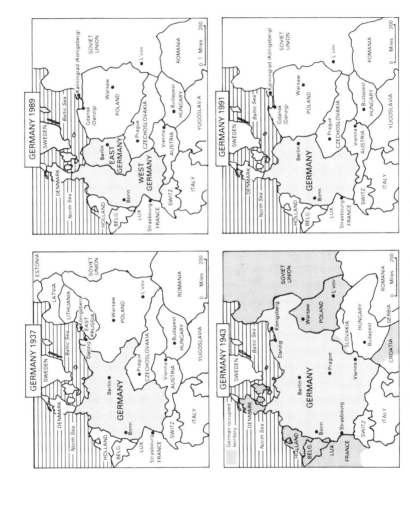

Figure 8.2 Maps showing changes in Germany's borders, 1937–91

Source: Based on *The Economist*, 2 September 1989

40 million inhabitants in 1939. By 1950, there were between 50 million and 52 million people because of the influx of refugees, mostly of German origin, from Poland, Czechoslovakia, Hungary, Russia, and East Germany itself. The task of reconstruction was daunting indeed. The division of Germany caused a generation of anguish, hostility, guilt and tension.

The bitterness of neighbouring countries compounded Germany's problems. The proximity of Germany's long border with an increasingly Soviet-dominated Eastern Europe did not help. The temporary division of Germany by the Allies, followed by the development of the cold war and 'Iron Curtain' across Europe, meant that Germany did not regain the unity, which had only been established seventy-five years earlier, for another forty-five years.

These painful experiences, living memories now to less than half of the German population, served to shape a modern but politically slow-moving fourth attempt to build a stable Germany. In attempting to do this, Germany faced the prospect of long-term, if not permanent, division into two states. In fact many young Germans had come to believe that such a division was not a totally bad thing.

Above all else, the past must not be recreated. Many had childhood memories of refugee camps. Some had Polish-speaking parents who, because they were Protestant and had German surnames, were sent to Germany after the war. These too had spent years in camps. Some had built their own houses with their families in the early 1960s under the government savings scheme. Some, who had been teenagers at the end of the war, were promoted very rapidly because the 'Nazi generation' was largely barred from holding public office. Some families had been divided by the partitions and remained hopeful of a united Germany. Most could not bear the burden of war and guilt and wanted a quiet time above all else.

Developments after 1945

The war and the effects of the war on the German political atmosphere had a great impact on German development. Firstly, Germany's industrial and economic strength, under the steadying and directed influences of a subdued political scene, quickly re-emerged. Secondly, the rebuilding of Germany's devastated cities largely followed traditional urban patterns. Thirdly, in the extreme post-war conditions, Conservative parties went to great lengths to *prove* their social commitment, and to harmonise the conditions of the German people, with an acute awareness of potential competition from socialist and communist neighbours. The division of Germany enhanced this pressure to perform. By the same token, the more left-wing Social Democrats fought harder *against* inflation and economic difficulties than they did *for* a programme of radical reforms. This produced the result that German Conservatives were 'above average in protective and distributive social welfare measures', while Social Democrats had little power to shift structures radically because of their need to stabilise production. 'Many issues of welfare policy were less controversial'; there was 'anticipatory obedience' to the major brakes on change, the Federal Bank and the Constitutional Court; and there was a great attempt at continuity, consistency and harmonisation of conditions, 'blocking the road for rapid innovation' (Von Beyme and Schmidt 1985). This unique situation only emerged gradually in the ten years following the war.

The structure of the post-war German state

The post-war West German state, founded in 1949 as the Federal Republic of Germany, was organised in a unique fashion, both to counter the fears of allies and to build internal stability and order. There were extraordinary checks and balances that made Germany stand out among European nations. Some of the most important characteristics were:

- a strongly federal structure with the *Länder*, the German regional states, both with their own government with independent powers and with representation in the federal government;
- the survival of the main social and economic structures, some dating from the time of Bismarck, such as the social security system;
- the absolute division of final responsibility between three main institutions; parliament, the federal bank and the constitutional court under which the 'ordinary' courts operate – with parliamentary powers divided between the Bundestag and the Bundesrat;
- the wide dispersal of national and state activity through several thousand government and non-government institutions, right across the Republic, in stark contrast to the massive dominance of capital cities and governmental structure in the other countries in the study;

Table 8.2 Population of principal towns in West Germany on 31 December 1986 (thousands)

Bonn (federal capital)	291	Dortmund	568
Berlin	3,400*	Stuttgart	566
Hamburg	1,571	Dusseldorf	561
Munich	1,275	Bremen	522
Cologne	914	Duisburg	515
Essen	615	Hanover	506
Frankfurt/Main	592		

Sources: Wirtschaft und Statistik 1987; *The Economist Pocket Europe* 1992.
Note: *Total Berlin population.

- the development of strong regional cities – Cologne, Frankfurt, Hamburg, Stuttgart, Munich – although this could change somewhat as Berlin regains its full status as the capital city (see Table 8.2);
- an emphasis on the 'social market economy' with an extraordinary focus on directing wealth towards socially unifying policies, such as overcoming the gross housing shortages and organising near universal employment and training programmes, while building the 'German economic miracle' and avoiding heavy, state-directed development;
- the detailed regulation of every level of activity by parliament – a German pattern that also dated from Bismarck, 'The Bundestag in the 1980s is one of the busiest parliaments in the world' (Von Beyme and Schmidt 1985).

In less than eighty years, Germany had undergone four shattering transitions. The last was the most far-reaching but led to remarkable post-war continuity, steady growth and a low international profile.

In 1990, the two parts of Germany were again reunited, making way for huge international changes and a fifth internal transition.

Chapter 9

The development of German housing from the Industrial Revolution to the Second World War

Until the Industrial Revolution, which began in earnest in the 1860s, German housing had been provided largely on a self-build basis by the predominantly rural population. The explosion of industrial population and output had huge repercussions on housing. There was a massive influx of rural people and large migrations from distant German enclaves such as Eastern Prussia, far inside Poland and the former Soviet Union, coupled with a rapidly rising birth-rate and spiralling death rate.

In the Ruhr, cities doubled in size every ten years with the total population growing from 870,000 in 1871 to 3.5 million in 1913. The numbers living in cities of more than 100,000 multiplied sevenfold in forty-five years from 2 million in 1871 to 14 million in 1914.

The development of early housing provision in the Ruhr towns and cities showed a strong and early preference for small buildings, with two to four units per house being common. The earliest industrial housing was literally self-built by workers but acute shortages arose very swiftly as industrialisation escalated. The need to move rapidly after jobs, the huge rise in densities, and the arrival of workers from much more distant German territories than the Ruhr hinterland, led rapidly away from individual units and ownership to two new forms of rented provision. Employer-provided housing became all-important, ensuring a stable and 'tied' workforce as well as a healthy (or healthier) one. Tenancies depended on employment and, increasingly, the ability to employ depended on the ability to provide housing. The company became a major provider of housing, at least in the industrial cities. Maybe a quarter of a million housing units were provided this way before the First World War in the Ruhr alone (estimate based on population figures and housing conditions described in Van der Cammen 1988).

The other main source of housing was the boom in speculative building by private landlords. The demand for renting was extremely high, the shortages were chronic and continuing, the birth-rate was soaring so that immigration only accounted for 20 per cent of the bursting urban population.

The dominant housing form in the Rhineland was low-rise, three-storey flats in small buildings, often called *Dreifensterhaus.* In the more eastern cities, such as Hamburg, Hanover, Berlin, and Leipzig, *Mietkasernen* tended to dominate.

MIETKASERNEN

Although in the Ruhr and Rhineland small houses predominated, other cities built large blocks of twenty or more flats to try to accommodate the workforce. *Mietkasernen,* as the four-storey tenements or 'renting barracks' were called, were built so that workers and their families could rent room by room off common landings. Subletting became very common. Wollmann writes of 'unparalleled housing misery in cities' (Wollmann 1985: 133). Conditions were such that the need for employers to provide better housing grew, creating a continuing flow of company dwellings. The building boom barely kept pace with the inflow.

THE FIRST CO-OPERATIVES

Social housing organisations pre-dated both the emergence of the German state and the advent of the Industrial Revolution. The birth of limited dividend, co-operative housing organisations was inspired by the Rochdale Pioneers. In 1848, the Gemeinnützige Baugesellschaft was set up in Berlin. The early co-operatives were founded as a utopian self-help way out of urban housing problems. The ideal German model was to build small, integrated communities of houses with vegetable gardens, strongly influenced by Owenite ideas from Britain. As the Industrial Revolution gathered speed, co-operatives became a practical vehicle for housing improvements among workers and the main foundation stone of social housing development.

The first state support came in the form of a legal framework for co-operatives and limited dividend housing companies in 1889. This form of organisation gave a limited return to investors, ensuring that the housing was used social purposes and not primarily for profit. Under the impetus that this provided, about 125,000 co-operative housing units had been built by 1913, about the same number as were built by British model-dwelling companies and philanthropic trusts.

Germany had the advantage of providing the framework for new forms of sanitary housing *simultaneously* with her industrialisation and first burst of prosperity. The self-help co-operative organisation dominated in Germany at this stage and the partly self-financing, privately-oriented 'dividend' system made social landlords part of an interwoven web of provision of immense variety. None the less, subsidies played a growing role and governments were drawn into housing development, if in a more indirect way. In spite of the state backing for limited dividend companies, direct intervention was extremely restricted. But the pattern slowly emerged by the turn of the century of local authorities providing low-interest loans and land, on condition that the limited dividend and co-operative companies operated 'in the public good', without profit apart from the limited dividends. Table 9.1 shows the growth in this form of housing organisation up to the First World War.

Thus, in the incredibly fast-moving fifty years from 1865 to 1914, Germany both became a unitary state and established her modern urban housing structures.

Three tenures developed, which were to flourish into the 1980s: private renting, including company lets; co-operatives and limited dividend housing companies, which were to play an increasingly important role in the industrial regions until mismanagement and corruption scandals brought their future into question in the early 1980s; and owner-occupation, which survived largely as a rural and semi-rural form of housing, without becoming a dominant urban form as in other countries.

HOUSING DEVELOPMENTS BETWEEN THE WARS

There was an immediate shortage at the end of the First World War of 2 million dwellings (Gesamtverband der Wohnungswirtschaft[1] 1989). The average amount of housing space per person was estimated to be three square metres, little more than the size of a modern single bed and only one-fifth of the housing space available in 1945 (Wollmann 1985). There was serious threat of revolution. State involvement in housing was almost inevitable in such an acute crisis.

Three main strategies were adopted by the Weimar State from 1919. Firstly, rents on all existing buildings were strictly controlled. This rent

Table 9.1 Different forms of non-profit housing associations* in Germany, 1889–1918

Company form	1889	1893	1909	1918
Co-operatives	28	101	–	est. 1,402
Joint-stock-companies (public limited companies)	27	–	–	58
Limited dividend companies (private limited dividend companies)	–	–	20	73

Source: Rahs 1982, pp. 9, 16–17, 26.
Note: *Gemeinnütziger Wohnungsunternehmen.

control remained in place with few variations until 1961. There were legal limits on rent increases, but rents were allowed to rise steadily, in line with costs. Rents for older pre-war property lagged behind rents for newer property because of the lower original costs, but in both cases costs included the cost of repair and the cost of depreciation. This made rent control less arbitrary and restrictive than in France or Britain.

The second strategy involved the requisitioning of empty property and the nomination of homeless households by local authorities to rented property. This ensured the fullest possible use of the stock for those most desperate and least able to gain access. It also heavily involved the state in the use of private property.

Thirdly, new building was heavily subsidised and targeted at both middle- and low-income groups. The collapse of the German economy, straight after the First World War and again after the 1929 world recession, meant that large bands of the German population, a majority in fact, became vulnerable and unexpectedly dependent on state aid, including many middle-class households. State subsidies took various forms, involving a combination of grants, interest-free and low-interest loans, mortgage guarantees and tax concessions.

GROWTH OF LIMITED DIVIDEND COMPANIES

For a while, the government was uncertain whether to channel subsidies directly through state organs or to use private institutions with the help of public subsidies. The latter prevailed. As a result, limited dividend housing companies, with the help of local authorities, played a growing role.

Local authorities were encouraged to take shares in the housing companies and to help set them up where they did not already exist; to ensure control over access through a system of nominations and the fullest possible use of subsidies. In some cases, the companies were entirely publicly owned. The municipally sponsored limited dividend companies were favoured over the more independent co-operatives because they could be more easily directed towards targets which the government wanted to meet, and expanded very rapidly, creating a 'mixed economy' in the German urban fabric. Most housing was regulated; some of it was provided through publicly controlled institutions; local authorities were directly involved in the administration of the housing programme and in arranging access.

Two-thirds of all new housing in the inter-war years was publicly subsidised. This was in part a result of the collapse of German financial markets and uncontrolled inflation. Owners of property paid a special tax which helped to fund new building. According to German official statistics, only traced from 1927 and missing in some of the Nazi period, German limited dividend companies built in the region of 580,000 units in the years

1927–39.[2] They were major builders in the pre-war Nazi period.

Co-operatives which had dominated state-supported housing before the First World War continued to grow but were increasingly outflanked by limited dividend housing companies, sponsored by employers, local authorities, churches, and trade unions. By 1939, there were over 4,000 non-profit housing associations – three times more than in 1918 (Rahs 1982). In that period they had built well over 1 million dwellings, although figures are confused by incomplete records, Nazism, the war, and territorial gains and losses. Table 9.2 shows the volume of house-building between the wars.

NAZI TAKE-OVER

The co-operatives experienced a further setback under the Third Reich, when their governing boards were taken over by the Nazi state. Some co-operatives actually dissolved themselves rather than accept this (Gesamt-verband der Wohnungswirtschaft 1989). Trade union-sponsored housing companies, the forebears of the giant Neue Heimat and major builders between the wars, were also commandeered and their property seized.

As part of the onslaught on mass unemployment in the mid-1930s, the Nazis recruited young workers into building squads for houses as well as roads, thus raising the levels of housing production and cutting unemployment. It is easy to see the popular appeal of such a strategy in a period of chronic shortage and unemployment. At the same time, of course, the housing programme helped Germany's industrial recovery with demand for materials, tools, transport and machinery. It gave the state a strong hold on housing activity and led to the creation of detailed legislated housing standards for building and management. While on the political front economic development helped create the German capacity and will to regain her lost territory and international status, on the housing front it meant that between 1929 and 1939, about 2.5 million new dwellings had

Table 9.2 Net additions to housing stock in Germany, 1919–39

	Net additions	Average per year
Total stock in 1918 14,000,000		
1919–29	1,966,381	178,762
1930–36	1,691,818	241,688
1937–39	804,715	268,238
Total additions	4,462,914	
Total stock in 1939 18,462,914		

Source: *Statistisches Jahrbuch* 1929–89.

been added to the German housing stock.[3] Because of the huge population upheavals and the redrawing of German national boundaries, figures are fairly confused until the late 1950s. However, by the time war broke out, some of the most chronic shortages had been alleviated.

In Germany there was no inter-war slum clearance programme. The relative newness of her stock and her urban populations meant that almost all new building represented strict additions. There was strong support for the construction of houses, although no records have been found that show even an approximate breakdown of the inter-war stock into houses and flats.

Thus, in the inter-war years, stemming directly from the earliest state intervention in housing in the 1880s, the foundations of modern German housing policy were laid: rent regulation and rises in line with costs to cover repair and management as well as capital costs; nominations of needy households to all types of housing, coupled where necessary with a form of 'billeting' (*Einquarterung*); strong public support for limited dividend housing companies and co-operatives; the involvement of many different kinds of organisation in the provision of housing; even-handed treatment (compared with Britain) of 'social' and private landlords, involving consistent and unchallenged support for private landlords; strong government regulation of housing. This framework had been built, under constant political pressure, in response to the needs of Germany's booming urban population.

In 1940, a new social housing act was passed by the Nazi government, extending existing legislation and establishing the legal framework for non-profit housing that was to last until 1990 – *Wohnungsgemeinnützigkeitsgesetz*.

NOTES

1   GWW – Gesamtverband der Wohnungswirtschaft – German Federation of Housing Associations.
2   It is impossible to give accurate figures after 1935. All pre-war figures are estimates based on information from the Federal Ministry of Housing and the GWW.
3   See note 2.

Chapter 10

Post-war housing

GERMAN HOUSING DEVELOPMENTS AFTER 1945

The German housing structure in 1945 was in nothing less than ruins. Much private ownership had simply vanished in rubble. By 1945 there were six people per dwelling compared with 3.6 in 1936. This represented a loss of maybe 5.5 million dwellings (estimate based on German government and Gesamtverband der Wohnungswirtschaft (GWW) figures), although the vast population and property upheavals of the early post-war years make the accuracy of any figures questionable.

The Allied Control Council, which ran German affairs after 1945, temporarily confiscated all private property. Local authorities were given the job of helping organise shelter for the homeless. They nominated families to rooms. Every surviving or potential property owner was drafted into the emergency housing programme. Over half the new housing was directly subsidised.

PRIVATE 'SOCIAL' LANDLORDS

Private landlords were desperately needed in the post-war crisis. Both private rents and the availability of private property were tightly controlled. In exchange for help with building or rebuilding, landlords were required to charge an agreed 'social' rent and to house only households who had been given a certificate of eligibility by the local authority. This was based on an income limit, which was set to determine need: 60 per cent of the population were eligible because housing problems affected such a wide band of the population.

From the extreme post-war situation, the West German social housing system emerged into a new and more broad-based approach. Any landlord, whether private owner or limited dividend company, became a 'social' landlord if they accepted nominated households in exchange for financial help. This 'social' housing was then regulated for as long as the subsidies lasted. Not only were landlords hugely diverse, but local authorities played a key role in ensuring that huge masses of needy housholds were housed. The Allies pushed every institution to the limit in co-operating with the central council, but were only temporarily involved in local systems. The new government role gradually emerged as a broad umbrella system, covering an immensely varied and dispersed housing network.

EAST GERMANY AFTER THE WAR

As German reunification became more and more problematic, and a Communist government took over in East Germany, the system in that part of the country evolved very differently, although still broadly based on the three pre-war sectors – co-operative companies, private landlords, owner-occupiers. In addition, the state began to build directly. The fact that so much property in East Germany was abandoned by people who moved westwards made the situation very different from that in post-war West Germany and ownership patterns much more confused. The state took a dominant and directive role, though it expropriated less than might have been expected, in anticipation of German reunion. From 1950 onwards, East German housing conditions developed within a completely different economic framework from that of West Germany. Therefore, the discussion on pp. 109–49 applies only to West Germany.

NEW BUILD PROGRAMME

The first post-war German housing task was rebuilding what could be rescued from the rubble. The second task was to launch a massive new-build programme. As national political life re-emerged, the *Länder* governments were given full powers to put in place whatever programmes they could, using the most responsive local structures to hand – the local authorities and local landlords. Housing was a regional and local responsibility. The federal government, with considerable American aid, orchestrated generous subsidies, ensuring that they went to help the people who were not housed at all and determining the minimum acceptable housing unit that could be built to ensure help for the maximum number of households.

In 1949, 223,000 new properties (net additions) were completed. By 1950, this had jumped to over half a million. Although there are doubts about the accuracy of figures relating to the immediate post-war period, and although there are inconsistencies in the available information from official German sources, it is important to indicate the basis on which new policies and initiatives developed. Table 10.1 shows the situation in 1950, covering two-thirds of today's Germany.

Table 10.1 Population, households, and housing stock in West Germany, 1950

Population ('000)	Households ('000)	1-person households (%)	Stock ('000)
49,850	16,650	20	10,082

Source: Emms 1990.

HOUSING DEVELOPMENTS IN THE 1950s

In 1950, under the first Housing Act, the post-war framework for sub-sidising any builder or landlord who agreed to minimum dwelling size, cost rent levels (after subsidy) and access for people with local authority eligibility certificates, was established. The emergency control of the housing market, which lasted until 1960, paid dividends both in volume of housing production and in low production costs for German goods – low labour costs partly resulting from low housing costs (Wollmann 1985). In turn, the rapid growth produced wealth that could be ploughed back into expanding housing programmes. About 6 million new housing units had been built by 1960, 2 million of which were owned by the limited dividend companies and a further 1 million of which were 'social housing units' built by private landlords.

PRIVATE RENTING DEVELOPMENTS: INDIRECTLY SUBSIDISED BUILDING

The majority of German housing since the war was provided without direct subsidies, nearly 11 million dwellings in all by 1990. Generous tax incentives and a secure, stable, rented market appear to be the two main reasons why private renting remained the most significant tenure, unlike all the other countries in this study.

Private landlords were encouraged to invest in property for rent on advantageous terms through incentives:

* They could write off depreciation of the capital value of the property at a steep rate over the first eight years of ownership.
* Rents in Germany were calculated in relation to the *actual* costs of ownership, management and maintenance. The full costs could be offset against tax on all sources of income.

These measures helped keep rents down by reducing costs significantly. Controlled rents in periods of acute shortage prevented landlords from profiteering – or restricted their ability to do so, thereby reducing the need for direct government intervention. But rent controls were not punitive and they were relaxed as soon as possible. While the profits to be gained from being a landlord were sometimes restricted, landlords were able to gain a secure capital asset, providing a steady return on investment over time and, in some periods such as the 1960s, private renting became highly profitable. High property values, a continuing though reduced shortage of rented housing, and a relaxation of controls ensured this return. Because of multi-party support for private renting, landlords flourished with little fear of political intervention.

Rents were set more and more loosely while rents on existing tenancies could only rise roughly in line with costs. Over time, rents ratcheted up as new properties were produced at a rapid rate with new rents set to cover new and higher costs. Later the system of comparable rents for existing tenancies was introduced (see p. 122).

Thus, through tax advantages, favourable loans and subsidies, and a regulated but stable, private rental system, private landlords continued to be attracted into the business of letting property and to stay in it because of guaranteed returns over time. Anyone on a secure and taxable income stood to gain in the short and long run from investing in rented property. For this reason, doctors, lawyers and other professionals, as well as businesses, could be drawn into investing their surplus earnings and savings in rented housing.

DIRECTLY SUBSIDISED PRIVATE RENTING

Over the early post-war years, government subsidies to private 'social' landlords met the difference between the controlled or registered rent and the real cost of producing and running the dwelling. Assistance to private landlords took the form of low-interest or concessionary loans to cover the early period of repayment when interest charges would be steepest. In return, the landlord agreed to take a nominated tenant with an eligibility certificate. The landlord also had to agree to charging a cost rent for the duration of the period of subsidy. Subsidies for cost rents were said to encourage landlords to maximise costs. This may have been true in big cities with a large social housing sector, but generally costs for social housing were lower (Pfeiffer 1989). At the same time, the government was guaranteed a return on the public investment through nominations and the landlord was attracted into letting the unit to a needy household. Private landlords renting social units enjoyed a tax advantage denied to non-profit organisations in that they could deduct the depreciation from their taxable income. This system of 'social housing' through private landlords is still broadly in place, although the actual level of the subsidies has changed over time and the volume of money targeted at 'social' private renting has also changed.

OWNER-OCCUPATION SINCE THE WAR

After the war, Germany put the lion's share of its effort into the production of small rented flats in major urban areas, where displaced households were crowded into chaotic remnants of pre-war housing. Individual property rights were over-ridden by huge social needs.

Quickly, the idea of galvanising individual savings and encouraging people to provide housing for themselves emerged and, in 1956, a house-building law was passed, encouraging owner-occupation through grants, low-interest loans and guarantees. Special savings and lending associations *Bausparkassen*, somewhat akin to giant credit unions, were encouraged. Through them, individual savers received a low-interest payment on their savings, but having saved about 40 per cent of the cost of a house, they could borrow about 60 per cent in addition at a low interest. The system of low-interest payments on savings and low-interest loans balanced itself out. Sometimes second mortgages were needed to make up the amount. Subsidies were available for those with eligible incomes – about 60 per cent of the population. A strong motive was the need to give refugees who had lost all property in the war the chance to build a new life and a sense of belonging, with a stake in the new Germany (Wollmann 1985). It was also a way of mobilising individual savings and drawing banks and builders into more diverse housing activity.

However, saving up such a large portion of the total cost was a lengthy business and people could not normally consider buying until their family responsibilities were reduced and their incomes more stable and higher. Owner-occupation was so expensive that it was only attractive if buyers could afford to pay for something significantly bigger and better than their previous accommodation.

LAND PROBLEMS FOR OWNER-OCCUPATION

With Germany's high population density, individual house-building usually involved finding land outside the cities. The existing urban housing stock was almost entirely rented out profitably by private landlords, so until recently there was only a small market in second-hand homes for owner-occupation. The strong powers that local authorities had to acquire land meant that land for individual owner-occupation was hard to come by and expensive. Local authorities favoured high density 'social' rented developments because of high demand. The acute shortages, lasting into the 1960s, and the emergency housing controls, lasting for fifteen years after the war, meant that absolute priority was given to producing the maximum number of small, utilitarian units.

COMBINING OWNER-OCCUPATION AND RENTING

Many new owner-occupiers built a house with a rental flat attached, as special tax incentives applied to those who were prepared to be landlords while becoming owner-occupiers. If a rented flat was *within* the house, there was an effective double incentive, with tax concessions and the promise of income from rent to repay part of the very high owner-occupier costs, much higher in Germany than elsewhere. This was a very attractive proposition, both financially and in terms of the quality of housing an owner could thereby afford. The result has been that 8 million one- and two-family houses were built between 1949 and 1986, nearly half of the massive 17 million additions. Figures are not available to separate single-family houses from two-family houses but until the late 1970s, the combined owner-occupier/rented unit was possibly more common than the single family house. Much of the new post-war private-rented sector was also built in blocks of under six units, often with a subsidised owner-occupied unit within for the landlord or landlady. The special tax concessions for building two-family houses led to some abuse and many such units quickly became single family houses again.

GROWING POPULARITY OF OWNER-OCCUPATION

Forty per cent of the population became owner-occupiers by 1990. There were a number of reasons. Firstly, the post-war German population regained wealth extremely quickly. As it aged, more and more people could afford to – and aspired to – buy more housing. While rented housing remained popular, over time people wanted to acquire bigger and better housing. They were encouraged to vacate relatively cheaper urban rented housing to make room for the more needy and younger households constantly queueing. Government policies to galvanise savings and encourage people to build both released existing units for lower-income households and created new units for the better-off.

Secondly, the huge influx of dispossessed Germans from eastern territories were eager workers, savers and often builders too. The West German government was anxious to incorporate them.

Thirdly, the intrinsic appeal of a single-family house in suburban or semi-rural areas became increasingly sought-after as congestion and environmental pollution gathered speed. German urban planning was functional and based on economic rebuilding. The miracle of recovery exacted a high price in dense townscapes, followed by rapidly extending urban sprawl. There was growing dissatisfaction among the increasingly affluent middle class with regimented blocks of flats, limited space and tight controls.

After 1960, the government sought to make owner-occupation available to wider and wider groups. Limited dividend housing companies were encouraged to build for sale, and Neue Heimat, the now demised trade union-owned housing company, built over a million houses for sale in the 1970s. The idea of owner-occupation began to gain popularity, as in other European countries. However, the tax incentives to invest in private-rented housing were always higher than the tax subsidies for owner-occupation. As a result, a private tenant indirectly received more housing subsidies than an owner-occupier and renting continued to dominate.

LIMITED DIVIDEND COMPANIES

The voluntary limited dividend companies and co-operatives had suffered great setbacks immediately before and during the war, either having their property seized and their boards dissolved or being taken over. There were less than half the number of non-profit companies in 1945 that had existed in 1939 (Emms 1990).

They were urgently restored or new ones were formed to help the government's emergency housing programme. They had survived and been used as a vehicle of public intervention, often closely linked to local authorities and *Länder* governments. They were more tightly regulated than private landlords and were viewed as a permanent source of low-cost, affordable rented housing for eligible and nominated households. Much of the early post-war housing was provided through them.

Principles of non-profit housing

Strictly speaking, limited dividend companies were not non-profit since investors received a 4 per cent interest on capital, whether they were member-based co-operatives or were owned by private institutions or public bodies.

The founding principles of social housing companies, enshrined in German law in 1940 under the *Wohnungsgemeinnützigkeitsgesetz*, were: that they would operate only for the *public good* (*Gemeinnützigkeit*); that they would charge *cost rents* rather than economic or market rents;[1] that dwellings would be no larger than 120 square metres so as to be able to accommodate the majority of households on moderate incomes; and that they would have as their sole and continuing purpose the *building and management* of housing (though related facilities such as shops, nurseries and community facilities could be provided). The companies were required to have a *continuing building programme*, which became an increasing burden as subsidies were reduced, costs rose, and management difficulties multiplied with the onset of industrialised building. This law remained largely in place until 1990 when the special status and privileges of limited dividend housing companies were abolished.

Subsidies

The subsidies, in the form of initial grants, low-interest loans and subsidies for commercial loans, meant that over the thirty to forty years of repayments the housing companies had significantly reduced costs to enable them to help households of modest means. These special subsidies were also available to other landlords if renting to eligible households. Private landlords could then also become 'social' landlords. The subsidy programmes were introduced in a situation of chronic shortage and the housing companies burgeoned, targeting a wide income band. The scale of operation became an increasing problem under the impetus to build. Over 7 million households were eligible for social housing after the war. The companies, along with other 'social' landlords, housed households below the income limit with eligibility certificates issued by the local authority. The limited dividend companies grew particularly fast in large cities because local authorities and *Land* (regional) governments relied more and more heavily on them for their housing programmes.

Rents in most cases were below private market rents. An exception to this was the large-scale modern developments of the 1970s, which were normally funded under a special programme with large, predetermined rent increases built in (Rahs 1992).

Structure

There were four main parties to the development of limited dividend housing companies: central government; the *Länder* governments; local authorities; and the housing companies themselves. It worked as follows. *Central government* provided the legal and financial framework giving grants for the *Land* government programmes. The *Land* governments (regional governments), with primary responsibility for housing, developed housing strategies, identified resources, including land, as well as channelling finance to the companies. The *Länder* or regional governments played a much greater role in Germany than other countries in this study, not only in ensuring delivery of government policy, but also influencing it through the Bundesrat and developing it within each state. Thus, the more urbanised *Länder*, often dominated politically by the Social Democratic Party, gave a heavy emphasis to publicly subsidised housing programmes, directing assistance from regional sources to supplement government aid towards social housing companies. The *Länder* also fostered the building of company-owned 'new towns', particularly in the city states of Hamburg, Bremen and Berlin, but also in the Ruhr area, around Cologne and in Baden-Württemberg.

The *local authorities* were the direct planners and co-ordinators of development. They were responsible for registering households requiring housing, which gave them a clear idea of the level of demand and need. They issued eligibility certificates to households within the income limits, making them eligible for social housing by either non-profit or private 'social' landlords.

In post-war Germany, the local authorities also had responsibility for registering rents in the private-rented sector and kept track of the level of empty property in order to nominate particularly needy households to both private landlords and housing companies. As a result, they actively monitored shortages and need. They had planning powers and power to acquire land – they could liaise directly with the social housing companies over needed developments. Many large cities had a major share in companies. In this way, they were able to influence developments directly. The location, size of building sites, distribution of dwelling sizes, rent levels – all were heavily influenced by the local authority's role. However, they did not build directly.

The *limited dividend and co-operative companies* provided the main vehicle for social housing development. The 1,240 *member-based co-operatives*, owned a total of about 1 million units. A majority owned fewer than 500 units. Although these started out as self-help organisations it was not a condition of membership to require rehousing or be a tenant. Individuals wanting to invest spare money in a good cause for a limited return of 4 per cent could become a member of a co-operative by buying a share. Many co-operatives originated in churches, trade unions and voluntary bodies. The co-operatives largely operated on a small, local scale, had good quality stock and managed it meticulously within the non-profit law. Co-operative housing was generally popular. On the whole, the smaller co-operatives have been able to resist government pressure to become more market- and profit-oriented, partly because of their membership base, partly because of their strong founding principles of self-help linked to 'the public good'. The investors in co-operatives backed these principles with financial commitment. However, the larger co-operatives acted more like limited dividend companies and became much more bureaucratic.

About 540 housing companies were *private limited dividend companies* and sixty were public limited companies. The companies owned an average of 5,000 units each, but the fifty largest companies, mainly the big city companies in which local authorities and the *Länder* had a major or controlling stake, had an average of 20,000 units each. These fifty owned one-third of all social housing. They operated on much more bureaucratic lines, but were pushed by financial and legal autonomy into operating on semi-business lines. Although they were not allowed to make a profit in a strict sense, they could invest surpluses in profit-making activities. Any profits from this investment would be ploughed back into the activities of the non-profit investing company.

The limited dividend housing companies provided 25 per cent of all rented housing and 15 per cent of the total German housing stock. In the major cities, up to 50 per cent or more of all housing was provided by them. Between 1950 and 1960, over half of all new-build units were 'social'; of those, two-thirds were built by limited dividend companies. Their share of the total stock dropped from 30 per cent at their peak to less than 20 per cent in 1980 and 15 per cent in 1990. The loss of relative strength was caused by the great boom in private building and by problems in the social sector (see pp. 127–46).

The government, at the major turning points in German history – after the 1914–18 war, in the 1930s, after the Second World War, and again in 1990 – relied heavily on these housing companies, both as the most responsive and malleable instruments for a burst in house-building but also as the major vehicle for guaranteeing access to those most in need.

The social housing companies built about 1.5 million homes in the 1950s; nearly 2 million in the 1960s; and just under 1 million in the 1970s. A small number of these were sold to private owners and the companies also built directly for owner-occupation. By 1990, a total of about 4 million units were in the ownership of the companies. They built some housing *on behalf* of private investors.

While they were major builders in the early post-war period, private owners were increasingly favoured, when once the worst crisis was over.

THE ROLE OF LOCAL AUTHORITIES IN HOUSING PROVISION

Local authorities had a powerful role in all forms of housing provision, as well as in relation to the social housing companies. They helped maintain standards. With state governments, they could encourage house-building.

Access

Particularly in the years of acute shortage, local authorities played a vital role in the question of access. Homeless families and families in acute crisis (e.g. that could only be reunited if they were housed) were nominated directly to empty property. Large housing departments developed in the major cities to co-ordinate the huge demand, to vet applicants, to issue eligibility certificates to all those within the income limits for social housing, to put landlords and prospective tenants in touch with each other, and to nominate directly in cases of emergency.

Local authorities negotiated lettings agreements with non-profit bodies which helped meet rehousing needs and provided landlords with support. Cities such as Bremen have pioneered new forms of collaboration between local authorities and housing companies to try and meet new housing needs.

Funding channel

Local authorities acquired a major role in funding too. They provided grants and subsidies to owners derived from federal and *Land* (state) programmes; they matched government grants with their own funds raised through taxes. Their planning role enabled them to subsidise social housing, providing land at less than full market value, or give an *Erbbaurecht* (right to build) on local authority land.

Local authorities were also responsible for enforcing rent control and for keeping a register of all private rents within their area – a role they still play, although direct controls were progressively withdrawn after 1960.

Development role

Local authorities played a crucial 'commissioning' role, helping determine and encourage the scale of housing programme that was considered necessary. Local authorities did more than 'enable' – they exercised planning control powers; they had resources to acquire land to distribute to builders. In the heyday of mass housing, with support from the *Länder* governments, they orchestrated the development of large, new 'satellite cities', often with a number of social landlords.

Local authority links with social landlords

Local authorities worked directly with all kinds of landlords, also establishing new limited dividend companies to expand their role.

SUBSIDIES TO ALL THREE TENURES

Figure 10.1 shows the way subsidies and tax concessions filtered through all tenures. It illustrates the social role of all tenures and the possibility of limited dividend companies operating in the unsubsidised market.

'Free' housing meant housing with no outstanding subsidised loans and without local authority rehousing obligations, whether private, social or non-profit. 'Subsidised' social housing, conversely, meant housing in any sector with undischarged, publicly subsidised loans, to which local authorities often had nomination rights. Eligible households within defined income limits could apply for social housing and social landlords had to let 'social' units only to tenants with eligibility certificates, although the landlord selected the particular tenant. In addition, local authorities bought nomination rights to specific dwellings, to which they had the right to nominate a particular household which the landlord was obliged to accept. Thus a combination of rationing and rent controls, subsidies for social housing through private and non-profit landlords, grants and tax conces-

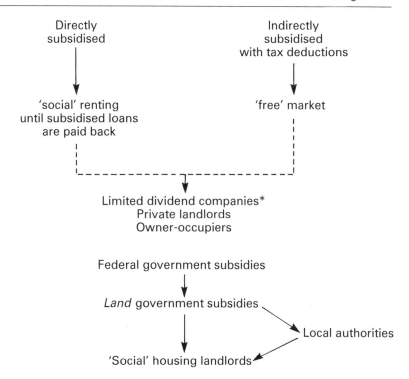

Figure 10.1 The channels of subsidy across housing tenures in West Germany

Note: *Until 1990, limited liability companies formed the one segment of the housing
market not to receive tax concessions. With the abolition of the non-profit status
(1 January 1990). they now receive the same tax treatment as private renting.

sions for owner-occupiers and tax incentives for private landlords, ensured
a rapid rebuilding of West Germany's housing stock (Tomann 1989).

IMMIGRATION

Germany, virtually uninterrupted in the post-war period, absorbed large
numbers of immigrants. Between 1950 and 1960, half a million East Europeans a year arrived, totalling 5.2 million in the decade; foreign workers
from southern Europe and Turkey began to arrive in the 1960s, totalling
half a million in 1968 and rising to 4.5 million by 1985. By far the largest
immigrant group was from Turkey, and by 1988 they formed 2.3 per cent of
the German population. This steady flow, coupled with the fragmentation
of German households into smaller and smaller units, kept up the demand
for high levels of building, in spite of the steeply falling birth-rate and
pessimistic predictions of absolute population decline. Increasingly, immigrants lived in older, inner, rented areas alongside other poor households.

Tables 10.2, 10.3, 10.4 and 10.5 show the size of the foreign population in Germany, its youthfulness, its concentration in certain inner areas of cities, and unemployment levels.

The continual pressure of immigration and the concentration of poorer immigrant households in declining areas fuelled social tensions, and created demand for renewed housing efforts.

THE 1960s

In 1960, the emergency housing system was dismantled. Rents and security were considerably freed up, particularly in the old housing areas. Social rental units from then on could be sold under certain conditions to owner-occupiers, although in practice this did not often happen as social housing

Table 10.2a Foreign population in West Germany, 1951–88 (millions)

1951	1961	1971	1973	1985	1988
0.5	0.7	3.4	4.0	4.7	4.5

Table 10.2b Total foreigners in West Germany by country of origin, 1986

Country of origin	Percentage	Country of origin	Percentage
Turkey	32	Poland	3
Yugoslavia	13	Spain	3
Italy	12	Asia (India, Iran,	
Greece	6	Lebanon, Others)	8
Austria	4	Others	19

Source: *Wirtshaft und Statistik* 1987.

Table 10.3 Age of foreign and native population in West Germany, 1988

	Age of foreigners		Age of West Germans
	Numbers	% of total	% of total
Under 21	1,200,000	31	20
21–45	2,200,000	48	65
45–65	900,000	19	
65+	125,000	2	15

Source: *Statistisches Jahrbuch* 1988.
Note: The foreign population is disproportionately youthful with very few elderly and a bulge in the 21–45 age group.

Table 10.4a Population of foreign origin in West German cities, 1987

City	Foreign population (%)	City	Foreign population (%)
Frankfurt	25	Düsseldorf	16
Offenbach	22	Cologne	15
Stuttgart	18	Mannheim	15
Münich	17	West Berlin	14

Table 10.4b Population of foreign origin in certain older inner areas of West German cities, 1987

City district	%	City district	%
Berlin Kreuzberg	28.8	Münich-Ludwigsvorstadt	40.3
Cologne-Kalk	31.4	Frankfurt-Bahnhofsviertel	80.0

Source: Wirtschaft und Statistik 1987.
Note: Foreigners were concentrated in old inner-city neighbourhoods.

Table 10.5 Levels of unemployment among foreigners in West Germany compared with national average 1978, 1980, 1986

	1978 (%)	1980 (%)	1986 (%)
National average	3.8	3.3	8.2
Foreigners	10.4	13.7	11.7

Source: *Bevölkerungsstruktur und Wirtschaftskraft der Bundesländer* 1978, 1980, 1986.
Note: Foreigners made up 8.5 per cent of the workforce, a higher proportion than their size in the population, reflecting their more youthful age structure. None the less, foreigners were disproportionately affected by unemployment, particularly in the early 1980s.

was generally built in blocks of flats, while owner-occupiers wanted single-family houses.

After a decade of building nearly 600,000 units a year, the most acute shortages were over. There was a fairly strong shift away from public subsidies towards private initiative. However, market pressures made it necessary in 1965 to reinstate the regulation of social housing rents and to reserve those units for households with modest incomes (still broadly defined to include about 60 per cent of the population).

Housing allowances

In 1965, to compensate for rising rents and a freer system, a housing allow-ance system was established. The relaxation of rent controls, the ending of the emergency system, and the growing emphasis on owner-occupation and private renting were coupled with more and more households having difficulty in paying market rents. German households paid much higher proportions of their income than the British on rent, often 29 per cent, but poorer households might have to pay 30 to 40 per cent. Some sort of personal income or rent support became inevitable if rents were to move towards reflecting cost. Only a very limited proportion of households were eligible – 12 per cent of tenants and 13 per cent of social housing tenants. While low-income owner-occupiers were eligible for allowances, in practice very few received them and 90 per cent of allowances went to tenants. With the help of housing allowances, households were expected to pay up to around 20 per cent of their income to meet housing costs.

In 1967, a second housing subsidy system was introduced, with a higher income limit, lower subsidies, and a greater emphasis on owner-occupation. This pushed up rents. Direct subsidies to owner-occupation were also improved.

The shift to private market solutions produced its own problems. In the late 1960s, new housing pressures began to emerge that created stress in poor areas through:

- rising rents in old, cheap areas as rents were freed of controls and demand for low-cost housing continued;
- increasing conversion of private rental units to owner-occupation;
- *Fehlbelegung* (mis-renting), as existing tenants' incomes rose above the eligibility level for social housing, preventing newcomers on lower incomes from getting in;
- the steady fall in production, particularly of low-cost social housing:

Although there were by then 4 million social rental units, pressures at the bottom intensified.

REGULATING PRIVATE RENTS

In 1969, security of tenure was introduced and a loose form of rent regulation, *Vergleichmieten* (comparable rents), was brought in in 1972. At each new private tenancy, the landlord could set rents at the market level; for existing tenants, rents could only rise in line with comparable rents. The existing rents of five other similar properties in the area were used to establish comparable rents. This level was verified through the town hall, which kept a register of all local rents (*Mietspiegel*). Over time, rents roughly kept pace with inflation and in areas of shortage, outstripped it. There were many ways of making the system bend in favour of rising rents. While rent rises caused hardship for lower-income families, they helped encourage continued investment in private rental units.

The production figures continued at a very high level until 1965. In the boom years from 1950 to 1973, 13.5 million units were built.

NOTE

1   Costs reflected repayments on capital after all subsidies had been allowed, plus allowable spending on management and maintenance.

Chapter 11

Housing in the 1970s and 1980s

HOUSING BALANCE IN THE 1970s

The renewed emphasis on production resulted in a record 714,000 new units in 1973 in both 'social' and 'free' markets. This boom coincided with the oil crisis and a collapse in the housing market. For a time there were 500,000 unsold, low-cost, owner-occupier units in Germany and many empty units on larger estates. The proportion of publicly subsidised units dropped to one-fifth of the total.

The intense German post-war effort had produced by 1974 a rough balance between the number of households and units, with 16 million new units and about 10 million preserved and restored older units. Production dropped steadily from 1974, although it did not fall below 350,000 units a year throughout the 1970s.

The rapid rise in costs after the oil crisis led to the government changing the subsidy system and increasing social rents in order to reduce the burden of subsidies. Rents were to rise in line with other sectors up to a limit. Tenants' security was enhanced at the same time.

URBAN RENEWAL

Most private-rented housing was built in inner cities and much of it dated from pre-war or the early years following the war. Rents for this type of property remained low and conditions and standards were increasingly outmoded. The poorest households tended to live in these areas. Increasingly, foreign workers and their families became concentrated there. As the housing shortage diminished, urban renewal programmes were mounted, following the *Städtebauforderungsgesetz* of 1971 to upgrade inner-city rented property. Five hundred and sixty inner-area renewal schemes received public support. Renovation grants were made available, through which landlords could improve older rented property to modern standards. Inner areas became sought after again as the penalties of commuting mounted. Owner-occupation in 'gentrified' old property became attractive.

The low rents for older property became less attractive to many landlords than conversion to owner-occupation. Many lower-income households were displaced as older properties were demolished or converted. Only about 25 per cent of converted properties went to these groups. The boom in outer estates helped this development, as poorer households were displaced outwards. The supply of new private-rented units continued, but there was an accelerating loss of low-rent, older private units. The urban 'gentrification subsidies', which urban renewal programmes quickly turned into, put increasing pressure on poor households. The policy was eventually amended because of this in the late 1980s to make conversion more difficult (Wullkopf 1992).

OWNER-OCCUPATION

Owner-occupation in the 1970s took a major upward turn, under the urban renewal schemes.

Many of the owner-occupier landlords who had built two units – one for rent – in the early post-war period incorporated their rented flat into the house as their outstanding loans were repaid and their debts reduced, making a bigger and better-quality unit for themselves. Many rented flats disappeared from the market in this way. Condominium-type developments, owner-occupation in purpose-built blocks of flats, also became more attractive. The sale of private-rented flats to occupiers was allowed and could appeal to buyers in a situation of shortage where market rents might continue to rise steeply.

There was steady government support for new-build, owner-occupied units throughout the 1970s. From 1976, more owner-occupied units were subsidised than were rented units (Hills *et al.* 1990). Owner-occupation thus became more attractive to a significant minority of the increasingly affluent, ageing, and security-conscious German population.

There was also some growth in second homes, almost entirely owner-occupied. For people renting small flats in cities relatively cheaply over a long period of time, buying a country house for holidays and retirement, through savings, could be a sound proposition.

'FREE' HOUSING

By the end of the 1970s, the movement towards market rents for all landlords was almost universal, although Berlin had lingering controls.

The emphasis was on decontrol with the allocation of dwellings according to the landlord's choice, rather than through local authority nominations. Social landlords could free their rents by the early repayment of government-subsidised loans, as a result of which they were released from controls. This meant that the number of 'social' rented units in the private-rented sector dropped significantly in the 1980s as repayments reduced outstanding loans.

The result of the freeing of controls was that, overall, private rents in Germany rose faster than the cost of living. The major urban areas, but particularly the booming areas of Frankfurt and southern Germany, had much faster growing rent levels than elsewhere.

One effect was to increase the numbers of housing benefit recipients and its overall cost, although Table 11.1 underlines the fact that the majority of Germans in all social groups met their direct housing costs in full. Many argue that a combination of high wages and the large rented sector led to competition between landlords for tenants, and that the relatively free market encouraged small individual investors to continue to rent out units. It also shows the concentration of housing benefit among tenants within social housing, and among immigrants.

Table 11.1a Housing allowances and their distribution in West Germany, 1988

Recipients	%
Tenants	93
Owner-occupiers	7
Elderly	35 (1981)
Unemployed	17 (1981)
Percentage of all tenants who received housing allowances	12
Social housing tenants who received allowances	25
Owner-occupiers who received allowances	1.5
Immigrants who received allowances	40

Note: Housing allowances reduced the proportion of income a household had to pay in rent from 41 per cent of income to 24 per cent; 32 per cent of housing allowance recipients also received 'social assistance', a much wider income support for those in need. By 1988, 1.9 million households received housing allowances.

Table 11.1b Recipients of housing allowances and costs in West Germany, 1972–86 (DM billion at 1970 prices, adjusted by GNP deflator)

	Recipients ('000 households)	Spending
1972	1,278	1.05
1975	1,849	1.20
1980	1,622	1.10
1985	1,572	1.26
1986	1,950	1.67

Source: Hills *et al.* 1990.
Note: Billions are in thousands of millions.

MASS HOUSING

A major influence on the changing focus of government intervention was the growth in large, peripheral, social housing estates in the 1970s.

The social housing companies had acquired an important civic role in city areas, carrying out many of the major new town developments that became so fashionable throughout Europe in the 1960s and 1970s. On the whole, it was limited dividend companies rather than co-operatives that were involved in mass construction, partly because they were often closely associated with local and state governments, partly because they were bigger and more prepared to undertake large-scale developments; co-operatives remained smaller, more independent, and more quality conscious. The housing companies carried forward most of the major large-scale, outer-estate developments. These in Germany, as in France, could only be located on the edge of towns, or even right outside them, because of the already densely rebuilt urban areas. It was still unusual in Germany after war devastation to demolish inner-city housing on anything like the scale contemplated in Britain. Inner-city land was also more expensive and this provided a strong incentive to build further out.

Municipally sponsored housing companies were obvious tools for achieving an important government goal, the creation of large numbers of modern homes for people on limited incomes in the areas of greatest demand. The urban densities and levels of overcrowding among the poorest households, the rapid population and household growth, the development of modern building techniques and the German origins of the modernist architectural movement with Gropius and others around the Berlin architectural school (Wolfe 1991), all ensured that Germany would build its share of 'mass' housing with brutalist or functionalist design and a 'machine' approach to living, particularly if it was to house workers. There were, however, significant local pressures in the decentralised West German system that opposed schemes and prevented a repetition on a similar scale in Germany of either the French *grands ensembles* or the British slum clearance estates.

The fashion for 'mass housing' and factory-style housing construction arrived later in Germany than elsewhere but for a while it was popular in social housing circles. There was an attempt to break out of the urban strait-jacket. In the late 1960s and 1970s, 250 outer estates of more than 500 dwellings, constructed mainly in large, concrete blocks, became the dominant form of 'social housing'. German outer estates were very large, with two-thirds of units on estates of over 2,000 dwellings each (Gibbins 1988). The 600,000 industrially built flats of 1965 to 1980 made up half of all non-profit units built in that period. Even though they comprised only 10 per cent of the total social housing stock, they eventually saddled the housing companies with significant, long-term costs, serious management difficulties and a tarnished public image. Today, *Grosswohnsiedlungen* are renowned for their lack of social provision, poor transport links, segregated population and disproportionate numbers of unemployed, one-parent families and dependent households. Twenty-five years ago, however, they offered for the first time good-quality flats (internally they were usually attractive) to people who had never before had such a chance. The greater social problems reflected in part the attempt by social housing organisations to open up opportunities to those in greater need.

There were many reasons for the adoption of mass building. There was a dominant belief that it would provide new solutions to urban congestion and the difficulties in satisfying demand for higher quality. There was also a belief that it would provide quality for the masses who, although now largely housed, occupied very minimal dwellings put up in great haste after the war. It provided housing companies and government with a vehicle for rehousing previously excluded disadvantaged social groups. It combined Germany's industrial strength with the ambitions of the social housing organisations and state and local governments.

THE STYLE OF THE OUTER ESTATES – *GROSSWOHNSIEDLUNGEN*

Some outer estates were built using conventional building techniques. None the less, the idea of producing at high speed the components for prefabricated building, copied from French and Danish ideas, was appealing. Panels measuring three metres by three metres and lifted by giant cranes to ten or more storeys were an enticing advance over bricks three inches by nine inches, 700 of which would constitute a single panel, each individually laid. On this basis, tenants would 'get more for their money'. This was an important rationale for politicians, builders, and housing professionals. Builders, of course, could see the scope for large profits.

Gropius's architectural modernism could come into its own in a way that had not been possible in the 1930s. In his memory, one of the biggest estates in Berlin was called 'Gropiusstadt'. The estates were generally planned for 10,000 or more people, in many cases double or treble that. They were conceived of by the main urban authorities with the resources and population to think big, such as Hamburg, Bremen, Berlin, Düsseldorf, Frankfurt, Cologne and Münich. With Germany's dispersed regional structure, most major regional centres saw the need for, and recognised the usefulness of, large new estates.

In many cases, the projects were too big to be undertaken by a single housing company. They were co-ordinated by the local authority, the *Land* government, or by a major housing organisation with town planning and urban development subsidiaries, such as Neue Heimat. Many companies

Plate 11.1 German caretaker on a bicycle

fell in with the 'new town' schemes on the promise of growth, prestige, resources and public favour. Fourteen giant estates of over 5,000 dwellings each were built, with blocks of thirteen storeys or more.

Problems began to emerge by the mid-1970s, often before the estates were completed. People had begun to move in long before the last building phases were complete. Amenities, facilities and transport were often added in the last phase or after completion, slowing down the process of 'settling in'. Over half the outer estates had a higher than average turnover of tenants.

Early unpopularity, slack demand, high costs, lettings difficulties and economic recession led to cuts in the planned size of many developments. This in turn led to the cancellation of some amenities. The estates often ended up smaller, poorer and more cut-off than was planned. The 'new town' concept was realised in very few cases.

Some of the biggest 'new town'-style estates, such as Neu Perlach in

Münich or Chorweiler in Cologne, had a significant number of condominiums or single-family houses built for owner-occupation. However, they were almost entirely publicly subsidised – 87 per cent of all units and 89 per cent were rental flats (Greiff and Ulbrich 1992). A majority in 1988 were still subject to lettings agreements with local authorities and therefore housed many poorer families.

SOCIAL PROBLEMS

Social housing was built for people in need. Throughout the 1960s and 1970s, the concentrations of need grew. Table 11.2 illustrates the growing proportion of disadvantaged or dependent households that social housing companies housed. However, the concentration of dependent households in the outer estates was much more extreme. A quarter of all claimants lived on estates built since 1970, yet they represented only 16 per cent of the social housing company stock and a tiny proportion of the total stock. The proportion of young people under 18 was very high – 32 per cent of all residents were under 18 on larger estates, 50 per cent higher than the average. The concentration of foreign families was also higher. In Hamburg, a city with a high proportion of social housing, the concentration of foreigners on outer estates gradually rose as Table 11.3 shows. However, this often reflected in part a rise in their numbers in the population.

The figures shown in Tables 11.2 and 11.3, which caused alarm in many housing circles, reflected a number of positive changes in the way social housing organisations operated. Firstly, their lettings policies became more

Table 11.2 Proportion of disadvantaged or dependent households housed by social housing companies in West Germany, 1988

	Rented housing in general (%)	Social housing (%)
Large households (4 persons +)	17	19
Manual occupation	43	53
Incomes below 2,000 DM per month	54	64

Source: Gibbins 1988.

Table 11.3 Rise in proportion of foreigners living on outer estates of Hamburg, West Germany, 1973–83

Year	1973	1975	1978	1980	1982	1983
%	6.3	6.8	7.6	9.0	9.7	9.9

Source: Gibbins 1988: 108.

relaxed as the housing shortage eased off. In the mid-1970s there was even a surplus in many areas and access to the new outer estates became relatively easy for previously excluded groups, particularly foreigners. Secondly, housing allowances became more generous in this period, enabling lower-income families, who had previously been forced to live in older, often overcrowded flats, to move into newer social housing. Thirdly, partly as a result of the better supply, many cities did away with *Obdachlosensiedlungen* (estates for homeless people) which were exclusively for people in great difficulty.

These three factors led to much greater acceptance of poorer and more disadvantaged households, including minorities. Inevitably, as much of the new supply was in outer estates, growing numbers of poorer households ended up there, but in the 1970s there was also a conscious aim to make the new estates more available to lower-income households who had previously been ineligible.

These new lettings policies created familiar problems. Estates became stigmatised as the concentration of 'problem' families grew. The level of dependence on social security meant lower incomes and greater economic problems. When unemployment rose in the 1980s these areas were seriously affected, in part because of their social composition. The social problems of large estates were characterised as the 'five A's – Arme, Alte, Arbeitslose, Ausländer, Alkoholiker' – 'the poor, old, unemployed, foreigners and alcoholics'.

As the scale of building and management operations expanded, the existing systems became clearly inadequate. Because so many of the largest estates were owned by publicly sponsored housing companies they tended to be closely tied to local authority lettings systems and to form part of larger and more bureaucratic housing organisations. Therefore lettings were not always carried out as sensitively as necessary. Repairs costs mounted rapidly and far outstripped the government guidelines.

Rent losses rose through a combination of lettings difficulties, poverty, and management problems. The increasing numbers of empty units created a financial crisis for the housing companies, illustrated by the spectacular collapse of Neue Heimat, but common throughout the housing companies owning large estates.

Remedies did not get under way till the mid-1980s.

Chapter 12

The Neue Heimat crisis

In 1982, Germany experienced a serious breach in the fairly continuous evolution of limited dividend companies since the war as the major providers of social housing.

Neue Heimat, much the most powerful limited dividend company with no less than 150 subsidiaries at home and abroad, builder and owner of 400,000 rental units throughout Germany, as well as developer of over 1 million low-cost owner-occupier homes, was caught out in spectacularly fraudulent activity and discovered to be on the verge of bankruptcy. The collapse of Neue Heimat had deep repercussions in German housing organisations and internationally.

The story of Neue Heimat in many ways epitomised the problems of mass housing. Many of the outer estates of the 1960s and 1970s were built by Neue Heimat and its growth and experience served as an international model. The following section looks at what happened in outline (see Fuhrich *et al.* 1983 for a fuller account).

SCOPE FOR SCANDAL

Neue Heimat was 'the largest non-profit housing organisation in all of Western Europe' (McGuire 1981), a model of innovation, co-operation and social responsibility according to Fuerst (1974). The very advantages of the German system, its public–private structure, its many channels of development and its flexibility, provided scope for possibly the most extraordinary social housing scandal of the post-war era.

UNION ORIGINS

Neue Heimat Hamburg was founded in 1954 by the German Trade Union Federation as an amalgamation of dispersed housing activities, sponsored by different trade unions throughout the Republic. Union-sponsored limited liability companies had been developing rapidly since the 1920s. The aims of the new organisation were: to help house poor families; to provide enough houses for all; to 'lobby the Government for an active housing programme in line with social principles', to improve amenities and community facilities; to create balanced communities by housing better-off people alongside poorer groups. These wide aims embraced political and social as well as housing objectives.

BEYOND NON-PROFIT

Neue Heimat led the way in ambitious involvement in town planning and urban renewal as part of a wider housing role. It bought large amounts of land for building, then applied for planning permission for pathfinding new developments. Soon, new activities were launched alongside the primary non-profit social role to which housing organisations were restricted under German law. In 1962, Neue Heimat did a deal with a private company to allow it to co-operate in private development. It established a subsidiary, 'Neue Heimat International', to spread its development programme throughout Europe in partnership with organisations in host countries. It also set up a research organisation which operated on a consultancy basis.

In 1963, under a new and highly ambitious president, Neue Heimat began to lobby for changes in non-profit regulations, mainly to reduce restrictions, so that it could operate more freely in the expanding field of new towns, industrialised mass building, and service supplies. By then, Neue Heimat owned 130,000 properties and was a highly successful non-profit company, maybe the most successful. The rules it sought to escape provided a shield for non-commercial undertakings, but also restricted the scope of non-profit organisations. Many considered the rules over-bureaucratic and over-protective.

PROFITABLE NEW TOWN DEVELOPMENTS

In 1969, 'Neue Heimat Städtebau' was founded as a profit-making subsidiary enterprise to build new towns and carry out urban renewal. At the same time, 'Neue Heimat Kommunal' was formed to develop large-scale communal facilities, shopping centres, schools, universities, industrial units and so on. Special government exemption was granted in order to allow Neue Heimat to move from its direct housing role, which was imposed by law on all limited liability companies. Neue Heimat was encouraged to extend its operations by this government support, but a big problem for Neue Heimat was its inexperience in profit-making commercial undertakings.

The group of Neue Heimat companies ended up carrying out all stages of property development from research, programming and planning, to land acquisition, contract supervision, and then renting, selling, leasing and administering the property. Large numbers of international experts were recruited, who ended up selling their services to governments and private enterprises through the research, urban development and communal facilities subsidiaries. A wide range of financial institutions became involved in funding Neue Heimat's enterprises, including the six leading German banks.

LOW-COST OWNER-OCCUPATION

Neue Heimat seized eagerly on the government's encouragement of low-cost owner-occupation. In the 1960s, it built a tenth of all German units for owner-occupation; by the end of the 1970s, it was building half.

Neue Heimat's most tangible aim was to produce in volume, cheaply and fast, to lower the price of housing. It was nailed on its own success in achieving this, as the glut in social housing and cheap owner-occupier units in the late 1970s was a main cause of Neue Heimat's financial difficulties.

By 1980, 2 per cent of the German population lived in Neue Heimat property, an extraordinary achievement (Rahs 1986). Neue Heimat's rate of expansion far outstripped the capacity to manage, to finance, or to secure risks. Its success in non-profit, regulated housing was not matched by commercial know-how in competition with private real estate firms. Table 12.1 illustrates the pace of growth.

SPECULATIVE VENTURES

Neue Heimat became involved in some extremely ambitious projects. It built a new, privately-owned town for the city of München, called Neu Perlach. When the new town was formed, a private estate agency, Terra-finanz, owned by the president of Neue Heimat and several local managers, bought the required land from local peasant farmers at a very low price. The company, Terrafinanz, then sold the land for many times that price to Neue Heimat. Neue Heimat paid the price, using its cushion of government subsidies and its special status to pull off the necessary financing. The owners of the 'transaction company', insiders to Neue Heimat at the

Table 12.1 Rate of expansion in housing units managed by Neue Heimat in West Germany between 1960 and 1984

	1960	*1984*
Units under Neue Heimat management	134,380	397,398
Asset value	1 billion DM	19.3 billion DM
Number of staff	2,347	4,231

Source: *Deutscher Bundestag* 1987.

highest level and therefore privy to its affairs, were the gainers of huge windfall profits at the expense of the small farmers and future tenants.

Neue Heimat built a new university complex for the state of Lower Saxony. It bought up land in Brazil and Mexico, with a long-term plan to carry out ambitious developments which never got off the ground. These ventures brought huge losses as land values collapsed and property markets slumped after the oil crisis.

The quality of Neue Heimat's building was also questionable. Neue Heimat property was often recognisable by its experimental design, the use of cheap materials, and the large, ugly, modern blocks. Often the least popular areas in the largest outer estate developments, such as Chorweiler, were those put up by Neue Heimat (Rahs 1989).

COMPLEX, ILLEGAL STRUCTURE

The structure of Neue Heimat and its subsidiaries was extremely compli-cated as Figure 12.1 shows. It lent itself to corruption. The government report, following the investigation of fraud, argued that it was constructed in such a way as to facilitate corruption (*Deutscher Bundestag* 1987). The staff of Neue Heimat (non-profit) and Städtebau (profit) were intertwined – both companies were run by the same managing director; the regional divisions of both organisations were heavily dependent for staff, money, programmes, etc. on the single head office of Neue Heimat; the share-holders in *both* companies were identical, a totally *illegal* state of affairs; managers, staff and boards of both companies formed a single union. Neue Heimat was the umbrella body for both non-profit and profit-making activities. They did not have separate meetings, keep separate records, or account for their activities to official inspectors, financiers or government. Their auditors were shown to have a direct financial interest in the fraud and profiteering that was going on. For example, Neue Heimat tried to gain 'benevolence' by giving donations to interested parties. By 1979, 25 per cent of Neue Heimat's funds were distributed in this way (*Der Spiegel*, nos. 4/7/17 1986).

DER SPIEGEL REVELATIONS

In 1982, the German political scene and the housing world were rocked by revelations in *Der Spiegel* of corruption in Neue Heimat, the squandering of union resources, and the duping of major institutions (including the government, banks, *Länder*, and unions themselves). The main accusations were that public funds had been used to line the private pockets of individual managers, and that the non-profit status had provided a shield and cover for illicit activities including transferring assets to profit-making subsidiaries and bailing out private losses with public resources. The central

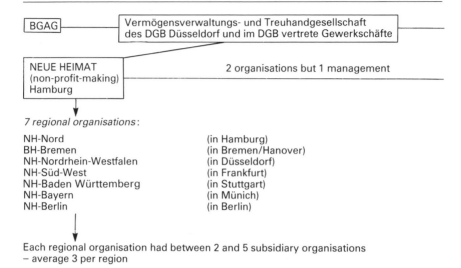

7 regional organisations:

NH-Nord	(in Hamburg)
BH-Bremen	(in Bremen/Hanover)
NH-Nordrhein-Westfalen	(in Düsseldorf)
NH-Süd-West	(in Frankfurt)
NH-Baden Württemberg	(in Stuttgart)
NH-Bayern	(in Münich)
NH-Berlin	(in Berlin)

Each regional organisation had between 2 and 5 subsidiary organisations
– average 3 per region

Total: 20 non-profit-making organisations

Scale of NEUE HEIMAT (1982)

Around 150 organisations/companies
5,700 staff (3,500 at the time of sale in 1986)
6.5 billion DM turnover p.a.

Figure 12.1 Structure of Neue Heimat and its subsidiaries in West Germany,
        1982

(Property Trust Company of German Trade Union Federation – DGB – in Düsseldorf and unions represented in DGB)

NEUE HEIMAT STADTEBAU GMBH
Hamburg
(profit-making)

Regional organisations in German *Länder*

plus:

Neue Heimat Kommunal
Gesellschaft zum Bau öffentlicher
und sozialer Einrichtungen mbH
(for the construction of public and social establishments)

GVG Grundstücksfinanz – und
Verwaltungsgesellschaft mbH
(real estate – finance and management)

Mediplan Krankenhausplanungs-
gesellschaft mbH
(hospital planning)

Begebau Beratungsgesellschaft für
Gewerbebau
(advice for industrial units)

Neue Heimat Fertighaus
Vertriebsgesellschaft
(prefabricated house sales)

VBV Versicherungs-Betreuungs –
und Vermittlungsgesellschaft mbH
(insurance agency)

Baudata Gesellschaft für bau – und
wohnungswirtschaftliche
Datenverarbeitung mbH
(construction and housing data)

plus:

46 Building Contractor and Real Estate Companies

Further Education Institutes
(e.g. Berufsfortbildungswerk)

Acon Gesellschaft für Werbung &
Kommunikation GmbH
(Advertising and Communications)

Total (in Germany): *Around 60 organisations*

plus:

ABROAD

NEUE HEIMAT INTERNATIONAL (umbrella organisation for all international organisations)

France ⟶ Around 12 organisations
Austria
Belgium
Italy
Luxembourg
Monaco   several dozens
Switzerland
Finland
Venezuela
Brazil
Mexico   main ones
Israel
Africa
Others

Total (abroad): *60–70 organisations*

problem was the financial collapse of Neue Heimat and its subsidiaries, resulting from large-scale speculation in land and property, the changes in the housing market and incompetent management.

The leadership of the company coupled these miscalculations with a dishonesty that astounded the public because of its exploitation of the protection provided by the *Gemeinnützigkeitsgesetz*. A few examples will illustrate the scale of the dishonesty. The president, Vietor, founded a private heating company which supplied heating units to Neue Heimat properties at 30 per cent above the normal price. Other firms were established to supply kitchens, TV aerials, and other equipment. They were also more expensive than those supplied by other firms. Private beneficiaries acquired cut-price villas at the expense of non-profit activities.

BANKRUPTCY AND SALE OF STOCK

It became clear that Neue Heimat was in deep financial trouble. Neue Heimat acted more and more rashly, as the situation became more desperate, creating an unrescuable situation that drew the 150 linked organisations into the spiral of bankruptcy. This made them all share in the need to pass money around to stave off collapse.

Neue Heimat managers, after the bubble had burst and bankruptcy looked inevitable, took the view that government and creditors alike could not afford *not* to rescue them. This belief prevented the unions from taking drastic action early in the crisis. All the major institutions involved took a similar view, regarding the collapse of Neue Heimat as unthinkable.

In 1985, under a new director, Hoffmann, the company announced the sale of 100,000 units to stave off bankruptcy. Under pressure from the tenants, Social Democratic *Länder* agreed to buy out Neue Heimat in their areas. This secured the social housing for the tenants and let Neue Heimat off the hook. As soon as it became clear that Social Democratic regional governments would take over the properties, the proposed sale price was doubled (see Figure 12.2 showing ownership and sales to stave off bankruptcy). Later in 1985, the sale of the remaining stock of over a quarter of a million dwellings was agreed for a token price of one Deutschmark (40p) to a master baker in Berlin by the name of Schiesser. Schiesser was unknown to the housing world. The assumption was that the banks would support the sale. The outstanding debt on the property was 17 billion[1] Deutschmarks and Neue Heimat's creditors refused to support the deal.

UNION RESCUE

One year later, the unions bought back the Neue Heimat property at a further huge loss, having been leant on by banks, tenants, and the Social Democratic Party. The Social Democratic Party (SDP) was by now heavily

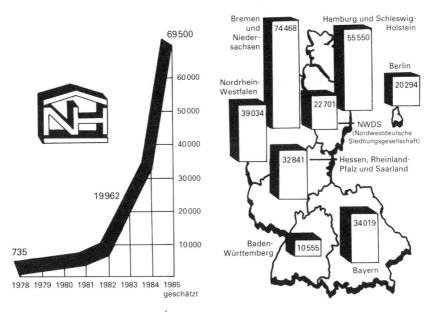

(a) Sales

Notverkäufe ...

Seit 1978 von der Neuen Heimat in der
Bundesrepublik verkaufte Wohnungen
(kumuliert)

(b) Stock in 1985

... solange der Vorrat reicht

Wohnungsbestand der Neuen Heimat
(insgesamt 289 462 Wohnungen,
Stand 22.8.1985)

Figure 12.2 (a) The sale of Neue Heimat properties in West Germany in an
attempt to stave off bankruptcy; (b) the distribution of Neue
Heimat's stock in different states of West Germany, 1985 (after sale
of 100,000 units)

Source: *Der Spiegel* nos. 4/7/17 1986

tarnished with the scandal as a long-time ally of both the unions and Neue
Heimat. The buy-back cost the unions 15 million Deutschmarks in fees and
other costs. Neue Heimat property was then sold on to a range of housing
organisations, some private and some sponsored by local and regional
governments. The entire stock was broken up and Neue Heimat no longer
exists.

GOVERNMENT INVESTIGATION

In 1987, the government appointed an investigation committee which
largely confirmed the well-rehearsed scandal. Losses on the operations of
Neue Heimat Städtebau and Neue Heimat International were the main

direct cause of the bankruptcy. Neue Heimat Hamburg began to show direct losses in 1981 (193 million DM), which almost doubled in 1982. Large speculative land holdings lost their value, housing units built for private ownership remained unsold, rent arrears and high levels of empty property created losses in the non-profit companies. In the climate of the 1960s and early 1970s, the investment gambles of Neue Heimat would probably have been successful and of themselves were not illegal.

The main causes of empty units were a big drop in demand in the late 1970s, supply broadly matching the number of households, unemployment, and inability to pay through stagnant incomes and slow economic growth. Neue Heimat's critical errors were identified as:

- resources being shifted from the non-profit company to Neue Heimat Städtebau;
- land speculation based on assumed rising values;
- over-building, particularly for owner-occupation, with 7,000 empty unsold units in 1981;
- complex financial arrangements with 115 credit institutions;
- cheap social housing developments with poor environments, materials, and construction which proved unpopular with tenants;
- a short-term surplus of housing in the early 1980s;
- reliance on union capital which was eventually withdrawn;
- poor maintenance, with less money spent on repair than comparable organisations – 9.5 DM per square metre per year in Neue Heimat, compared with 15.2 DM as the German average – repairs money was diverted into other activities.

THE ROLE OF THE UNIONS

It is hard to see how the German unions could have been duped by such large-scale misjudgements and, eventually, law-breaking. Workers' representatives had sat on company boards and had been involved in social housing since the early days, but there was no suggestion that they gained from the profiteering that went on. Quite simply, they were taken in by the heady expansion, along with government representatives, major banks, partners in Neue Heimat's speculations, and tenants. The unions were simply not able to manage or control their large real estate enterprises and they were not cautioned by their official advisers. If anything, they were encouraged by the incentives and the collaboration of government and banks in expansionist schemes. The German system, with its strong private orientation and major public subsidies, leant itself to this overgrowth.

The imperative to build, the greed of the building industry and those connected with it, the potential for sewing up deals, short-cutting the process of government and using public funds to promote private gain, made all parties vulnerable to misjudgement and corruption; the fast-changing world of large-scale developments provided cover for those willing to play for high stakes.

The 'ring-fencing' of housing accounts that protected German non-profit companies did not work in Neue Heimat's case because of its size and power, its illegal structure, and the willingness of its directors to put at risk the non-profit activities.

The government investigation's recommendations paved the way for the abolition of non-profit status for all social housing companies. Their main recommendations were as follows:

* profit and non-profit activities should not be mixed;
* regional housing organisations were preferable to national ones;
* there should *not* be an obligation on housing organisations to build;
* maintenance money must *all* go on maintenance;
* no asset transfers that penalised existing tenants should be allowed;
* controls should be tighter and sanctions against misdemeanours should be stronger.

POLITICAL FALL-OUT

The government published the investigation report in the middle of the 1987 election campaign and used it to great advantage. The Social Democrats put up a feeble defence of the unions, while at the same time the unions withdrew campaigning support from the SDP in retaliation against its attack on union incompetence. The two immensely powerful organisations completely failed to rise to the challenge of the crisis and undermined each other.

AFTERMATH

Vietor, the Neue Heimat ex-president who led much of the profiteering, died before the investigation finished its work. Many of the other main actors were enjoying long prison sentences at the time of writing.

The outcome for Neue Heimat tenants, staff and property was in practice a rescue. Several *Länder* governments created new companies to run Neue Heimat's huge holdings. But the repercussions of the scandal and collapse of Neue Heimat will last as long as the property.

It is a deep irony that government protection, coupled with risk-taking enterprise in a 'free' market, should have brought to ruins such a major social housing organisation. Its link with unions, its commitment to help the poor, its ambitious building programme, all made its problems significantly worse. As a private landlord, its profiteering would have had fewer repercussions, whether successful or not. Linked as Neue Heimat was to government and the social housing movement, its collapse in such ignominy undermined the public support for government-sponsored social housing.

The National Confederation of German Trade Unions in part owned Neue Heimat. The Conservative government spared no pains to discredit Neue Heimat, the unions, and the SDP. In so doing, the limited dividend structure became the sacrificial lamb. As the scandal unfolded – with stories of luxury villas for managers, land deals in Brazil, and tenants being charged more than once for services feeding the popular imagination through exaggerated media coverage – so government support for limited dividend companies shrank. This fitted with the government agenda to withdraw from subsidised housing and create a freer market.

At the point when empty units were becoming a serious problem on the unpopular, new, giant social housing estates outside the big cities in the early 1980s, it was politically and economically expedient to cut the social housing programme and to call into question the special tax and legal status of the housing companies.

Thus, political interests, business mismanagement, market failure, international economic change, personal corruption, and intense media exposure all combined in an unusually concentrated way to bring about the collapse of Neue Heimat.

NOTE

1   The US billion (a thousand millions) is used throughout.

Chapter 13

Changes in the 1980s

A number of important measures were introduced in the 1980s which changed the German housing system radically. The first major change came in 1983 when the government announced plans to free up rents even further, including limited dividend company rents, so that they moved closer to market rents. The second change was to encourage social landlords, including the companies, to redeem their public loans ahead of time. A significant number did this as their capital debts shrank and there was a rapid acceleration in the number of 'free renting' units. This seriously reduced the local authorities' scope for nominations (Kreibich 1986). The third move was to encourage the sale of company property to raise money for repair and to stave off mounting financial problems. This had only limited success, as most social housing was in flatted blocks. The fourth move was to refuse to engage in a rescue operation for Neue Heimat, thereby facilitating an end to the special legal and financial status of housing companies. The fifth change was a shift in favour of subsidising owner-occupation, which grew significantly in the 1980s.

ABOLITION OF 'PUBLIC GOOD' LAW

The government believed that putting the social housing companies on the same footing as private landlords would achieve many benefits – greater efficiency, greater investment, less 'mis-renting', better standards of repair, less bureaucracy and complacency, less government financial involvement, less red tape.

In 1988, the government announced plans to end the 'public good' law protecting social housing companies – the *Wohnungsgemeinnützigkeits-gesetz*. They would no longer be required to build; they would be allowed to diversify their activities; they would be liable to the same taxes as private landlords and profit-making companies but would also be allowed the same tax privileges. In the housing stock for which they received no subsidy, they would be allowed to charge market rents for new tenancies and they would be free to operate for profit on the same basis as other landlords. An advantage to the government was a potential for additional tax income. Housing co-operatives continued to enjoy tax exemption if they only rehoused members.

These changes were made law in 1988 and implemented in 1990. There was nothing to stop the social housing companies from continuing to operate on a non-profit, 'public good' basis and the co-operatives and municipally owned companies were less affected by the legal changes than privately sponsored companies. However, the pressures to raise rents and to provide for better-off more stable rent-payers, and the temptation to maximise profit, if only to expand the enterprise and produce more units (one of the government's aims), would be strong. Meanwhile, housing companies were forced to remedy problems in their outer estates.

SPECIAL HELP FOR OUTER ESTATES

In 1983, money first became available to experiment with improvement projects. In 1985, the Federal Ministry of Housing initiated research into the situation on the outer estates. At the same time, these estates became eligible for federal renovation subsidies under the urban renewal programme. By 1988, eleven large estates had been renewed through the *Städtebauförderungsgesetz*. In 1987, a new building law introduced improvement measures for large estates.

In all, between 1983 and 1988, 24 million Deutschmarks were set aside by the federal government for these programmes. But in Germany, with its decentralised federal structure, it was generally the *Länder* and local authorities, together with the housing organisations themselves, that raised the money to tackle problems. Special financial measures were introduced by state and local governments to help the housing companies regain viability. One key to attracting and keeping rent-paying tenants was cutting rent levels in unpopular estates. These had often risen above private rents because of the funding arrangements and costs (Emms 1990), and acted as a major deterrent to new tenants. Although housing allowances helped some households, even in social housing areas a majority of people paid the rent in full and all had to pay a proportion.

A special surcharge for high-income tenants, if they were 20 per cent above the eligibility limit for new tenants, had been introduced in the 1980s in some *Länder* to compensate for the high costs of repair and controlled rent levels. The surcharge was levied on people's incomes rather than on the housing unit, causing a serious distortion in rent levels with people paying different amounts for similar units. This surcharge was held partly responsible for the steep rise in vacancy and turnover levels in newly built, high-cost new town developments. It also squeezed further the population of outer estates. An increase in management problems followed on from the high turnover, particularly on the unpopular large estates where rents were already high and where 'mis-renters' were particularly needed because they helped retain a social mix and were considered 'good tenants'. Stable, better-off, long-standing tenants were valued by the housing companies as enhancing their attractiveness and viability but were driven out by the levy on 'mis-renters'. In some areas like Hamburg and Bremen, the levy on 'mis-renters' was done away with to encourage more economically secure households to stay. Housing companies increasingly tried to redeem their public loans to reduce their social lettings commitments, so that they became free to let a higher proportion of their stock to better-off tenants.

Other strategies were adopted. Sales of flats were initiated to fund essential repairs and to reduce costs. Management and maintenance money was increasingly focused on repair, on environmental upgrading and on building security. By 1988, half of all housing spending was targeted at repairs and modernisation.

MORE SENSITIVE MANAGEMENT

Housing companies generally adopted a more responsive, more competitive management strategy to entice tenants in and persuade them to stay. A key was to change the relations between landlord and tenants to a more service-oriented approach. There was growing acceptance of the need for tenants' representation, local meeting places, a local management office and direct contact (City of Cologne 1989). Tenant-housing company committees were set up in some areas. The surroundings cried out for upgrading. Intensive planting and better management control were two of the most commonly sought improvements. An effort had to be made, both to integrate the areas into the city and to integrate within the estates the variety of landlords, services and enterprises involved. Changing the estates from single-function, rented dormitories into varied, multi-use city areas was a strong theme. Some cities and companies developed tenant liaison structures, employment and training initiatives and social support. Finally, the access policies had to be changed to create more diversity. There was acute awareness of the contradiction between growing chronic shortage and lettings difficulties.

NEW TOWNS

The size of German estates and the 'new town' approach offered some advantages. Many facilities, schools, police stations, employment centres, nurseries, social organisations and cafés existed and, with efficient management, appeared to flourish. Shopping centres were done up and let at below-market rents to encourage commercial activity. The estates often had busy centres and mothers with children were often around in the after-noons because of the short school hours in Germany (children normally come out at around 1.30 p.m. up to the age of 11).

In a comparative study where tenants on German and British large estates were asked about social relations, neighbourliness and social activity, tenants on German estates reported much greater satisfaction with social facilities, links in the community and general friendliness (Hillier *et al.* 1987, Couch 1985). Thus, although German outer estates experienced problems, these did not seem that difficult to put into reverse when companies were given both support and new incentives to operate more efficiently. In fact, as the 1980s advanced, empty units on even the most unpopular large estates disappeared. The estates themselves were often better maintained; the dwellings were in demand, as the number of needy households rose; and by 1991 (Bochum Conference 1991) housing companies experienced long queues for their flats, although many long-term problems in the *Grosswohnsiedlungen* remained.

As the acute housing crisis of 1991 created the need for new-building, estates were being developed again, albeit with fewer units per estate and more social and environmental care. The lessons of the 1960s, 1970s and 1980s were at the forefront of new plans (Wullkopf 1992).

NEW SHORTAGE

The 1987 census revealed a crude shortage of 740,000 dwellings after nearly a decade of government belief that there was a housing surplus. Household formation had outstripped new building. There were big rent increases because of the shortage in high demand areas. Rents often rose above the housing allowance maximum, causing actual homelessness.

The huge surge in demand for housing through rapid immigration, the prospect of reunification, the steep fall in the supply of new housing, and renewed economic growth, created a general pressure to house people. Local authorities faced the problem of nominating growing numbers of needy households to a declining supply of 'social' units as many landlords moved out of subsidy. Their legal hold on limited dividend companies was greatly weakened by the abolition of *Gemeinnützigkeits* status. The economic problems of non-profit housing companies were subsumed by the growing needs.

It was hard to drive forward the expanded programme without social landlords. This underlined an inherent contradiction in the role of social housing companies. They were extremely useful vehicles for large-scale and emergency programmes. But the scale of their developments and their bureaucratic ties with government created management difficulties, experienced to a much more limited extent by private landlords. It remains to be seen whether they can become flexible enough to operate as private 'social' landlords and engage as fully as is required in new government programmes.

CRISIS OF CHEAP HOUSING

Not only is Germany currently in crisis over its deficit of low-cost units; the number of dwellings owned by non-profit housing bodies to which local authorities could nominate needy households was set to drop from 4 million units in 1985 to 2.5 million in 1995.

By the 1980s, early post-war housing units had actually cleared their long-term public debts altogether. They then became 'free-renting', unsubsidised units, though the companies could still only charge *cost* rents rather than market rents. Over 1 million units belonging to the non-profit housing companies became 'free renting' in the 1980s. Rents rose on these units and access again became difficult for very poor households.

There were therefore growing conflicts over the social role of housing companies in spite of the desire to see them operate as private landlords. The government clearly saw this as a way out of the burden of non-profit housing yet local authorities required cheap and accessible housing for newcomers.

The housing crisis reached the top of the agenda, as immigrants of German origin flooded westwards from eastern Europe. Figure 13.1 illustrates this. The government quickly shifted gear to encourage housebuilding. Most subsidies were directed towards individual ownership – 70 per cent of all 'social' housing in 1988 was for owner-occupation.

An expansion of low-cost private renting became urgent. Special grants were made available to private owners willing to convert basements, cellars, attics, barns, garages, shops, disused factories – almost anything with walls and a roof – into rented accommodation as long as existing building standards could be adhered to. The grants were strictly limited to the production of *additional* housing units and were geared to encouraging specifically low-cost rented housing. The units were counted as 'social' units and had to be made available only to eligible households. These grants were not available to help owners extend or improve existing housing units.

URBAN RENEWAL DECLINES

The subsidised conversion and amalgamation of older inner-city units into elegant homes for affluent owner-occupiers had seriously added to the housing pressures; the subsidies that encouraged the process were progressively changed. City governments in pressured areas, such as Frankfurt, applied standards for conversions more rigorously in order to make the process more expensive, less attractive to would-be gentrifiers, and

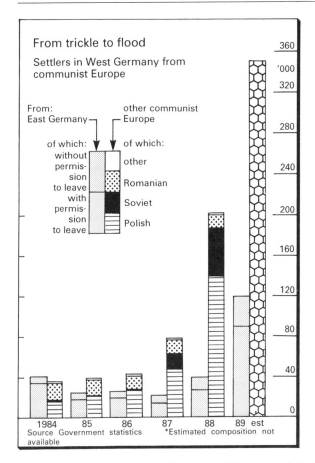

From trickle to flood

Settlers in West Germany from communist Europe

From: East Germany — other communist Europe

of which:
without permission to leave
with permission to leave

of which:
other
Romanian
Soviet
Polish

1984   85   86   87   88   89 est
Source Government statistics   *Estimated composition not available

Figure 13.1 Settlers in West Germany of German origin, 1984–9

Source: *The Economist*, 16 September 1989

Note: The second column showing 'other communist Europe' represents, almost entirely, ethnic Germans.

more difficult to carry out (*Frankfurter Allgemeine Zeitung*, 16 December 1989). Coupled with the new grants for the creation of additional rented units within the existing stock, the stricter enforcement of the law aimed to preserve the supply of private-rented accommodation, slowing down the rate of conversion to owner-occupation.

A RENEWED ROLE FOR LOCAL AUTHORITIES

The relaxation of rents and the drop in supply of 'social' units made the local authorities resume a more active role. With the problems of immigration and reunification, local authorities received special funds to 'buy back' nomination rights to 'free' rented units and to mount the new programme of conversions. They retained their all-important role in relation to access and development.

The following measures were announced by the government towards the end of 1989 in an attempt to stem the housing crisis, some of them in direct contradiction with the liberalisation policies of the 1980s:

- DM 2 billion per annum for social house-building;
- temporarily until 1992, increased emphasis on tax subsidies, in order to promote the building of new housing with controlled rents and allocation commitments;
- the promotion of ownership through a new building society financial programme;
- a Federation–*Länder* programme for student housing;
- tax subsidies and low-interest loans to promote the creation of new housing units in existing housing (e.g. basements);
- loans by the Kreditanstalt für Wiederaufbau for local authorities and other housing providers to provide temporary housing;
- temporary relaxation of the tenancy law in order to encourage lettings of holiday/weekend housing;
- changes in the planning and building law to reduce the time to get building approval;
- financial assistance to *communes* by the Kreditanstalt für Wiederaufbau for the development of new building land;
- expansion in building capacity through the mobilisation of unemployed building workers and workers from related fields, many from eastern Europe.

(Hasselfeldt 1989)

In 1990, an emergency target programme of 150,000 low-cost rented units was announced. In the same year, Germany expanded its population, size and housing stock by 25 per cent through unification with East Germany, creating a totally new situation. Emergency centres and camps again appeared, as Germany faced the problem of 800,000 people in temporary accommodation or on the streets (Greiff and Ulbrich 1992).

Times are changing rapidly in Germany. The future remains unclear. For the moment, there is almost a panic atmosphere as attempts are made to house yet another colossal movement of German people – the fifth in 120 years.

Chapter 14

Housing change in East Germany

In 1945, most of the East German housing stock dated from before the First World War – about 4 million of the 5 million units. Cities like Dresden had been blanket-bombed, and war damage was devastating. The immense political tensions and uncertainty over the future of East Germany hampered the building programme, so that by 1960 less than half a million new units had been produced. The strong military and political involvement of Moscow in the affairs of East Germany added to the problems. Stalin accepted the inevitability of reunion and therefore prevented major departures from the existing ownership pattern. Industrial and military development received priority over urban investment. As a result, over 3.5 million homes in East Germany were still privately owned at reunification in 1990.

Since 1960, East Germany built nearly 3 million new housing units, over 40 per cent of the total stock. Over 1.5 million pre-1919 dwellings were demolished in the process, leaving 2.4 million pre-1919 dwellings at reunification. Most of these were extremely run-down, barely having been modernised since the war. Some older inner areas of Leipzig are virtually intact in their pre-war condition. In some cases they had been abandoned by their owners, who fled to West Germany. About half a million are now being reclaimed by their former owners (in all over 1.5 million property claims have been lodged). As a result of the large stock of old, unmodernised houses, over one-third of East Germans live without basic amenities such as hot water, bath, shower or indoor toilet.

CO-OPERATIVES

Co-operatives survived the war and were used by the East German government to develop publicly provided housing. The fact that members invested savings in shares, contributed labour and ran their internal affairs, was a great attraction in a situation of conflicting demands and insufficient resources. In the early 1960s, when new house-building gathered speed, co-operatives accounted for well over half of all new building. The co-operatives continued to build new housing up till 1990, contributing about a quarter of the new-build stock of over 3 million put up since 1960. However, they were increasingly controlled by the state, which imposed mass design, volume building, and state allocations, creating many ugly, monotonous and over-large estates. The co-operatives remained smaller than state housing bodies, with anything from 500 to 10,000 members. But in spite of signs of resistance in Berlin and elsewhere, they became very much an arm of the state apparatus (Chamberlayne 1990).

STATE PROGRAMME

In an attempt to tackle the outmoded housing conditions, the East German government built a growing share of new housing directly through state bodies. By sacrificing the co-operative requirement of organising members and galvanising savings, this appeared a quicker route. With no member input, standards fell and allocations were entirely state-run. State housing could be administered through existing state systems to a minimal level for costs that were largely hidden within the all-pervading state apparatus. Rents were extremely low – only 3–4 per cent of income – and the acute shortage of modern housing ensured demand for poor quality, state-provided, mass estates. Very few repairs were carried out and the post-1960 buildings were often characterised by weather penetration, poor insulation and sanitation problems.

Direct state building came to dominate, comprising three-quarters of all new construction by the late 1960s. Most new developments were on large estates. One hundred and twenty-five very big developments of over 2,500 units were built between 1960 and 1985, over 1 million industrialised units in overwhelmingly repeated grey concrete blocks. These estates make up one-seventh of all East German housing. Half the population of East Berlin lives in massive developments, the biggest having between 35,000 and 60,000 units. They totally dominate the city landscape. In all, there are seventy giant estates in the Eastern *Länder*, with an average of 11,800 units each. Another fifty-five have between 2,500 and 5,000 units. Modern mass estates were let almost entirely to active workers directly by the state and, according to a recent government study (Bundesministers für Raumordnung, Bauwesen und Städtebau 1991), their populations were relatively stable, social problems, such as were experienced in the West, not having arisen prior to reunification.

PRIVATE BUILDING

In the 1970s, the proportion of private building grew again because of the failure to satisfy demand and the need to galvanise individual effort. Self-build home ownership became gradually more important. From 1970, about one in ten new units were built by individuals on a self-help basis, attracting individual investment, labour and commitment. In all, at least a quarter of a million new dwellings have been built this way. Figure 14.1 shows building developments.

The housing situation at reunification in East Germany is summed up in Tables 14.1 and 14.2, based on estimates provided by the West German Federation of Housing Associations. Overall space standards were much lower than in West Germany (18 square metres per person compared with 35 in the West) and many households were forced to share (Greiff and Ulbrich 1992).

REUNIFICATION

Reunification has caused major upheavals in the housing market in East Germany. Property rights, housing standards, repair requirements, rent levels, social landlord structures, housing allowances, and nominations have all been officially integrated into the West German system. In prac-

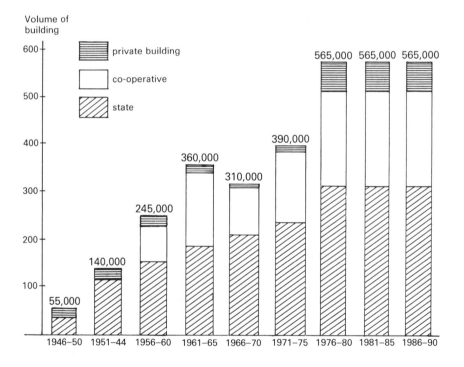

Figure 14.1 Volume of house-building in former East German territories showing proportion provided directly by the state, 1946–90

Sources: *Statistisches Jahrbuch* 1988, Chamberlayne 1990

Table 14.1 Distribution of ownership of housing in East Germany, 1991

Tenure	%	No. of units
Private renting	25	1,750,000
Owner-occupation	25	1,750,000
State ownership	35	2,500,000
Co-operatives	15	1,050,000
Total		7,050,000

Sources: Estimate based on information collected by Monika Zulauf 1990, 1991; Gesamtverband der Wohnungswirtschaft 1991.

Table 14.2 Distribution of property by age in East Germany, 1991

Date of construction	Pre-1919	1919–39	1945–60	1960–90	Total
No. of units	2,400,000	850,000	800,000	3,000,000	7,050,000

Source: Zulauf 1990, 1991.

tice, this is proving much more complex than anticipated and a chaotic state of flux prevails (Gesamtverband der Wohnungswirtschaft 1991). The immediate aim is to cope with homelessness and integrate the systems. A longer-term aim is to bring East German standards up to Western levels to prevent further floods westwards.

The exodus to the West in 1989 and 1990 initially created unheard-of slack in the market. But this was followed by a flood of West Germans laying claim to their pre-war property in the East and a commercial boom following the conversion of East German currency into West German Deutschmarks. There has also been large-scale immigration to former East German states from other parts of Eastern Europe. The situation in 1991 became economically and politically unstable, with rising inflation, large-scale unemployment, and protests by East Germans over government failure to deliver results that even approximated to the pre-unification hopes. In major cities there is a serious housing crisis, attracting a massive federal reinvestment programme. Poor standards and low rents are considered such a deterrent to private housing investment that significant tax and subsidy incentives have been reintroduced specially to cope with the impact of reunification (Wullkopf 1992).

SOCIAL HOUSING AND CO-OPERATIVES

All state-provided and socially owned housing was converted by law in October 1990 into local authority-sponsored limited dividend housing companies and housing co-operatives. These are now governed by federal

German law. Rents were rising steeply, from 3 per cent up to 12 per cent of income. The aim of covering management and maintenance costs in full and all service costs had already produced a fourfold increase by late 1991; however, administrative and service costs were so high that this had not yet been achieved.

The very high level of unemployment in East Germany makes housing allowances, which have been introduced, even more important than in West Germany. Housing allowances in the Eastern regions can cover service charges for heating, water, rubbish and waste disposal, as these are significantly higher in the former East German states due to serious inefficiencies (Wullkopf 1992).

EMERGENCY MEASURES

In 1991, there was a growing shortage of housing in Berlin and landlords were obliged to rent out empty property. A system of emergency nominations, equivalent to billeting, was put in place, although some housing organisations were ignoring the local authority nomination system in the prevailing confusion.

The West German nomination system for households in need of a home was introduced and all co-operative and local authority-sponsored housing (i.e. all housing that was previously funded by the East German state) was supposed to be allocated through local authority housing vouchers. No income limit was imposed for access in East Germany because of dislocations caused by the property claims of West Germans and because of the movement westwards from Eastern Europe. Both pressures created an extraordinary housing situation. Everyone in need of a home was made eligible. Emergency vouchers attempted to cope with extreme cases but many had to be accommodated in short-stay and makeshift accommodation. In practice, controls were very weak in the chaotic conditions (Gesamtverband der Wohnungswirtschaft 1989).

The German Federation of Housing Associations (Gesamtverband der Wohnungswirtschaft) was given the task of reorganising the East German social housing stock of 3 million units; implementing rent rises and ownership transfers; protecting members' shares, and ensuring the independence of the co-operatives. The possibility of introducing the right to buy was mooted (Gesamtverband der Wohnungswirtschaft 1990).

The transition for the German Federation was enormous, as it acquired responsibility for almost double the number of units virtually overnight on reunification – from 4 million to 7 million.

STANDARDS

The needs in East Germany are intense. Its building industry is outmoded and poorly trained; its industrialised building is of very low quality; space standards are minimal; its housing management systems are under-resourced and overmanned. Investment in housing dropped significantly in the year before reunification. About 1.8 million units had no inside toilet and 1.3 million no bath or shower. In all, 5.5 million East Germans lived on outer estates with serious construction and repair problems. The gap between East and West German housing conditions was so great as to hamper development.

The management problems for the East German co-operatives and new local authority-sponsored companies became suddenly much more complex. Not only did they have to adjust to independent legal status and revolutionised financial arrangements, they were also expected to handle community and neighbourhoods' housing issues. Previously, the East German state, with its ubiquitous secret police, had exercised iron-like control over all aspects of local as well as national life. The secret police, according to one observer, had substituted for estate management. The West European warden-caretaker did not exist. Therefore local problems of control and enforcement quickly mounted (Clark 1991).

Yet solving the housing crisis through West German money and know-how was difficult too because of the emergency housing situation in the West, which also demanded extraordinary resources. The aim became to help East German firms develop rapidly, mopping up unemployed labour, developing necessary skills and adapting to Western ways simultaneously.

DISMEMBERMENT

The dismemberment of the former East Germany was so sudden and so dramatic as to drain confidence from the East and exaggerate differences. Not only have many properties and businesses been claimed by former owners; educational and social institutions have been newly subjected to West German controls and standards, leading to closures and job losses in many cases. Overmanning was a serious problem because of the commitment to full employment, and East Germans have found the swamping with consumer goods and open frontiers more of a burden than anticipated as more and more young moved West and more and more activities were expropriated by West Germans. One illustration of the problems of adaptation is the unmanageable increase in rubbish from Western consumer packaging as shops have opened up, stocked with Western goods. The East German system simply could not cope.

Reunification is a reality but it may take years for the dreams of a united, economically strong and socially stable Germany to blossom. Meanwhile, the 17 million inhabitants of East Germany experience a humiliating exposure to an alternative system, within which they have so far had great difficulties in competing.

Chapter 15

Summary of the current German housing pattern, and conclusions

SUMMARY OF THE CURRENT GERMAN HOUSING PATTERN

The German housing system is based on three premises:

1   Private owners are the logical providers of the majority of housing, either through individual ownership or through renting out to others – 85 per cent of former West German housing is privately owned; about half the housing in the new Federal States (former East Germany) is privately owned.
2   Limited liability companies and co-operatives (the German 'non-profit' social sector) are important organs for state assistance and for social provision but should fit into the overall pattern of private provision – increasingly they are treated as other private landlords; one-fifth of the housing stock in United Germany is owned by them.
3   Subsidies should be used to help a broad band of lower-income households, regardless of tenure – the 'social housing' sector includes 30 per cent of owner-occupied housing and about 10 per cent of private-rented units.

A majority of all households are eligible for 'social housing'. Social housing means housing for which subsidised loans are currently being provided and for which households with eligibility certificates issued by the local authority can apply. Units are no longer classed as 'social housing' when the subsidised loans have been repaid and there are no longer lettings obligations, although lettings obligations may last for a further eight years. With non-profit housing companies, subsidised loans usually last for thirty years, but they can be repaid earlier thereby 'freeing' the unit from the requirement to accept nominations; low-income and first-time owner-occupiers can receive subsidy if they are eligible.

Some characteristics of German housing provision affect the delivery of housing services: the intense regulation of finance and of standards of repair and access leads to more even and better general conditions than in Britain (Couch 1985). The role of the social housing organisations and the local authorities is much more complex in Germany because they are part of a much more varied set of arrangements.

Housing costs much more in Germany – particularly owner-occupied housing. The hours of work required in order to earn sufficient money to purchase a house are three times higher than in Britain (Nationwide Building Society 1987).

German housing has been greatly influenced by the huge changes that have constantly overtaken her development, not least the latest upheavals in East Germany, leading to the collapse of Communist control, the development of a democratic system in less than six months, and the shrinkage of time-scales for the reunification of Germany from ten years, optimistically dreamt about in Autumn 1989, to an immediate reality in 1990.

Owner-occupation has not been singled out for special treatment. Many current and past government reports ignore it altogether. Government housing information is kept according to subsidy programmes, which generally apply across all tenures. The focus of concern is twofold: the number of housing units required and provided and the cost in terms of public subsidy and tax concessions. Whether they produce owner-occupied, private-rented or non-profit housing company units is less important than the volume, location, cost and suitability of production. However, owner-occupation has become more important in recent years.

The variety of channels of production creates a more flexible and responsive housing market. Figure 15.1 shows the large volume of unsubsidised housing, the changing positions between private and non-profit social landlords, and the high level of total output. Over time, the role of limited dividend companies and the level of subsidised social rented housing shrank.

By a gargantuan effort, Germany had produced 20 million post-war dwellings by 1988. The total West German stock reached well over 27 million by 1991. Table 15.1 shows the evolution in population and housing between 1950 and 1990; Table 15.2 shows the distribution of the stock by tenure and subsidy; Table 15.3 shows how production was split fairly evenly between multi-storey flats and smaller one- and two-dwelling units, shifting increasingly towards smaller units.

Ownership is highly diversified because the rented stock – 60 per cent of the total – belongs to private or semi-private institutions. Even the housing companies in which local authorities or regional state governments, the *Länder*, form a majority of the shareholders (about 30 per cent of all companies) operate under business regulations, must balance their books as independent entities, and meet their full costs out of income. Rents must be charged on that basis. This gives the companies, simultaneously, considerable independence and autonomy, and exposes them to real competition from comparable private landlords as well as from other housing companies.

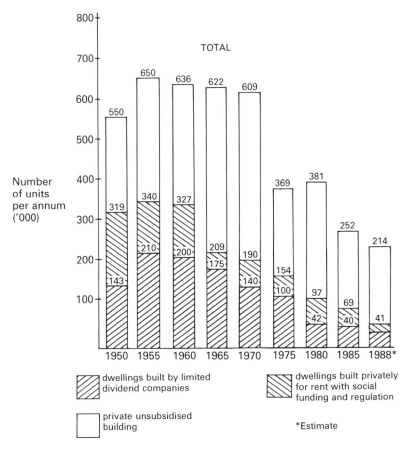

Figure 15.1 West German house-building, 1950–88: proportion of 'social'
housing against unsubsidised building and that built by limited
dividend companies

Source: Gesamtverband der Wohnungswirtschaft 1989

The owners of rented accommodation range from churches, local
authorities, trade unions, insurance companies, small businesses, management companies and industrialists, to professional individuals such as
doctors, lawyers, dentists and financiers. In some cases, private investors
were responsible for funding mass housing estates (City of Cologne 1989).

Although most of the developments described here relate to the former
West Germany, the same system has now been installed in former East
Germany and the housing problems are likely to merge. One major effect
of reunification, however, has been the reintroduction of federal building
subsidies, and the extension of tax and grant incentives for building and
conversion and special measures – on the one hand to decontrol and

Table 15.1 Population, households, and housing stock in West Germany, 1950–90

Year	Population ('000)	Households ('000)	1-person households (%)	Stock ('000)
1950	49,850	16,650	20	10,082
1961	54,733	19,399	22	16,002
1970	60,176	21,991	25	20,807
1980	61,431	24,811	30	25,406
1985	61,024	25,336 (1982)	32 (1982)	26,076
1987	62,399	26,100	33	26,595
1990	62,640	27,600	–	27,100

Sources: Emms 1990 (up to 1985); German Census (for 1987); Greiff and Ulbrich 1992 (for 1990).

Table 15.2 Distribution and structure of West German housing stock by tenure and subsidy, 1978, 1982 and 1987 (percentage of stock of dwellings)

	Owner-occupied	Housing associations: Not social housing	Social housing	Private landlords: Social housing	Not social housing	Total stock (million dwellings)
1978	37.5	4.0	10.5	7.2	40.8	22.6
1982	40.1	4.0	10.4	45.5		23.2
1987	42.0	3.7	9.1	3.5	4.17	26.3

Source: Hills *et al.* 1989.

Table 15.3 Housing completions in West Germany 1949–86, showing construction of multi-storey flats and one- and two-unit houses

Period	1949–56	1957–66	1967–76	1977–86	Total
Multi-storey flats	1,897,000	2,790,000	2,838,000	1,113,000	8,638,000
One- and two-unit houses	1,340,000	2,430,000	2,221,000	1,997,000	7,988,000
					= 16,626,000

Source: Gesamtverband der Wohnungswirtschaft 1989.
Note: An estimated 3 million units in addition to the above were approved for building in the period.

privatise former East German housing as rapidly as possible, and on the other hand to involve local authorities in the new programmes required to tackle the extraordinary problems. Table 15.4 summarises the tenure pattern in both parts of Germany.

CONCLUSIONS

One of the strongest influences on post-war German housing was the imperative to build. This led to an 'all hands to the wheel' policy, with help for all producers willing to house people. The government made it attractive to private investors and owners to build, and gave housing companies an incentive to grow by providing generous subsidies. There was perhaps some sense of competition between the limited dividend companies, which by law had to operate 'for the public good' and subsidised private landlords who wanted to make money out of social need. The Germans appear to have had less difficulty in straddling this divide than other countries. Politically, it was a matter of emphasis rather than substance. Social Democratic state governments favoured the more public route of the limited dividend companies and co-operatives, but certainly not to the exclusion or even detriment of 'social' private landlords; while the Christian Democratic emphasis was somewhat reversed, the substance was similar, with a preference for private landlords but frequent recourse to social landlords.

While there were compelling reasons for supporting limited dividend housing companies through public subsidy programmes at times of crisis in German history, there was never a compelling argument for turning local authorities into direct landlords. Both the private orientation of the German economy, and the widespread European pattern of building through autonomous housing bodies, influenced German social housing. The strong co-operative origins of the social housing movement also created pressures for a non-profit rather than a state social housing system.

Table 15.4 Distribution of tenure in Germany based on 1987 Census (W. Germany) and Institut Wohnen und Umwelt 1989 (E. Germany)

	Former West German States	*Former East German States*	*Total*
Private landlords	12,000,000	1,700,000	13,700,000
Limited liability companies and co-operatives	4,000,000	3,600,000	7,600,000
Owner-occupiers	10,500,000	1,700,000	12,200,000
Total	26,500,000	7,000,000	33,500,000

Source: Greiff and Ulbrich 1992.

These traditions are still alive today and prevented the absorption of co-operatives into the now largely private-rented housing market.

The German economy grew very rapidly, with an extremely low rate of inflation. The result was roughly twice the available state resources for social programmes such as housing, compared with Britain (Couch 1985). This not only affected the volume of housing production (number of units) but also the quality of housing, with higher standards of insulation, better equipment, central heating in over 75 per cent of all units, and a much better standard of basic building. This, however, did not necessarily apply to space standards or urban layout.

A crucial influence in the development of housing provision has been the linking of subsidies to performance, rather than to ownership. Therefore, Germany used and subsidised directly any builder, private, public or semi-public, that would build.

Traditionally, rural dwellers largely built and owned their own housing. This legacy survived in the pattern of owner-occupation today. In communities of fewer than 50,000 inhabitants well over half of the houses are owner-occupied, whereas in major cities of more than half a million people under 20 per cent are owner-occupiers. In the inner areas of major cities such as Dortmund, the proportion may be as low as 2 per cent (Couch 1985). The more rural *Land* governments also tend to favour owner-occupation, while the more urbanised *Länder* favour renting. The government itself has over a long period given greater tax incentives to landlords; these were gradually extended to owner-occupiers. There is thus a much greater balance of incentives between tenures, and in recent years the emphasis has been on 'free market' private provision.

A huge volume of modest housing was built after the war. There were on average two rooms fewer per dwelling than the average British home. German rents have been charged by the square metre, helping to create demand for smaller units. The subsidy system also strictly limited the size of dwellings in an attempt to ensure maximum production.

Although a majority of Germans live in small flats, the amount of housing space per person has risen sharply from 14 square metres per person in 1950 to 36 in 1985 – a two-and-a-half-fold increase. The number of people in a dwelling more than halved from five in 1950 to 2.2 in 1990.

While most German housing was built in small blocks of flats in cities, increasingly single-family or two-family houses became popular and a majority of recent production has been in that form. As needs have risen through the process of reunification, so policies have reverted to the encouragement of building in estates, albeit with more careful guidelines for amenities and environmental needs.

The main success of German housing policy was to keep a large and varied renting system. Social housing was frequently moved between front-

and back-burner as new crises developed. Currently it is enjoying something of a comeback with intense demand through rapid immigration and new shortages. But its low-rent, fast-build approach produced problems similar to elsewhere, if on a smaller scale – inadequate standards; inefficient, clumsy decision-making, and concentrations of disadvantaged households. Therefore, whenever possible, the government tried to reduce social housing and expand private provision. Although owner-occupation has been much slower to expand than in other Western countries it has enjoyed growing popularity which is likely to continue.

The influx of East Europeans created demand that could not immediately be met, either through rapid access to owner-occupation or through rapidly rising 'free market' rents. At the most extreme, there are now 130,000 single people who are literally homeless in the united Germany (Greiff and Ulbrich 1992). Therefore, the government pumped money into new social building programmes specifically targeted at East Europeans in West Germany (Tomann 1991). To the extent that this succeeds, it will create demand among poorly housed West Germans and in the Eastern states of Germany, for similar treatment. The government is therefore likely to continue its umbrella role, encouraging all sectors, damping the sharpest effects of the free market, and building up the role of 'social' landlords again, albeit within a freer, more market-oriented pattern (Gesamtverband der Wohnungswirtschaft 1989). This mixed outcome is a response to more extreme conditions in Germany than in other countries because of the peculiar needs of reunification and recent immigration from Eastern Europe. It underlines the difficulty of relying on the market, the possibly inescapable role of the state in sponsoring housing provision, and the tendency for crises of shortage and social need to push forward the role of social housing providers at least for the foreseeable future.

Part III
Britain

All history is unfinished history, and just as we have had more than one yesterday, so we can, if we choose, have more than one future.

Asa Briggs, *A Social History of England*

Britain was the cradle of the Industrial Revolution, of scientific invention, and of many forms of state provision. Yet Britain's role in the making of post-war Europe and in tackling housing and urban problems has shown few signs of leadership, glamour or flair.

Britain's economic problems are well known. In the 1890s, Britain had more registered shipping tonnage than the rest of the world put together (Briggs 1983). Today, the British merchant fleet barely survives. In 1958 when the six continental states formed the European Economic Community, Britain refused to join. Her per capita income was one-third higher than in those founder states. By 1979, it was lower.

Britain is the only country in Western Europe to have one-quarter of its housing stock directly owned by government. While this situation is changing rapidly and likely to go on doing so, Britain has uniquely large, bureaucratic and politically contentious public landlords, a structure that marks her out and may reflect or shadow her decline and conservatism in other areas of activity.

Britain is more urbanised and more densely populated than any other country in this study. Britain has a higher proportion of single-family houses than Germany, France or Denmark. That combination makes Britain – and British housing in particular – a richly fascinating area of study. This study aims to show the achievements and the problems, recording a housing history that in many respects is second to none, 'a Rolls-Royce effort' (Donnison 1987).

Although Britain is considered and discussed here as a single state, there are four distinct geographical and administrative countries of the United Kingdom – England, Wales, Scotland and Northern Ireland. The study covers mainland Britain, but not Northern Ireland. Where figures or information refer to only one part of Britain, this will be highlighted. Many statistics are available only for the United Kingdom as a whole. This is made clear in the text. As Northern Ireland comprises just over 2 per cent of the total population, the figures for the United Kingdom give a clear picture of general conditions in Britain.

Whether Britain is like a 'late afternoon tea party', an 'engine that has run out of steam', a 'complacent and arrogant but faded grandee' – or something more positive – may remain uncertain until her place in the European Community, on which her future surely depends, becomes clearer. Much will be determined by her internal political stability, the modernisation of her communications, educational and industrial networks, and the productive re-use and adaptation of her long-established infrastructure and urban patterns – particularly her valuable and now partly under-utilised cities and housing stock.

The story of British social housing sets out to show the almost inevitable drift towards 'Thatcherism', coupled with a strong and sometimes surprising defence of her remarkably resilient social institutions.

Part III.1 The cotton mills owned by Robert Owen at New Lanark, 1825

Chapter 16

Background

BASIC FACTS

Income levels are lower in the United Kingdom than in the continental countries. The UK and France are very close in average household size, birth-rate, proportion of racial minorities, and levels of unemployment. Early urbanisation led to lower rates of house-building since the war than those of northern European countries. Table 16.1 gives some background figures.

HISTORY IN BRIEF

Early urban growth

Three major invasions, by Romans, Danes and Normans, failed to conquer large areas of the British Isles but laid the foundations of agricultural prosperity, trade, and early urban settlements that made Britain ripe for industrialisation in the late eighteenth century. A tenth of the British people lived in towns by the fourteenth century; London was by far the largest city in Europe by 1600. The Reformation, the Parliament of

Table 16.1 Facts about the population of the United Kingdom, 1991

Population	57,200,000
Population per sq. km	234
Average number per household	2.6
Under 15s	19%
Over 65s	15%
Non-white minorities	6%
Population in urban areas	89%
Unemployed	10.3%
Percentage of total expenditure per head on housing and energy	15%

Sources: *Economist Pocket Europe* 1992; *Eurostat General Statistics* 1992; Office of Population Censuses and Surveys 1992.

Commoners, a constitutional monarchy, the insular pragmatism of British government, the eventual union with Scotland in 1707, all helped pave the way for a huge leap in economic, scientific and social progress in the eighteenth century.

Industrial Revolution

The Industrial Revolution happened in Britain eighty years before anywhere else in the world for complex reasons: efficient agriculture; inland waterways; cheap and highly developed shipping; plentiful and accessible raw materials (coal, iron, tin, copper, salt); political stability; a developed urban and social structure; and contact between the classes (Barraclough and Stone 1989, Briggs 1983). There were many other factors, such as the ignominious slave trade that deeply tarnished any credit Britain drew from that part of her history which put her so firmly in the lead. Huge gains at a huge price may not in the long run have proved to be gains at all.

The first inventors, entrepreneurs and builders irrevocably changed the landscape of Britain and then the world. Factories, mines, and towns experienced explosive growth. Output of coal alone grew twentyfold in the nineteenth century. Hundreds of thousands of miles of canals, rail tracks and roads were built to carry people and goods at ever faster rates and in accelerating volumes (Briggs 1983).

In 1800 there were fifteen towns with over 20,000 people; by 1891 there were sixty-three. Local authorities first emerged with some constitutional strength in 1835, set up to control conditions. In 1834, the Poor Law was amended with the intention of creating the 'Work or workhouse' ethic. Co-operatives and Friendly Societies were the Victorian poor's answer to rapid change and harsh conditions, while private charity – of rich to poor and among the poor themselves – kept the state at bay and made it a somewhat neutral arbiter rather than orchestrator of urban conditions (Thompson 1990).

By 1860, Britain was the richest country in the world, with twice the wealth of France and three times that of Germany (Briggs 1983). This 'convulsion of prosperity' fuelled imperial ambitions and encouraged export to embryonic and underdeveloped areas of the world. By the late nineteenth century, Britain no longer led at the frontier of industrial invention – the motor car, the radio and much later, computers (pioneered in Britain but developed and applied elsewhere), were to fuel the American and German growth.

Social unrest

By the turn of the century, social unrest and tensions began to emerge after a long period when economic rather than political change had largely contained dissatisfaction and kept an uneasy peace between the owners and productive masses (Thompson 1990).

The pace of incremental legislative and social reform was quickened by growing militancy among trade unions, which mushroomed from 1.5 million members in 1896 to 4 million by 1914. As British wealth and dominance were threatened by the rise of Germany, so was her internal cohesion threatened by those who had not fully shared in the dominant structures of British society. The Suffragettes, the Irish Home Rule and Republican movements, militant trade unionists and the newly formed Independent Labour Party, all underlined the fragility of Victorian grandeur, now almost a generation behind and rapidly becoming a thing of the past. The early Welfare State was born in this turbulent period before and immediately after the First World War, and the beginnings of national insurance took over where Friendly Societies left off.

Plate 16.1 Working class by-law housing in 1900

The First World War

The First World War overtook the lives of 8 million men, at least three-quarters of a million of whom died in the mud trenches of northern Europe (Winter 1986). Britain paid a huge price in order to contain Germany. Virtually every able-bodied adult up to the age of 40 was conscripted by 1916. Food was rationed, railways were nationalised, supplies controlled and women drafted into men's jobs. Two-thirds of the workforce was government-regulated (Briggs 1983). The war itself failed to establish lasting peace and did irreparable damage to Britain's leading position. At the end of the war, the return of 5 million demobilised soldiers created fear of revolution.

A stormy interlude

The government maintained the apparatus of state control and state provision that had helped the country through the most serious threat to her sovereignty for 900 years in order to introduce progressive housing and educational legislation. But labour unrest culminated in the General Strike of 1926, and the great Wall Street crash of 1929 led to chronic depression and nearly 3 million unemployed. The inter-war years proved a brief and stormy interlude, still overshadowed by the horrors of the First World War, while menaced by the gathering clouds of the Second World War – a 'bitter society' (Briggs 1983).

The Second World War

The Second World War was maybe Britain's 'finest hour'. But Britain was seen like the biblical David, a declining island people literally isolated in an alien and Fascist-dominated Europe. No wonder at victory she regarded herself as occupying the moral leadership of the world, having lost her pre-eminence in virtually every practical field.

Women did not give up their jobs after the Second World War. There were chronic labour and skills shortages as the economy recovered and women could no longer be so readily spared for the home. Some wartime changes became permanent.

Returning soldiers formed new families and both marriages and births boomed, while family break-up, greatly heightened by war conditions, affected many more than ever before. The population grew, and so did the number of households. In 1931, there were 10 million households; by 1951 (there was no census in 1941), there were 13 million (Halsey 1989: 363).[1] The total population rose in the same period from 46 million to over 50 million. Should they 'rise on ladders or wait in queues?', asked Winston Churchill, wartime Prime Minister.

Labour's 'fair deal'

The post-war Labour government offered a strongly egalitarian 'fair deal' – universal social security, child benefit, a free national health service, free secondary education for all, a commitment to full employment – a vote-winning vision first presented to the coalition cabinet by William Beveridge during the war. By 1951, thanks to Labour's efficient, rapidly implemented universal welfare system, it could be claimed that primary poverty (lack of bare necessities for survival) affected only 3 per cent of the working class, a huge drop from 33 per cent in 1939, although these figures have been frequently challenged (Abel-Smith and Townsend 1965).

British urban society absorbed 2 million immigrants from New Commonwealth countries in Africa, Asia and the Caribbean, during the post-war boom. From 1970 her population grew only very slowly, her main cities declined, and her economic position in the world tumbled.

The comfortable cushion of past growth and imperial preference locked the British economy into outdated patterns of management, production and investment, while the swelling ranks of trade unions – 11 million members by 1970 – attempted to guard and defend their idea of a rightful share of what had been achieved. Britain's decline, half concealed by post-war growth, seemed terminal to many European and American observers, as they built anew, even though in actual purchasing power Britain remained rich and close to her European partners (OECD National Accounts Statistics 1989).

Thatcherism

The Conservative victory of 1979, bringing a woman Prime Minister to prominence for the first time in European history, marked the sharpest turn of events since the Industrial Revolution began. The Welfare State was called into question, British manufacturing and trade suffered greater losses than in any previous period, and the consensus was broken on historic sacred cows, including welfare and housing.

The full significance of the Thatcher years is perhaps most sharply and poignantly highlighted by the re-emergence of significant destitution and begging on the streets of central London. The numbers of unemployed fluctuated between 2 and 3 million for the whole decade. There was little resemblance between Victorian stability, prosperity and international pre-eminence, and the harsh increase in polarisation, relative poverty and neglect of infrastructure that marked out Britain so singularly from her partners in Europe at the beginning of the 1990s.

NOTE

1   Household figures are for England and Wales only.

Chapter 17

Early housing developments

PITCHED ROOF HOUSES

British housing history roughly mirrored her economic and political progress, starkly highlighting the social divisions as well as the wealth and the industry of the nation. Early growth in town-based crafts and trades, dating from Roman and early medieval times, encouraged villages and towns that stand today in much the same shape and with some of the same houses as they did 500 years ago. Britain's housing heritage offered an easily replicated housing model to private builders, making two-up and two-down, single-family dwellings with pitched roofs the most common, most durable, and most lastingly popular form of housing.[1]

Those who were employed on the land often lived in squalor, as the rural population expanded faster than the supply of rural housing, and the solid cottages of earlier centuries survived to house larger households in primitive conditions (Burnett 1980: 36).

EARLY URBAN GROWTH

The thousands who were made destitute by the enclosure of agricultural land in the eighteenth and nineteenth centuries, a pre-condition of modernisation, either made shift in hovels or moved into growing cities, attracted by the great freedom, wealth and opportunity as much as coerced by rural change and poverty (Briggs 1983). The agricultural revolution fed the Industrial Revolution as much as rural upheaval fed the growth of cities.

The rate of urban house building kept pace with or outstripped the huge explosion in town populations in the first half of the nineteenth century (Burnett 1980). Many skilled workers and artisans were well housed by the mid-nineteenth century. The average mid-Victorian working family in skilled employment occupied four rooms! The major industrial growth areas were virtually new towns. Manchester, Newcastle, Glasgow, Liverpool and Birmingham went through incredible explosions of population, and cheap, terraced housing was quickly put up by factory owners for their workers and, later, by builders for sale to private landlords.

NEW LANARK – A UTOPIAN CO-OPERATIVE

Enlightened industrialists began to reject the brutalisation of the workforce that went with child and female labour, provoking a growth in legislation to control conditions and early rationalist, philanthropic experiments. Robert Owen's renowned New Lanark mill in Scotland was probably the best known, most utopian and most far-reaching in its impact. Between 1800 and 1829, he converted an ailing textile mill into a prosperous business and thriving community by adopting a revolutionary combination of benign but tight management, social welfare, and education – providing quality housing for his workers, embryonic sheltered homes and pensions for retired workers, a school, a play and entertainment centre, and adult education. He restricted child and women's labour early in the century; he stressed supervision and encouragement rather than punishment and repression. He believed strongly in the power of education and the benefits of co-operation. Robert Owen regarded human nature as universally reformable. He held the strongly utilitarian view that owners' self-interest lay in providing 'moral conditions' within which workers would become allies of the owner, producing more and better goods (Owen 1970).

His ideas were echoed in many early continental housing initiatives. The Rochdale Pioneers formed the first workers' co-operative in 1823 on these principles, founding a world-wide movement.

BACK-TO-BACKS

As urban pressures mounted, back-to-back housing became common, particularly in the north. This took the form of long rows of two- or three-storey, two- or three-room dwellings that economised on bricks, tiles and land by being built literally back to back, with a party wall on three sides of each dwelling. The communal water supply was often impure; there were no drains; and closets were shared by whole streets. Back-to-backs made the removal of refuse and waste water difficult, but were the most economical form of nineteenth-century urban workers' housing. They were soon banned in many towns, but Leeds Council insisted on preserving the form right up to 1909 thereby earning, into the modern era, a vilified image for Victorian terraced housing. Twentieth-century enthusiasm for slum clearance was fuelled by the legendary back-to-backs.

SLUMS

More and more people were drawn into cities in the middle of the century, and maybe a million starving Irish peasants swelled the numbers of inexperienced and often ignorant new city-dwellers. The cost of transport for unskilled workers limited their access to housing outside the immediate industrial areas, while the already dense development of the cities meant that poor families simply had to double up in existing lodgings. Cellars, sheds, lean-to's, attics – all were brought into use until by the 1860s poverty, squalor and rough living were commonplace. About 1 million households (nearly 6 million people) lived in shared accommodation in growth areas – almost one-third of the total population. Severe outbreaks of cholera and a rising death rate in many cities led to serious attempts at reform.

Much of the densest overcrowding was in London, the international metropolis that attracted and sustained a huge unskilled proletariat, both disconnected from its rural roots and weakly connected with the industrial maelstrom. The infamous 'rookeries' became the haunt of new arrivals and of the failed 'dregs of Victorian affluence'.

THE POOR DISPLACED

The poorest inner areas were under constant assault from city developers, hungry for land for railways, commercial centres, banks, warehouses, schools and hospitals. Removing the chaotic slums was easier than either improving them, regulating them or finding real alternatives. The result was even greater overcrowding for the newcomers, the casually employed and the casualties of *laissez-faire*. 'Tens of thousands are crowded together amidst horrors which call to mind what we have heard of the middle passage of the slave ship' (Mearns 1970).

OCTAVIA HILL

Prominent Victorians, many of them women, led pioneering reforms in health, housing, education, child care, and prisons (Lewis 1991).

Octavia Hill, the most famous housing reformer, opposed demolition of poor dwellings, believing that clearance and rebuilding at great cost would exclude the very people who most needed help and were made homeless by the clearances. She advocated sound management, renovation, repair, and strict enforcement of rent payments and tenancy conditions on those normally considered too feckless to shoulder responsibility.

She persuaded private investors to buy rented property for her and her fellow workers whom she trained in housing management. John Ruskin was her first benefactor in 1863. Although she only took over, restored and managed 15,000 properties in fifty years of effort, she inspired many twentieth century developments in tenant-oriented housing (Power 1987a). With her support, the Artisans' and Labourers' Dwellings Act of 1868 was passed, the first significant control on slum landlords. But she strongly opposed the building of blocks of flats as oppressive and hard to manage.

MODEL DWELLINGS

The failure of private landlords to meet bare needs on a purely commercial basis for the very poor and casually employed led philanthropists such as George Peabody to invest their wealth in housing. Lord Shaftesbury persuaded him to set up the Peabody Donation Fund in 1862, investing his private fortune in a trust that would build, charge 'affordable' rent, and provide a 3 per cent return that would be reinvested in housing.

In 30 years the Fund had built 5,000 solid, utilitarian flats in dense blocks 'to ameliorate the conditions of the poor and needy of this great metropolis and to promote their comfort and happiness'. Other trusts and model dwelling companies followed suit.

The model dwelling companies and philanthropic housing trusts opted to build flats rather than the traditional terraced housing because they believed they could concentrate more dwellings on scarce urban land by stacking them five floors high, offering modern sanitary facilities because one pipe could serve five flats compared with one house. Flat-building combined local authority action to clear the worst slums from central areas, with the need to provide the maximum number of low-cost units for the poor, and a way to pay for expensive central land.

But rents were out of reach of the very people they were displacing from the cleared rookeries. To make matters worse, slum rents rose for the poor as more slums were cleared.

Need for legislation

Victorian growth depended on cheap casual labour. The 'survival of the fittest' literally described their fate. The death rate in many towns rose higher than in the countryside, and employers often sought new rural migrants because they were stronger than urban residents debilitated by poor conditions (Royal Commission 1885). The process was self-fuelling, as Lord Shaftesbury described it, with young rural migrants coming to town full of hope and promise, being quickly broken by the long hours, the shocking conditions, bad diet, and insecurity.

Surprisingly quickly, attempts were made to regulate conditions in towns, for two main reasons. Firstly, the growing middle classes were directly affected by epidemics, such as cholera, and were concerned to see better conditions out of fear of chaos. Secondly, as machinery advanced and urban life became more complex, skill and stability were required of the workforce. This was only possible if housing conditions improved.

The role of government grew in recognition of the need to regulate the free development of towns in support of individual and private effort.

In 1835 the Municipal Corporations Act established modern city government. Lord Shaftesbury introduced Acts to control and encourage regulated lodging houses for the poor. But the first laws to tackle housing conditions directly were the Torrens and Cross Acts of 1868 and 1879. At last local authorities were empowered to remove slums and obliged to replace them with regulated housing. They could no longer simply clear poor dwellings but had to ensure that housing was built on the cleared land. Joseph Chamberlain, as mayor of Birmingham in 1873, pushed the pace by enforcing sanitary laws, providing water and gas, paving streets, and improving houses in poor areas. Birmingham became a model city.

Royal Commission

Local authorities were helped in their growth by a Royal investigation of conditions, headed by the Prince of Wales. The Royal Commission on the Housing of the Working Classes sat in 1885, hearing 'the great and the good' document and denounce the appalling housing squalor among the poor. It confirmed that:

> the evils of overcrowding, especially in London, were still a public scandal and were becoming in certain localities more serious than they ever were. The existing laws were not put into force, some of them having remained a dead letter from the date when they first found place in the statute book.
>
> (Hall 1988: 19)

There were too few landlords and, while conditions for the very poor were shocking, rents were high. Thus the Royal Commission on the Housing of the Working Classes in 1885 found that 'It was common practice in London for each family to have only a single room, for the rent of which nearly half of them paid between 25% and 50% of their wages' (Royal Commission 1885: 17). Overcrowding increased in the areas near philanthropic housing developments, such as those of the Peabody Trust in Covent Garden (Royal Commission 1885), as the very poor were squeezed into receding areas of cheap housing. Clearance and rebuilding did not of itself solve the problem of poor housing.

By-law housing

Local authorities, under their new powers, expropriated unfit property, enforced standards on new building, and planned urban development. By-law housing was the product, built in endless rows of brick-built, two-storey terraces, neatly laid out with alleys between the backyards to allow carts to pass through to empty sewage buckets. The vast majority of by-law housing was owned by small landlords – city dwellers who shaped and elected the new municipal politicians. Local authorities became increasingly powerful as they began to levy local property taxes (rates) on their electorate in order to carry out works to improve their towns (Thompson 1990).

By-law housing represented a great advance in standards and it spread at a pace that stayed significantly ahead of demand, so that by the beginning of the twentieth century less than one household in thirty occupied only one room, and over half had five or more rooms (UK Census 1911: vol. 8). Even in crowded London, under 14 per cent of the population occupied one-room dwellings and 46 per cent had four or more rooms. By-law housing reinforced the English attachment to single-family houses for all classes, even the poor, albeit in monotonous and uniformly stultifying inner suburbs. By-law housing was often put up by so-called jerry builders in cheap terraces with no foundations. But much of it was still standing three generations later, having survived the huge slum clearances.

Local authorities or trusts?

By 1890 local authorities were working with trusts and companies, as well as building model dwellings directly with cash from the rates. It was not until after the First World War that council housing, as it is now known, became significant. Local authorities had directly built only 20,000 houses for people in need by the First World War, while philanthropic landlords and model dwelling companies had built 120,000 flats. Local authorities did not expect or intend to continue as landlords, although the combination of rates income, slum clearance, public health and rebuilding powers helped determine their future role. Meanwhile, imaginative housing experiments that were to influence local authorities throughout the twentieth century emerged from the combination of benign industrialists, co-operative ideals, and public regulation.

Ideal villages

The village of Port Sunlight, built by William Lever, whose soap factory gave birth to the first major multi-national firm, Unilever, became a national monument. The strong pall of cheap carbolic soap from the famous soap factory hung over the workers' houses for a century.

Bournville, built by the Cadburys in Birmingham, and New Earswick, by the Rowntrees in York, had a strong Quaker ethos behind them, reminiscent of Robert Owen's 'Vision of a new society' – 'a happy home for many generations of children where they will be brought up amid surroundings that will benefit them spiritually, mentally and physically' (J. Rowntree in Wagner 1987).

Well-paid, well-treated chocolate and soap workers were housed in beautifully built factory villages, while the very poor and the very rich of the wider society remained as far apart as ever. The model communities aimed to mix the social classes; the houses were varied; there were gardens and open space, community halls and adult education centres. There was a strong emphasis on teetotalism and on workers' allotments. Both Cadbury and Rowntree emphasised co-operation and common interest between the workforce and owners.

The Garden City Movement

The Garden City Movement, immeasurably significant to twentieth-century housing, was also born of the dream of ideal communities. Ebenezer Howard, like Robert Owen, was a utopian. Unlike Robert Owen and the Quakers, he was not an ambitious and successful industrialist but a modest clerical worker who abhorred the conditions of workers in exploding cities. In 1898 he published a remarkable little tract, *Tomorrow: A Peaceful Path to Real Reform* (Howard 1898). Freedom, co-operation and community would grow in harmonious surroundings that united the values of the countryside with those of the town – the Three Magnets, as he called them. With the precision of a watchmaker he designed new garden cities outside the chaotic urban areas. Enclosed in a green belt of agricultural land, each city was funded through investors buying farm land cheaply and earning a return as the value of the garden city rose, attracting in new industry for the growing workforce. Transport would link the settlements to the main city. Profits, over and above the interest on investments, would be ploughed back into the new town for social provision and further development. The whole development would be self-financing, self-governing and self-fuelling. The idea would spread because it offered benefits to everyone – the investors, the new citizens, the industrialists, the government. The common good would be advanced without an increase in tax or subsidies.

Garden cities would be locally managed havens 'to banish despair and awaken hope in the breasts of those who have fallen; to silence the harsh voice of anger and awaken the soft notes of brotherliness and good will' (Howard 1898). It was a dream of moral, emotional and organisational power, befitting the age of explorers, eccentrics and inventors.

The Garden City ideal drew on the Quaker villages. It also drew on the anarchist tradition of European writers like Kropotkin, with his allotment movement, and the American frontier and back-to-the-land ideals of Walden and Thoreau. Letchworth, the first garden city, was designed by Raymond Unwin, the architect of Rowntree's New Earswick, and begun in 1903.

Neither industrialists nor government were at the helm of the Garden City Movement: it was led by a group of utopian planners, relying on moral force. The idea was incredibly hard to implement in practice. Existing cities had their motor and their magnet. New garden cities proved unattractive and risky in a period of international tension and national turmoil. It was easier to attract settlers to colonies abroad than to new colonies at home. Part of the problem was that the better-housed had little incentive to move and the slum dwellers themselves were too insecure and poor to form the front line. It took Ebenezer Howard till 1912 to get his first garden city to produce dividends, and the string of garden cities he dreamt of never materialised.

Garden suburbs

In spite of these problems, the idea caught on and, like a forest fire, swept in many different directions. The idea of satellite cities, independent yet linked, was subsumed into an all-embracing fervour for suburban housing development. Garden suburbs replaced garden cities and, in Britain and Europe, the strong outward movement of more affluent classes was converted into planned satellite housing areas over the following decades for more modest households.

Groups of better-off tenants organised co-operative developments owned by public utility societies, using many of Howard's ideas. These were a cross between housing associations and co-operatives and were able to borrow money at low rates of interest when the government passed its earliest planning legislation, the Housing and Town Planning Act of 1909. By the time of the outbreak of the First World War, 100 utility societies had been formed and many suburban developments were established on co-operative garden city lines (Hall 1988). Although the remnants of these initiatives still exist in places, they were largely buried by the war and by post-war development.

Famous developments, such as Hampstead Garden Suburb, set up in 1907, represented the triumph of unstoppable city sprawl over controlled, harmoniously planned, mixed urban communities, and the green belt. The garden city ideal, difficult to implement in practice, hugely influenced the later developments of both peripheral estates of social housing and of new towns. The 'informal, reposeful and natural' ordering of homes into 'day to day co-existence which would sooner break the estrangement of the classes', remained a powerful idea (Hall 1988).

CITY HOUSING

Cities themselves were barely affected by garden city developments before the First World War. The mass of poorly housed gradually spread out into the by-law housing that was the norm, while new waves of rural migrants continued to fill the slums. But the worst of the cellars, back-to-backs and rookeries were at least outlawed, if not abolished.

Scottish cities, with their tradition of tenement flats rather than houses, had many smaller dwellings and there overcrowding was much more serious, with two-thirds of all households occupying only one or two rooms – double the rate for London (Thompson 1990, Vol. 2, Ch. 4). But housing conditions had steadily improved. By the turn of the century, only one in ten households lived in one room, compared with a quarter of the population forty years earlier (M. Daunton, in Thompson 1990; Vol. 2, Ch. 4).

The Victorian and Edwardian eras, the heyday of British urban expansionism, foreshadowed the main patterns of twentieth-century housing development. Single-family houses dominated. Blocks of flats were uncommon, except in Scottish cities, and from the outset they were less popular and harder to manage than houses (London County Council minutes 1885–1964). The rapid development of suburbs, from by-law two-up and two-downs to elaborate garden city satellites, led to the incremental decline from the turn of the century of inner-city areas.

Private landlords were already in retreat under the impact of demolitions. The lower demand for rented housing from better-off households in inner areas, coupled with the burden of property taxes, caused a drop in property values from which private landlords were never to recover. Yet at the outbreak of the war they owned 90 per cent of the housing stock and rented it out to tenants more cheaply, relative to wages, than in France or Germany (M. Daunton, in Thompson 1990, Vol. 2, Ch. 4: 198).

By the outbreak of the First World War, Britain's terraced streets, condemned in slum clearance proposals and later glorified in Coronation Street, were a monument to an era of unparalleled prosperity, peace and social stability (Briggs 1983).

NOTE

1   See M. Daunton in Thompson (1990) for discussion of Scottish preference for tenements.

Chapter 18

After the First World War

In the Great War, the entire machine of government cranked up to take every ounce of labour and production it could muster into the massive war effort. Building came to a standstill and acute shortages quickly surfaced, particularly in Glasgow, where shipbuilding had boomed because of the naval threat from Germany and where overcrowding was double the London level. Landlords, squeezed before the war by local taxation and drop in demand, faced the tempting prospect of rapid rent rises. Workers – many of them, for the first time, women – were, however, vital to the survival of the nation, stripped of almost all its young men on the war front. Widespread unrest over rent rises led to rent strikes. The government passed the 1915 Act, introducing tight rent controls virtually across the board for the first time. The 1915 rent controls were singled out as the act that most clearly shaped twentieth-century Britain into the 'fundamentally stable, unprotesting, unenterprising and conservative society of the late twentieth century' (Thompson 1990).

The 1915 Act disrupted private property rights and free market provision; it led to the rapid retreat of the private landlord and the emergence of council housing; it paved the way for direct state subsidies to owner-occupation, leading to residential immobility by tying up savings in bricks and mortar (M. Daunton, in Thompson 1990, Vol. 2, Ch. 4).

Private landlords never recovered; no lobby or political grouping argued in their favour. The alternatives appeared to make more sense: good-quality, controlled and regulated council housing; and owner-occupation for respectable, stable earners of all classes.

The problem for private landlords lay in the shortage, and hence the need for controls, resulting in low rents that drove away investment and prevented repair of existing housing. Only the state, which imposed the controls, could break the vicious circle. In Britain, direct state housing seemed the answer. In other countries, compromise solutions emerged, combining controls, subsidies to private landlords, and the encouragement of semi-private housing organisations as buffers between the state and the private sector. Why these developments failed to emerge in Britain is not clear, since newly built dwellings for rent were exempt from controls from 1919, and the subsidies available for owner-occupation from 1923 were also available for private renting.

BOOM IN STATE HOUSING

Britain's extraordinary burst in state housing activity marked her out from Europe or America in a similar period. Many small landlords had tried to exploit shortage, poverty and the war situation. But far more important was the war effort itself, which galvanised every sector of society to a single state-orchestrated purpose. It was hard to conceive of the state doing other than directly controlling an expansion in house-building straight after the war when conditions were so acute.

Renting was almost universal, so although subsidies to owner-occupation were introduced shortly after the ending of the war, the bulk of new housing would be for rent. This required a landlord structure that the state could build on quickly and readily. Millions of poorly co-ordinated and reluctant private landlords, now triply burdened with rates, lower values and rent controls, were not an obvious vehicle. The philanthropic trusts and public utility societies required a return on investment but the state was reluctant to subsidise these private bodies directly on a large scale, although there were attempts in the 1920s to encourage public utility societies to build again.

The fear that 5 million returning servicemen would unite with militant women, who had manned the factories and led the rent strikes in mass unrest, created an urban housing crisis. Four years in the trenches had earned nothing less than 'homes fit for heroes'. The threat of revolution appeared real (Swenarton 1981). The emergency post-war housing programme under the Addison Act of 1919 therefore provided generous funding for local authorities to build, although it did not rely exclusively on them. A combination of weakened private landlords, rigid rent controls, depleted national resources, chronic shortages, and political unrest, led to ever more government-directed housing programmes through local authorities.

COUNCIL HOUSING

Local authorities represented the urban masses in a way that central government could not hope to do. They provided an important power base for the working-class majority, who were the main beneficiaries of any urban improvements. The local electorates had a strong vested interest in supporting embryonic council housing.

Councils could obviously play a critical role in channelling and containing discontent, in co-ordinating plans and in simplifying housing problems by providing it direct. The hope was that they would raise stand-ards and remove the profit element, thereby providing cheaper, better quality property to let. The government would provide cheap money. Local authorities were willing to build, provided that it did not cost the local taxpayers more. They saw council housing as a powerful vote-catcher. The government found local authorities more willing partners in improving conditions than private landlords, including philanthropic trusts.

COTTAGE ESTATES

Thus, in 1919 the Addison Act was passed, announcing a target of 500,000 new council homes in five years, to be built on generous garden city lines. All over the country, local authorities planned cottage estates of houses with gardens, to be built on the edges of towns at the low density of twelve homes to an acre. The old by-law pattern was condemned as dead-ening and wasteful, with 40 per cent of the land eaten up by tarmac roads.

In the new cottage estates, the roads took up only 17 per cent of the land, whereas open space took up over half. The houses were often gable-fronted with a genuine rustic air. Arches led through the middle of terraces into back gardens and allotments. People were moved from insanitary and often shared dwellings in crowded and dirty inner areas to what many described as palaces. But the open government subsidy meant that costs rose rapidly; labour and materials shortages made building slow and expen-sive; local authorities were not as efficient in planning the supply lines as small private builders responding to very local demands. Therefore in 1921, after only two years, the programme was halted and was replaced under the Chamberlain and Wheatley Acts of 1923 and 1924 with lower subsidies, less generous standards and a bigger local authority cash contrib-ution.

In 1924, private landlords were in heavy retreat. Shortages had increased, with half a million more sharing households than in 1911 (Halsey 1988: 367); rent controls were still considered necessary and local authorities were nowhere near their target of half a million homes. The result was a massive burst of lower-quality, cheaper council housing, rented to lower-income tenants at rents above the controlled private rents. Many families encountered real hardship in meeting the rents and, in Stockton-on-Tees, the death rate rose as families moved out to new estates (M. Daunton, in Thompson 1990, Vol. 2, Ch. 4). New council estates lost their garden suburb character and increasingly became new enclaves of poverty. The late-1920s' financial crash and the Great Depression made matters worse.

Local authorities did not plan their estates, as Ebenezer Howard planned his Garden City. Social facilities were rarely built in; industry did not move with the housing. In fact, part of the aim was to leave industry behind. There were reports of growing delinquency and social breakdown, and high levels of unemployment and poverty on the giant cottage estates outside London, Manchester and Liverpool.

Councils became the major providers of new rented housing. In all, well over 1 million council houses were built on cottage estates between the wars, although the Tudor Walters standards – inspired by Raymond Unwin the garden city architect – with the generous Addison Act funding were soon dropped. Only 100,000 flats were put up by councils in the inter-war period, mainly in London. Council housing was an overwhelming success in breaking the urban mould and in providing single-family dwellings for working people, although for decades much of it was only let to the stable and relatively prosperous working class. Now 70 years old in some cases, the majority of inter-war cottage estates are still intact, popular, and increasingly bought by individual sitting tenants.

SLUM CLEARANCE

The condition of the remaining controlled stock of private-rented homes plummeted over twenty years of neglect. The rate of household formation had greatly accelerated and the very poor could ill afford council rents. The number of sharing households continued to rise, from 1.7 million in 1921 to nearly 2 million in 1931 (Halsey 1988: 367).

Under the Greenwood Act of 1930, for the first time councils took direct responsibility for rehousing whole communities, for keeping rents within reach of the very poor, and for targeting their efforts at slum dwellers previously excluded from council housing. Responsibility for tackling overcrowding was added in 1935. Altogether, nearly 300,000 slum properties were demolished in the 1930s, but at least another half million were condemned (Halsey 1988: 368). One million council houses were built under the 1930s' slum clearance programme. About 4 million people were forcibly moved from slums in the period 1930–9, more than at any other period in this century (Halsey 1988: 385). It was a drastic and far-reaching policy that tackled some of the worst physical conditions, but often set up unforeseen social problems (M. Daunton, in Thompson 1990, Vol. 2, Ch. 4). The face of British cities was changed by the clearance of old inner areas and the construction of rings of council estates around virtually every town and city of any importance in the country. Few voices were raised against clearance, but entire inner-city populations were unsettled and much private renting was terminally blighted through demolition plans.

LOCAL AUTHORITY LANDLORDS

The Town Clerk of each council was responsible for co-ordinating inputs into housing management. While the slum clearance of the 1930s brought social needs to the fore, only a minority of councils had appointed housing managers by the Second World War. Politicians gave priority to production.

There was little sense of the landlord's obligations to council tenants who had no security of tenure. Welfare and repair needs were given low priority. Allocations and rent setting were often a matter of political expediency. Because council housing was largely built in cottage style, front-line custodial and janitorial services were not normally considered necessary. Only the rent collector, with a narrow and routine role, visited the estates (CHAC 1939). Problems therefore quickly mounted, with worrying signs of decay.

By the Second World War, council housing had become firmly embedded in the national psyche as the tenure that catered for slum dwellers. But it provided mainly suburban houses and it was vastly better than run-down slums.

THE RISE IN OWNER-OCCUPATION AND THE DECLINE IN PRIVATE LANDLORDS

Throughout the inter-war period, private landlords sold out into owner-occupation at a rapid rate. Nearly one and a half million dwellings were converted in this way between the wars.

The 1919 Act subsidised, for the first time, private building for owner-occupation and from then on private building boomed. Demand was enormous. Speculative builders were enticed by direct subsidies. The Conservative view was that owner-occupiers were a safer bet than private landlords. Making room for more tenants by helping some to become owner-occupiers was not seen initially as divisive and it was politically popular.

There was thus an unspoken political consensus on the advantages of owner-occupation. Much of the ethos that went with support for owner-occupation was embodied in the housing style of the country – single-family, terraced and semi-detached houses that were cheap and simple to put up and easy to run and sell.

The garden city and garden suburb movements further encouraged owner-occupation, and many of the early co-operative, public utility societies were bought out by their individual members.

Owner-occupation expanded enormously between the wars from 700,000 homes in 1914 to 3.5 million by 1939. At least one and a half million were built new, the rest were transferred from the private rented sector. Of the new dwellings, a quarter were directly subsidised by the state.

Building societies lent out over 2 million mortgages in the period. Loans

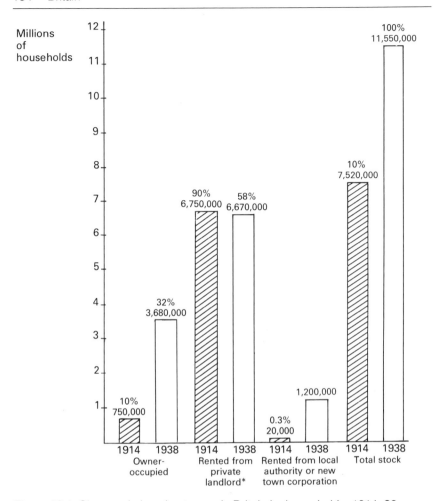

Figure 18.1 Changes in housing tenure in Britain by households, 1914–38

Source: Halsey 1988: 377

Note: *The increase in sharing prevented the decline in private renting from becoming steeper.

to wage earners, as opposed to white-collar workers, constituted up to nearly 40 per cent of all loans by the 1930s (Burnett 1980: 248). This built on the nineteenth-century tradition of workers saving to provide their own housing through Friendly Societies, which formed the original building societies. The early housing co-operatives in Britain were also formed so that individuals could end up owning their own home.

House prices fell as the supply expanded, and repayments were often little more than the rent for a better council property. Many workers found their way to purchase. It became possible for anyone in regular work to get a mortgage from the booming building societies if repayments took up less than 25 per cent of wages. The main stumbling block was the deposit, but gradually the amount required was lowered to 5 per cent of the cost. In the early 1920s you could buy a house for anything from £200 – two and a half years' wages for an average industrial worker. The pace of speculative building accelerated till in the five years before the war, over a million homes were constructed privately (estimate based on Burnett 1980: 246).

Thus, between the wars, owner-occupation became a popular tenure across the classes. In form, it was not radically different from suburban council cottage estates or traditional terraced by-law housing. The predominance of houses over flats, even in cities, made it easy for landlords to transfer their property and for builders to construct at low cost. Owner-occupied housing made up over 70 per cent of all new inter-war housing. The 1930s became famous for their endless suburban semis, invariably separated from council estates to ensure their status value; the occupants generally regarded council housing as not only inferior but a drag on property values because of the lower status of its occupants (Saunders 1990).

Owner-occupation went down-market as it became cheaper (Burnett 1980). It offered security, and attracted households who previously might have rented from private landlords. At the same time, lower building costs in the 1930s did encourage the construction of nearly a million new private-rented dwellings, including some by housing associations. These helped replace the rented dwellings converted to owner-occupation (DOE 1977a, Technical Volume). Rent controls were somewhat relaxed in the 1930s as shortages eased up but they were quickly reimposed when the Second World War broke out. Meanwhile, council housing went down-market, partly due to the drop in demand, as better-off prospective tenants could save and buy for a similar price.

By 1939, owner-occupation was a reality for a quarter of all households, a far more rapid rise than the rise in council renting, at far lower cost to the government.

When the Second World War began, Britain's modern tenure pattern was firmly set – rising, cross-class owner-occupation; expanding, increasingly targeted and lower standard council housing; a declining private landlord tradition.

The stock of houses in Britain had risen from 7.5 million at the outset of the First World War to 11.5 million at the outbreak of the Second World War. The ownership of these homes was changing, as Figure 18.1 illustrates.

Chapter 19

After the Second World War

DISRUPTION

The war had an immeasurable impact on home conditions. Building stopped, as in the First World War, but there was the additional burden of rebuilding bombed cities – London, Coventry, Birmingham, Liverpool, and Manchester all had flattened areas. Three-quarters of a million houses were destroyed or severely damaged. The evacuation of women and children during the early years of the war had highlighted problems of education, health, and poverty in the slums (Women's Group on Public Welfare 1939–1942).

The chronic housing shortage was coupled with deep changes in family and work patterns. Return to 'normal' life was slowed by the persistence of rationing and conscription, both of which lasted well into the 1950s. Britain's 'finest hour' was very much a legend that lived on in children's comics, radio programmes and the cinema, while real life was about doubling up in existing housing, putting up prefabricated homes, moving around in search of a new future. The population was unsettled by the war and looking for relief from six years of intense and sometimes heroic sacrifice. A Labour government was elected on the promise that the people's welfare would come first. The idea of the state building a better Britain, on the back of victory, seemed a logical follow-through to the war effort and a sure way out of a return to pre-war conditions.

REBUILDING

Squatting was rife and could only be countered with actual housing. Prefabricated houses were put on bomb sites and the rebuilding of damaged property began. An ambitious council building programme was launched, with a target of 240,000 units a year. The private building industry was in disarray because of the wartime agenda, whereas government and local authorities, which had played the dominant role throughout the war, were geared up. Council housing was set to grow rapidly again, as there was greater faith in the capacity of public bodies to respond than in the capacity of private ones to do so.

THE ROLE OF LOCAL AUTHORITIES

Local authorities in 1945 were already large landlords. They were therefore well placed to expand the housing programme. They were particularly strong as a political force in the major cities where housing problems were the most serious.

After the war, the Labour government considered setting up a National Housing Corporation to sponsor the construction of social housing on a mass scale. This idea was abandoned in favour of local authorities because Aneurin Bevan believed that local electors would have more direct impact on their local political representatives than they would on a government body (Foot 1975). They already had extensive planning powers, a major role in public health as it related to housing, and were the only bodies that could enforce closing and compulsory purchase orders on negligent private landlords. The scale of the problem seemed to require the action of these unitary but locally representative bodies that could ensure coherence – land acquisition, planning, slum designation, rehousing, redevelopment and building, political answerability and vote-winning.

The rebuilding was coloured by sharp disillusionment with suburban sprawl, popular as it was with home-owners. The 1930s' slum clearance council estates came under attack as mean and endlessly monotonous (Thompson 1990). There was the feeling that land was being gobbled up, city life was deteriorating, and everywhere there were semis or slums. There was too little colour or diversity and little mixing. Post-war Britain wanted to break the mould. Victory encouraged the notion that Britain could build its way out of battle.

By-law housing was classed as slums, often even where it was structurally sound. Inter-war semis were classed as 'boxes', whether saleable or not. Council housing was drab. Quality was all-important and housing developments would reflect the needs of different social groups as well as reinforcing their interdependence.

NEW TOWNS

New towns, the direct descendants of garden cities, came into vogue in the post-war euphoria. They offered mixed communities, a clean sweep, cheap development because of low land costs, a mass solution because there was space beyond the urban periphery, and an escape from slums and sprawl. They would provide a major addition to the stock.

The 1944 Abercrombie Plan for Greater London had proposed ten satellite towns. They favoured housing families in houses rather than flats –

'they fit the English temperament' (Hall 1988: 220). The New Towns Act of 1946 laid the ground for the fourteen new towns that were designated in 1950. They united the Garden City tradition, with state-driven public housing. They incorporated the concepts of the green belt and the relocation of industry. Development corporations were set up by statute that were unencumbered new bodies, bureaucratic and top heavy as they may later have become.

The new towns were low-density, widely spread housing areas, often embodying less adventurous urban ideas than their originator intended. They satisfied one kind of need, housing maybe three-quarters of a million households in twenty-two designated new towns, mostly in the south-east, over the post-war period. Houses were built for mixed social groups and for owners as well as tenants. But they were not built on the required scale, nor did they tackle slum problems directly. Nor did they help to rebuild the cities, a primary task. Later, the idea of greatly extending existing towns was adopted.

FLATS

The second strategy, partly determined by the new green belt policy, was to tackle the inner cities through massive, dense rebuilding. A reversion to the pre-war slum clearance programme became inevitable, as half a million dwellings, condemned as unfit pre-1939, were still awaiting demolition in 1951.

According to Abercrombie, replacing and rebuilding crowded slums, where sharing was common, at acceptable modern standards required both the new towns *and* high-density flats. Three-fifths of the existing population could be housed within existing cities, with a combination of houses and eight- to ten-storey flats (Hall 1988: 172). Anxiety over the use of agricultural land for building and continuing fear of uncontrollable urban sprawl reinforced the desire to build at high density.

Low-density, suburban housing could not satisfy demand. In order to provide a mix of social groups and of styles within council housing, special subsidies were introduced for the construction of some houses on expensive inner-city sites, as part of large developments, mainly of flats. The policy broke up the monotony of inner-city estates and helped retain more affluent tenants. But flats became the order of the day for city councils, particularly in inner areas. The government increased subsidies for inner-city land and for flats. The green belt, by limiting city growth, gave further impetus to rebuilding within the city.

Labour built 900,000 homes in the first five years after the war, short of its target of 1.25 million but a remarkable start to the housing boom. The houses were mostly built by local authorities, were to house all sections of the community, and were of exceptionally high standards. But post-war shortages led to huge difficulties and Labour lost the 1951 election in the face of major economic problems coupled with continuing chronic housing shortages.

When the Conservatives were elected in 1951, their pledge was to produce 300,000 units a year. By 1957 they had built, mainly through local authorities, over 1.5 million new homes. They did not question the role of local authorities since their imperative was to produce the maximum number of houses and demolish the maximum number of slums, a programme which unquestioningly required government action. Thus, together with Labour's effort, by 1956 over 2.5 million new homes had been built, three-quarters of them by local authorities or new town corporations. This more than doubled the council housing stock in just over ten years. Between 1951 and 1961, the number of sharing households was halved and the proportion of households that were overcrowded dropped from one in twenty to one in thirty-five (UK Census data 1951 and 1961).

RENEWED DEMOLITION

Bomb sites and pre-war slum clearance sites were increasingly rare. The crudest shortage had been met. A big head of steam was up in the building industry. The economy was booming and the post-war consensus encouraged a strong push towards equalisation of conditions. Universal health and education were not enough as long as Victorian housing conditions persisted. The very success of the Welfare State in reducing crude poverty generated demand across the society as a whole for the elimination of outdated housing conditions. The desire for a new start was married with the clean-sweep ideas of modernist architecture and in 1956 a major new slum clearance and demolition programme was announced, fitting neatly with both the political commitment to rebuild and the post-war technocratic approach to 'machine living'. Special subsidies were introduced that increased with every additional storey. The maxim of Le Corbusier, 'Existing centres must come down. To save itself, every great city must rebuild its centre' was being followed (Quoted in Hall 1988: 209).

Over 6 million dwellings dated from before the First World War. Britain had long been a highly urbanised society. Already in 1901, 77 per cent of the population of England and Wales had lived in urban areas (Halsey 1988: 326). The vast majority of town dwellers were in towns of over 10,000. Unlike other countries which urbanised much later and were largely able to build afresh for new city dwellers, Britain had to clear the backlog of old urban dwellings before she could build anew. If the urban infrastructure was to survive, much building had to be either recycled or replaced. The green field option had been largely expended on inter-war estates, vast areas of suburbia, and now the designated new towns.

SLUM CLEARANCE

In the ten years from 1955, 600,000 dwellings were demolished. By 1976, a further million had been destroyed through official slum clearance. The impact of adopting slum clearance was the very opposite of the aim. A clean sweep and starkly modern conditions were the objectives. Instead, throughout the post-war period, British cities were scarred with blighted, semi-abandoned streets of old, semi-derelict housing, demolition sites enclosed with corrugated iron, building works, filth and congestion.

Conditions spiralled as slum landlords were hamstrung by rent controls and clearance plans. Tenants lost all sense of belonging to their threatened neighbourhoods as they waited in long queues for their dream home to materialise.

The knock-on effect of the slum clearance programme was immeasurable. Local authorities lost rate income from the widespread blight. Landlords were deeply alienated by the process of compulsory purchase, which they increasingly contested. Shops and local industry were swept away in the process of clearance. Schools lost populations and other services declined. Property values plummeted and building societies 'red-lined' inner-city areas as too risky to invest in.

Meanwhile, local authorities were encouraged by successive central governments to declare more and more slum areas for demolition. Only in this way were the new housing targets realisable. The higher the building target, the greater the need for clearance. There were few other sources of urban land. But the larger the clearance plans, the greater the rehousing needs. Thus it appeared that slum clearance targets constantly outran rebuilding achievements. Table 19.1 highlights the growing scale of designation.

Table 19.1 Numbers of dwellings in slum clearance areas in the United Kingdom, 1939–67

Year	1939	1954	1965	1967
Officially designated slums	472,000	847,000	824,000	1,800,000

Source: Burnett 1980: 279.
Note: The criteria for declaration were very varied (see Power 1987a).

IMMIGRATION AND SLUM CLEARANCE

There was an unhappy coincidence of inner-area blighting and the arrival of immigrant workers. Immigrants were needed to man the growing public and private sectors – hospitals, transport, building. Throughout the 1950s and early 1960s, West Indian immigrants arrived at the rate of about

Table 19.2 Ethnic groups in Great Britain, 1990

Ethnic group	'000s
White	51,689
West Indian and Guyanese	461
Indian	786
Pakistani	462
Bangladeshi	108
Chinese	135
African	136
Arab	64
Mixed	308
Others	163
All ethnic minority groups	2,623
Not stated	509
All	54,821

Source: Office of Population Censuses and Surveys 1992.

30,000 a year, reaching a peak of 75,000 in 1961 before immigration controls were introduced. Asians followed a similar pattern, peaking at 47,000 in 1962. Families joined the settlers in large numbers, particularly from the Indian sub-continent, up to the mid-1980s. Most immigrants were from ex-British colonies which formed part of the New Commonwealth. They had British passports with full rights to British citizenship. Today, there are over 3 million British residents of minority ethnic origin, 1.25 million from the Indian sub-continent and over half a million from the Caribbean. Table 19.2 gives a breakdown of available figures.

Immigrants were discriminated against and were generally poor. They were rarely eligible for council housing (Rose *et al.* 1969). But they could, if they pooled their resources, buy into the run-down inner areas.

Cities like London, Birmingham and Manchester had several hundred thousand houses in clearance areas that were years from realisation because of the scale of the demolition programme and the difficulties in moving it forward due to rehousing problems. Immigrants bought cheap houses in those areas on a large scale from long-standing private landlords who were often waiting for a way to get out. Blight and strict rent control on existing tenancies made property cheap – houses with controlled tenants were almost valueless and immigrant landlords sometimes took over a half-empty house with a controlled tenant.

DECONTROL

From 1957 unfurnished accommodation could be let without rent control or security. Unscrupulous inner-city landlords exploited the new loopholes. There was also an explosion in furnished lettings to house-hungry and vulnerable immigrants at quite exorbitant rents. The phenomenon of 'Rachmanism' became a legendary part of British housing history, with a notoriety that has not yet faded. The 'red-lining' by building societies and the uncertain future of the property in inner areas meant that buyers, particularly immigrant buyers, had to pay exorbitant rates of interest to borrow 'shady' money (Holland 1965). They then let out rooms to friends and relatives at higher than average rents, very often on a furnished, room by room basis. Overcrowding, sharing and insanitary conditions became intense in the areas where immigrants concentrated. Racial conflict and actual riots broke out spasmodically from the late 1950s onwards.

The combined effect of slum clearance programmes and rapid immigration was immense. Conditions in major inner-city areas deteriorated even more rapidly. For the first time, the housing problem and rehousing programme took on a racial dimension that was ugly and unmanageable. The slow-moving clearance process was slowed even further as local authorities were unwilling to rehouse families of New Commonwealth origin. Areas could not be demolished until they were emptied.

There was therefore a second layer of blighting in areas of immigrant settlement. Old and run-down houses were occupied far more densely by the new, younger and often larger households than they had previously been. Repairs were more difficult to carry out. Property deteriorated even further. Inner areas around the edge of existing slum clearance areas were affected as the landlords there sold out under the increased demand for inner-city property from immigrants. This led to the declaration of further redevelopment areas in the late 1960s and early 1970s.

Much of inner London, inner Birmingham, Greater Manchester and the West Riding of Yorkshire, including many of the mill towns such as Bolton, Rochdale, Oldham, Bradford – areas of high immigration – were caught up in the spread of blighting. Other cities and towns experienced a similar process, if to a lesser extent.

DECLINING AREAS

Early industrial cities with lower immigration, such as Glasgow, Tyneside, Humberside and Liverpool, ran massive slum clearance programmes that ended up far outstripping demand or need. The post-war collapse of their industrial base was made worse by slum clearance, speeding up the loss of jobs and enterprise and greatly accelerating the loss of population from the cities (Donnison and Middleton 1987). In some cases, the very rationale for immigration – labour shortage – was undermined by the damaging long-term effects of widespread clearance, with its impact on jobs, services and economic activity as well as on housing (DOE 1977b).

REHOUSING PROCESS

The plan was to rehouse as many people as possible from a condemned area over a short period into an adjacent area of new flats built on a previously cleared site. The newly cleared site could then be built on to house people from the next slum clearance area, and so on.

It is easy to see why this system quickly broke down. Residents favoured slum clearance on the whole, because of the promise of a brand new house. But people's dreams were rarely fulfilled. They often wanted nearby rehousing but cherished the ambition of a house, not a flat. This made for constant delays in rehousing. New, available estates rarely matched all the needs of slum dwellers from adjacent clearance areas.

People had to be persuaded to accept flats. Several surveys showed their unpopularity (Power 1987a, Lambert *et al.* 1978, for example) and people were often pressured into accepting areas they had not chosen. There was, in practice, little choice. Matching the rehousing programme with the rebuilding and clearance programmes was far more complicated than had been envisaged. Delays, closed waiting lists other than for slum-clearance families, and spiralling inner-city conditions created a sense of coercion (Dunleavy 1981). Even then, the matching problems still led to long delays. Figure 19.1 shows the process.

The process had many perverse effects. It drew people *into* slum-clearance areas as the one sure way of getting rehoused by the council. A few weeks' delay could cost thousands of pounds in delayed contracts. The whole complex process could seize up. This encouraged some families to exercise their power, holding out for the best offers. More vulnerable and less sophisticated or stubborn families fell prey to the pressures and ended up in the least popular estates or in the next slum-clearance area. Houses, emptied for demolition, were often reoccupied if other houses in the area were not yet empty.

THE EXCLUDED

Many important categories of household were excluded to simplify rehousing. Single people and childless couples were generally not eligible. Newcomers were excluded in order to avoid 'queue jumping'. Insecure and furnished tenants were not eligible as 'transient', except if they could prove long-term residence – usually over five years, often up to ten. Clearance was often delayed for ten or more years through a vicious circle of partial emptying, partial refilling, and partial exclusion. By the mid-1970s, many of

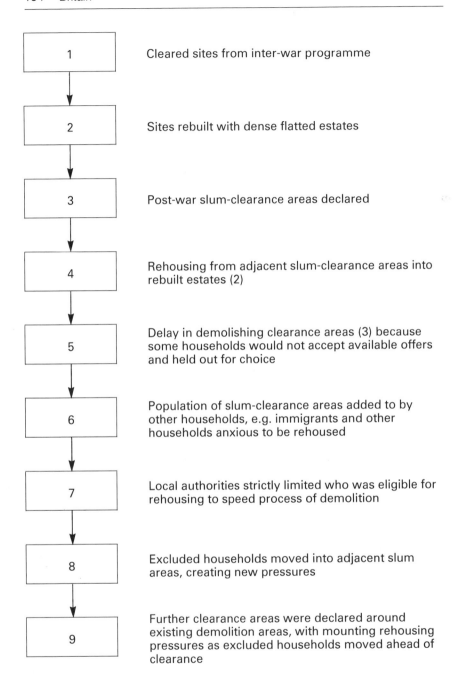

1	Cleared sites from inter-war programme
2	Sites rebuilt with dense flatted estates
3	Post-war slum-clearance areas declared
4	Rehousing from adjacent slum-clearance areas into rebuilt estates (2)
5	Delay in demolishing clearance areas (3) because some households would not accept available offers and held out for choice
6	Population of slum-clearance areas added to by other households, e.g. immigrants and other households anxious to be rehoused
7	Local authorities strictly limited who was eligible for rehousing to speed process of demolition
8	Excluded households moved into adjacent slum areas, creating new pressures
9	Further clearance areas were declared around existing demolition areas, with mounting rehousing pressures as excluded households moved ahead of clearance

Figure 19.1 The process of slum clearance and rehousing in Britain

the exclusions were abandoned as there were fewer and fewer cheap, private-rented areas to move on to.

RACE AND REHOUSING

Racial minorities were increasingly concentrated in and around designated redevelopment areas, but they were invariably excluded from rehousing. Councils delayed certain slum clearance and redevelopment areas in order to avoid dealing with areas of immigrant concentration (Islington Borough Council 1974). The delayed areas attracted greater and greater minority populations, in ever worsening conditions, as a solution to exclusion from better areas (DOE 1977b).

The blighted areas were used by councils to rehouse families from more advanced slum-clearance areas who had to be moved but were 'unsuitable' for new flats – generally so-called 'problem families'. 'Dumping' in redevelopment areas became common from the 1960s (Macey 1982). Trapped immigrant households, long-standing, controlled, elderly tenants unwilling to move, and 'problem families', were forced together (North Islington Housing Rights Project 1976).

Thus, the rehousing process proved slow, costly, cumbersome, coercive and divisive. It resulted in great social and racial segregation (Parker and Dugmore 1976).

HIGH RISE AND 'MACHINE LIVING'

For twenty years from the mid-1950s to the mid-1970s, mass housing was in vogue.

Most of the industrialised flat-building, and virtually all the high-rise[1] and deck-access building, took place within cities and towns on slum-clearance sites and only exceptionally on peripheral sites. Local authorities were virtually the only bodies to put up large-scale, medium- and high-rise, industrially built estates, although in the 1980s, private, inner-city developers increasingly adopted industrialised methods on city sites. About 1.5 million social housing units were built by local authorities directly, using industrialised 'mass' techniques (Association of Metropolitan Authorities 1984). Although in the 1960s urban local authorities increasingly adopted flats as the dominant housing form, high-rise flats formed only a very small proportion of the total. Even at the peak of their short-lived popularity in 1966 they made up only 26 per cent of new construction (Dunleavy 1981: 41–3).

The extreme damp of the British climate made thin, water-absorbing concrete walls and flat roofs a most unsuitable building form. Therefore industrialised building quickly threw up physical problems that were costly to remedy – damp, condensation and poor insulation.

Table 19.3 Local authority housing construction approvals[a] in England and Wales, 1953–75

Years	Houses	Medium[b] flats	High rise[c]	Total flats	Total
		(a) Thousands			
1953–57	535	174	42	216	751
1958–62	291	175	78	253	544
1963–67	370	226	182	408	778
1968–72	267	198	69	267	534
1973–75 (3 yrs)	181	126	7	133	314
Total	1,644	899	378	1,277	2,921
		(b) Percentages			
1953–57	71	23	6	29	100
1958–62	53	33	14	47	100
1963–67	48	29	23	52	100
1968–72	50	37	13	50	100
1973–75 (3 yrs)	58	40	2	42	100
Average, 1953–75	56	32	12	44	100

Source: This table is compiled from Tables 2.1 and 2.2 in Dunleavy 1981: 41–3.
Notes: [a] Size of building programme approved by government for subsidy.
[b] Medium = 4-storey flats.
[c] High rise is defined in this table and throughout the section as five storeys and above. High flat subsidies began at five storeys (Dunleavy 1981).

In the post-war period, under half of all social housing construction was in the form of flats. By the end of the mass housing era in 1979, with over 6 million local authority-owned dwellings, over 4 million were in the form of houses. Table 19.3 gives a breakdown of construction during the mass housing era, underlining the dominance of houses over flats and of low- and medium-rise over high-rise flats, apart from the mid-1960s.

HIGH RISE ABANDONED

A fatal gas explosion at Ronan Point in London in 1968 killed five people as one side of the council-owned tower block collapsed like a pack of cards. Almost overnight the high-rise programme was abandoned after only ten years. Fewer than 400,000 'high-rise' flats had been produced, only 13 per cent of the social housing stock. Industrialised methods continued to be used after 1968 'with the blocks laid on their sides', forming deck-access estates. The high-rise and deck-access housing was never popular with tenants. Tower blocks quickly earned notoriety in the Press and in the public eye, often proving difficult to let after Ronan Point.

However, over the ten years following the tower block explosion, large, medium-rise blocks were built all over the country. These eventually proved to be at least as unpopular and difficult to manage as high rise itself. Maisonettes, a peculiarly British attempt to provide two-storey dwellings within multi-storey blocks, often became notoriously unpopular due to noise problems and unconventional internal layouts.

The British experience of mass housing was quite distinct, with local authorities as direct providers on a huge scale, houses dominating over flats, even during most of the post-war flat-building boom, and mass housing being closely tied in with slum clearance. (See Dunleavy 1981 for a fuller discussion of high-rise housing.)

Plate 19.1 Ronan Point disaster marks the end of tower block building

NOTE

1   According to Patrick Dunleavy, all blocks of flats of five storeys or more could be classed as high rise. They were eligible for special subsidies as such.

Chapter 20

Council landlords

Local authorities were extremely effective at achieving the central goal of volume building, producing a massive 4.5 million units after the Second World War. They achieved the primary target of eliminating the worst Victorian slum areas and they also housed a significant number of other households in addition. Overall space and amenity standards rose as a result.

The problems with their role stemmed from four main factors: the style and quality of council housing; the management of building and rehousing processes and, in the long run, management of the stock; the ripple effects from the slum-clearance programme; the loss of private renting.

QUALITY

The *quality* of social housing after the war was forcibly cut, first by materials shortages, then by economic problems. But the real decline in standards came with the renewal of slum clearance.

The Conservatives favoured maximising private building and shifted the housing emphasis towards owner-occupation for those who could afford it, partly for ideological reasons but partly in order to limit the scale of local authorities' responsibility, thereby speeding up the slum-clearance process and targeting need. To do this more effectively, standards were cut and Harold Macmillan, the Conservative Housing Minister, promoted the 'people's house', a Corbusian 'workers' housing' concept writ large. Blocks of flats were pushed up, virtually devoid of design.

The facilities and layout of the dwellings themselves were a major advance. They were light and spacious compared with traditional two-up, two-downs or back-to-backs; they were usually centrally heated and equipped with modern amenities, satisfying to families who had coped with sculleries, tin baths in the living room, ice-cold upstairs bedrooms and serious disrepair.

The main problems stemmed from poor materials with minimal standards for wall thickness, roof cover, heating and lift equipment; and the almost total failure to provide an amenable estate environment or social facilities. Therefore the overall image of the estates was poor and the dwellings themselves often suffered from faulty heating, water penetration, noise problems and alarmingly rapid decay of access areas. The fact that medium- and high-rise flats were used for families with young children added to problems of noise, wear and tear, supervision, and, later, neighbour disputes, social problems and vandalism.

Therefore quality was a complex issue with some aspects of the new estates proving to be highly appealing, even seductive, and others exhibiting serious problems within the first few years of occupation. Tenants still in slums were ready to accept unpopular flats because of internal amenities, space, and modern conditions. But tenants already rehoused showed rapid signs of disillusionment and alienation. It took local authorities a long time to recognise the problem. It took tenants a long time to oppose clearance and flat-building.

MANAGEMENT OF THE PROCESS

The rate of building far outstripped the capacity to manage either the building or the political process. There were consortia of local authorities to maximise production capacity; contractors wined and dined politicians; government architects promoted uniform, off-the-shelf design; tenants were not consulted. The whole process acquired a heady, breakneck speed that spelt huge windfall profits for builders, major political success for vote catchers, and the rapid replacement of obsolete slums with the technically most advanced housing available (Dunleavy 1981).

Supervision was often poor; contractors sought to minimise input; essential elements, like the number of bolts holding concrete panels in place, were often below specification. Many parts, such as doors and ceiling panels, were of flimsy materials that were quickly damaged. Finishes were often omitted altogether. Therefore many estates had an extremely cheap and utilitarian look about them. Few came anywhere near to the promotional claims. Some parts never functioned properly from the start. Underground garages, heating and lifts were notorious examples.

To make matters worse, the process itself was often far more costly than envisaged. There were labour difficulties on many big sites; time-scales for completing developments slipped; supplies for giant contracts were hard to co-ordinate or control. Delays cost money.

Large local authorities were demolishing, building and letting several thousand dwellings a year. Slum-clearance sites normally involved at least 500 dwellings (Power 1987a). Many were over 1,000. The aim was to make them as big as possible to stay ahead of the rehousing needs. Local authority management structures became more and more complex with specialist professional staff to handle different stages in the process and their interlinking requirements.

RATIONALISING LOCAL AUTHORITIES

Co-ordinating a complex sequence of interventions with a large number of sites, all at different stages in the process, proved to be a task of daunting proportions. The amalgamation of local authorities, in 1964 in London and in 1974 in the rest of the country, was an attempt to rationalise and gain economies of scale in local government. It reduced the total number from nearly 2,000 to 402. As a result, British local authority landlords, some of them already very large, became unique in the scale of their housing management operation, with the average council landlord becoming five times larger overnight. Table 20.1 illustrates the vast scale of urban council landlords.

The complex interaction of specialist departments of the local authority, each with its own professional interest group, made the building process dominate over the management needs, as Table 20.2 illustrates.

Developing housing through unitary local authorities, rather than through single-purpose landlords, greatly complicated the task, since many functions overlapped, no matter how slightly, with many others. The possibilities for breakdown became immense and no one person, section, or profession was in full charge of the process. Meanwhile, each separate department carried other responsibilities for other services that divided their attention. Landlord–tenant relations were strained by these external pressures.

MANAGEMENT OF THE STOCK

Local authorities were not set up to become landlords. Their political role was often in conflict with their landlord duties. For example, local elections ran every four years, but slum clearance programmes ran for anything from

Table 20.1 Largest urban council landlords in Britain, 1991

Local authority landlord	Number of dwellings	Local authority landlord	Number of dwellings
Birmingham	109,898	Newcastle-upon-Tyne	41,286
Glasgow	142,000	Sheffield	75,211
Leeds	81,342	Southwark	57,577
Liverpool	58,256	Sunderland	45,699
Manchester	92,029	Tower Hamlets	41,200

Sources: DOE Housing and Construction Statistics 1992; Glasgow District Council 1992.

Table 20.2 The interaction of local authority professions at different stages in the process of redeveloping local authority housing in Britain, 1956–76

Local authority professional departments	Stages in development[a]	Years 1 to 7 in process of redevelopment[b]
Public Health Welfare Medical officers Lawyers Planners Valuers	Designation of redevelopment areas and compulsory purchase	Years 1–3
Surveyors Planners Architects Engineers Finance/accountants Education Health Fire services Building regulations Dept of Environment Public Health	Development of new housing plans	Years 1–4
Housing visitors Lettings officers Welfare officers	Rehousing	Years 2–5
Finance/accountants Engineers Surveyors Fire, water, electricity services	Demolition[c]	Year 5
Architects Surveyors Clerks of works Finance/accountants Building services – fire, water, electricity, gas Engineers	The building process[c]	Years 6–7

Notes: [a]Each local authority would have several areas in progress simultaneously. Each area might experience several stages in the development process simultaneously.
[b]Many areas took far longer than seven years.
[c]Contractors became involved at these stages.

seven to thirty-five years. Rent rises were seen as vote-losers, yet rents had to rise if repairs responsibilities were to be met.

By 1974, local authorities in England and Wales owned over 5 million dwellings. The oil crisis led to a sharp decline in council house-building through rising costs, high interest rates and public cutbacks. But it took several years for the cutbacks to show through because of the time lag in development. Many early 1970s schemes were completed ten years later.

While local authorities employed overall about one housing worker to every fifty dwellings, there were ever fewer estate-based staff. Even caretakers did not exist on many estates. Caretaking had originally had a strongly custodial and supervisory role on the early flatted estates, akin to the continental system. However, under the impact of scale developments, increasing centralisation, poor supervision and the constant dilution of town hall management, caretaking had in most places been reduced to little more than a minimal cleaning role. Caretakers were virtually alone on the front-line and their status dropped further and further. The actual number of caretakers on estates was cut progressively, and increasingly they were mobile rather than estate-based. Under this regime, tenants became increasingly dissatisfied (Parker 1983).

STANDARDS OF MANAGEMENT

The government had not given a unitary or regulatory structure to public housing services, unlike education, health or transport. There was no enforcement of standards or sanction against poorly performing local authority landlords.

Local authorities often ran exemplary parks, libraries, and schools. Many of the new towns, with their unified, modern structure set up with pristine clarity at the outset, were world models, including their housing departments (e.g. Stevenage). But generally local authority housing management was weakened through lack of definition; it was no one's specific responsibility. It grew in layers of reaction to problems and the division of functions was never resolved. At the same time, the size of the local authority stock expanded far more rapidly than the systems could cope with. The reaction of local authorities was to 'streamline' their systems, leading to the further removal of direct services, such as cutting rent collection, reducing caretaking responsibilities, cutting back on estate cleaning, withdrawing further into town halls. Estates deteriorated rapidly as a consequence of the remote, town-hall-based administrative system and the low level of direct services (Burbidge *et al.* 1981).

CENTRAL CONTROL

Central government in 1972 introduced radical changes in an attempt to force a link between standards, costs and payments.

The 1972 Housing Finance Act both instituted government control over council rents, attempting to push them up in relation to costs and incomes, and brought in a universal rent rebate system for those on low incomes. The aim was to increase revenue for repairs, to increase tenants' contributions in line with rises in income, to help fund the growing debt on the capital building programme, and to target help at those who were too poor to pay more. The Act was highly controversial, provoking rent strikes and a famous confrontation between central and local government at Clay Cross. It was repealed in 1975.

But the shift away from a universalist approach to council housing with low rents became permanent. Until then, rent fixing had been a local council responsibility and rents had nowhere near kept pace with incomes, costs, or inflation (DOE 1977a). Government from there on imposed more on local authorities, forced higher payments on tenants in work, reduced the commitment to council housing as a whole, and attempted to force improvement of the existing stock rather than continue endless new building. The shift to rehabilitation was in part an economic response, in part an attempt to tackle inner-city decline and racial tensions.

A MOVE TO RACIAL EQUALITY

A very serious situation had emerged in inner cities from post-war housing policy – devastated areas, bleak estates, exclusion, discrimination and sharpening social divisions. There were strong pressures to do something (GLC Minutes 1976). In 1976, the government passed the Race Relations Act, the first significant legislation aimed at ensuring equal treatment of all residents and outlawing discrimination on grounds of race, colour or creed. It applied particularly to housing and employment, although it embraced all areas of organisation. The Act was criticised for 'lacking teeth' but it provided a benchmark which has frequently been used in the fight for progress by black organisations. It gave Britain a legal mechanism against extremism, now sorely needed on the continent (*The Economist*, 7 December 1991).

It was followed by a path-finding housing law in 1977, the Homeless Persons' Act, which laid down a totally new obligation directly on local authorities to help house those households classed as in priority need and unintentionally homeless – i.e. not made homeless through their own decisions and action.

Priority need covered adults with dependent children, pregnant women, and vulnerable people such as the elderly, the mentally sick, or children leaving care. There had been an element of 'fair play' in the idea of housing first long-standing residents who had waited longest in very poor conditions (Macey 1982, Cullingworth 1969). But in the mid-1970s councils found themselves with a glut of difficult-to-let units. The Labour government was acutely aware of growing numbers of vacant council dwellings, amidst talk of demolishing some of the worst older council blocks. In 1977, the Greater London Council began to advertise old and unpopular flats in the main London newspapers on a 'Ready Access' basis to those who turned up first at County Hall, their municipal headquarters. The combination of the Race Relations Act and the Homeless Persons Act was to prise open council housing to minorities.

THE RIPPLE EFFECT OF SLUM CLEARANCE

Council developments had concentrated the most marginal households in the last slum-clearance and redevelopment areas. The last slum-clearance areas to be emptied housed the greatest concentrations of poverty and deprivation. Local authorities could not in the final round escape rehousing responsibility since there was nowhere else to push unwanted applicants, and actual homelessness was growing. They explicitly tried to rehouse the most disadvantaged people on unpopular estates, either on older and already problematic estates or on new, very large and unpopular estates. Racial minorities were disproportionately affected (Power 1977a, Parker and Dugmore 1976). This had a landslide effect and a bottom layer of difficult-to-let estates was rapidly created in most cities in the country as demand tailed off from the slum-clearance programme.

A perverse effect of the combination of slum clearance and rebuilding was that estates completed in the mid- to late-1970s were often the hardest to let, not necessarily because of their design *per se*, but because of their scale and the fact that slum clearance was over. The large 1970s estates were frequently under-occupied. Overcrowding, sharing, and lack of amenities became increasingly rare and, by the late 1970s, affected only a tiny proportion of households (Halsey 1989: 367). The proportion of sharing households dropped to 1.3 per cent. The earlier cramming and exclusion turned into a search for willing and suitable applicants (Burbidge *et al.* 1981).

DIFFICULT-TO-LET ESTATES

By 1974, the government had established that most large city authorities had at least three difficult-to-let estates; these tended to be the newest, largest industrialised estates and their problems related to the social make-up and management of the estates, as well as their physical structure (DOE 1974).

A full 'Investigation of Difficult to Let Estates' carried out in 1976 (Burbidge *et al.* 1981) provoked the government into action. In the course of 1976 and 1977, the investigating team visited thirty local authority estates, uncovering deep problems of social stigmatisation, management incompetence, and hostility between tenants and local authority landlords (Burbidge *et al.* 1981).

Intervention by government became inevitable, as local authorities were forced by the Homeless Persons' Act to open up their lettings to the most marginal households. The intensified welfare housing role evoked images of America's unlettable public ghettos. The logic for drastic action proved inescapable. But it was not until the end of the 1970s that anything very substantial was done to address the problems.

Chapter 21

Decline in private renting – rise in owner-occupation

The watershed for private landlords was the introduction and partial retention of strict rent controls from 1915. Until 1957, controls stayed in place. Many old properties were still rented out at pre-war levels. Not only could landlords not repair their property economically. There was little financial reward for being a landlord, as the rent income itself was low and prospects were poor. Many large landlord holdings were broken up into smaller-scale portfolios (Hamnett and Randolph 1988).

After the war, properties gradually lost their controlled tenants and new tenancies were at higher rents. If the accommodation was let furnished, new rents were not controlled and there was no security for tenants. This form of letting therefore grew, as landlords exploited both the shortage of private accommodation and the relaxation of control. Harassment, extortion, colour bars, all arose within the 'twilight' areas of private renting. The Milner Holland Report (1965) deplored the conditions of inner-city private tenants, particularly racial minorities in furnished, insecure lettings, and advocated urgent action to upgrade existing property rather than wait for the long hand of redevelopment.

There was repeated intervention to curb abuse. As a result, the supply of private-rented accommodation continued to decline.

Conversion to owner-occupation: gentrification

It was generally more profitable and a lot easier to make money by selling out into owner-occupation, which was strongly favoured through the tax system, especially after the abolition of schedule A tax on the capital gains of owner-occupation in 1963 and the introduction of a new capital gains tax on private landlords in 1965 from which owner-occupiers were exempt. These measures greatly enhanced the attraction of owner-occupation.

If older housing was structurally sound it became worth doing it up for owner-occupation. The process was called gentrification if better-off buyers moved into an area, forming a distinct identity alongside existing lower-income residents. It grew on the back of inner-city blight and as a reaction against slum clearance.

General Improvement Areas

Government policy also shifted gradually in the 1960s in favour of rehabilitation of older property.

Slum clearance seemed to be self-perpetuating and contaminating as the areas affected spread and as structurally sound but run-down property came to be included. From 1964, Improvement Areas offered government money to upgrade the environment of run-down, old areas, with improvement grants for owners to do up their houses. These measures were gradually extended and greatly enhanced the attractiveness of the old areas, fuelling gentrification. In some cases, long-standing slum clearance plans were withdrawn and General Improvement Areas were declared (Ferris 1972). Great profits were now to be made by selling out to owner-occupiers as demand for older housing soared. Rents on improved properties were regulated at below-market levels.

Available finance

Local authorities began to offer mortgages to new owner-occupiers to buy older property, previously hard to obtain because of the conservatism of building societies and the refusal to lend on properties considered risky investments through the 'red-lining' of inner areas. The option mortgage subsidy was introduced in 1969 for those on low incomes who did not benefit from mortgage interest tax relief because their incomes were too low. The subsidy reduced the interest charge, making repayments cheaper. Often young, would-be gentrifiers benefited. Teachers, social workers and other essential workers, were often on low enough incomes to qualify if they had children and only one earner.

The combination of these changes made buying old, cheap, run-down housing in potentially attractive inner areas near town centres appealing. High inflation made the value of investing in owner-occupied housing, with its steeply rising value, ever more attractive. If people could manage the initial years of big payments, within a few years their fixed costs would be overtaken by rising rents and incomes and they would gain a windfall asset.

Redevelopment plans were abandoned where pressures of gentrification and rising property values made compulsory purchase difficult and much inner-city terraced housing was freed from blight, paving the way for the further spread of owner-occupation.

In all, since the war, 4 million properties were converted, modernised and transferred from the private-rented sector into owner-occupation. Thus owner-occupation played an even bigger part, numerically, in the demise of the private-rented sector than slum clearance itself. This was a somewhat unexpected outcome, partly generated by the reaction against slum clearance, partly generated by the need for a flexible alternative to the tightly controlled council and private-rented sectors.

WINKLING

Private tenants meanwhile experienced increasing pressure from sales and gentrification, as well as council redevelopment plans. Speculative property dealers cashed in on the popularity of owner-occupation by buying into the declining, private-rented sector and selling on at big profits to gentrifiers. More and more private landlords were quitting while the going was good. The notorious practice of 'winkling' grew up, involving pressure tactics which included cash payments, to induce controlled tenants to move. The value of properties with controlled tenants was several thousand pounds less than identical properties without. It was therefore worth property dealers' while to pay off tenants to gain vacant possession. Up to £5,000 could be offered, depending on the area (Power 1973). Since controlled tenants were often elderly women on their own who wanted to stay put, the process of 'winkling' created massive adverse publicity. One outcome of the property speculation boom in owner-occupation and 'winkling' was renewed council intervention in 'twilight' areas, this time in the shape of Housing Action Areas (see p. 211).

HELP FOR PRIVATE LANDLORDS

In 1974, the government attempted to encourage private landlords to stay by offering full improvement grants at much higher values for each self-contained and modernised unit within existing houses. Landlords often did up the units and then sold each one off separately to owner-occupiers, in spite of increasing restrictions to try and ensure that converted flats remained available for letting. These conversion grants led to a multiplication of owner-occupied dwellings, two or three to a house in some cases. Far from stemming the decline of private landlords they accelerated the process by enhancing values, fuelling the ever-growing demand for owner-occupation. Without being able to charge market rents, private landlords did not get sufficient return on the value of their property. By selling, they could realise their asset and invest more lucratively.

NEW PRIVATE RENTING

If there were few incentives for private landlords to retain their property, there were even fewer incentives for new landlords. Very few new rented units were added to the stock, except through furnished lettings. Hotel and bed-and-breakfast accommodation began to expand and some owner-occupiers took in lodgers to help with the repayments. None of this had a significant impact on a rapidly shrinking market.

The continuing unpopularity of private landlords prevented any serious attempt at rescue. The whole process of decline appeared irreversible, as Table 21.1 shows.

Table 21.1 Number and percentage of rented dwellings in Britain owned by other than local authorities or New Towns, 1938–89

Year	No. of rented dwellings, other than local authorities and New Towns (in millions)	Total stock (%)
1938	6.6	58
1960	4.6	32
1971	3.2	19
1975	2.9	16
1989	1.6	8

Sources: Estimate derived from Holmans 1987; Hills and Mullings 1990.

A THIRD TENURE

The government, in an attempt to slow the drift towards monopoly local authority landlords and extend choice to tenants, embraced the concept of a third tenure – something between private renting and local authorities. Politicians were looking for a vehicle that combined the flexibility, private initiative and small scale of private landlords with the social responsibility and responsiveness to government needs of public landlords.

THE RE-EMERGENCE OF HOUSING ASSOCIATIONS

Housing associations and trusts had continued to provide rented housing from their Victorian beginnings. They had occasionally been offered carrots to engage in larger-scale, government-sponsored activity, but had generally been conservative, slow-moving and complacent (Emsley 1986).

By 1960, a number of things were becoming clear. The private-rented sector was shrinking too fast and its decline had proved hard to stop. When rent controls were relaxed, gross abuse appeared to force their reintroduction. The council sector was performing as planned but was monolithic and a costly burden on the state. It also excluded many households in need of a home – for example, those who could not buy but did not live in a designated redevelopment area. There were specific needs calling out for an alternative tenure: elderly people in run-down, rented accommodation, in need of more supportive housing; key workers, such as nurses and teachers; young, newly formed, childless households.

Unmet needs

Not only were many housing needs still unmet and cities crying out for a new direction, there was also evidence of growing disillusionment with councils themselves as they continued in their traditional role of housing established families in spite of social changes with more single people, more minorities, and so on (Cullingworth 1969). 'Big Brother' commanded the resources, yet often acted insensitively with little regard for community; councils actively discriminated by excluding virtually all newcomers, the life blood of every great city (Rose *et al.* 1969). Squatting organisations developed in areas like Notting Hill and Islington where redevelopment had led to boarded-up, unused property.

Advice centres were set up by Church groups to try and help young families into housing and provide legal advice to cope with harassment, unfair rents and difficulties with rehousing through the council. Action groups also emerged in redevelopment areas to organise better rehousing, to support excluded groups, and eventually in some cases to oppose demolition plans. Tenants got together to oppose gentrification where they saw their own future threatened (Ferris 1972).

The rebirth of housing associations

The Housing Corporation was set up in 1964 to support the growth of voluntary and charitable housing associations. The aim was to create a genuine mixed tenure. They let properties at cost rents with the help of government subsidies. There was a rapid growth, particularly in the acquisition of inner-city property for rehabilitation. But subsidies only covered part of the cost and rents were high compared with council rents or controlled rents (Smith 1971, 1989).

'Cathy Come Home'

The most famous spur to action was the heart-rending television documentary, *Cathy, Come Home*, in 1966, showing a young couple losing their private-rented home after having a baby. This film formed part of the new national Shelter campaign and it turned the scandal of private landlord abuse into a political issue that could no longer wait. Housing associations were called in to help and many of the advice and action groups registered as new housing associations, leading to their renewed growth as social landlords.

In 1974, the Housing Corporation was given a much greater role and grants to housing associations were made much more generous. About 90 per cent of the full cost of each unit was met by government directly. Rents were registered as fair rents, higher than council rents but far below market levels.

Under the impetus of the improved subsidies, housing associations became part of a more local, more direct and more responsive housing movement that restored old property, worked with local authorities, helped sitting tenants to stay in the area and even move back into their renovated homes in some cases. Their reputation rose far out of proportion to their contribution in numbers of units.

HOUSING ACTION AREAS

In 1974, Housing Action Areas were introduced to tackle some of the deep-set problems of inner decline by combining council action with community initiative in neighbourhoods of poor, run-down, old, terraced housing. Where private landlords could not or would not improve property, councils could sometimes compulsorily purchase for renovation. Councils were encouraged to work directly with housing associations which were locally based, geared to rehousing tenants within a local, rehabilitated stock and taking over from private landlords.

The local authorities created project teams that would make the improvements happen over five years. The schemes were strictly time-limited. The role of the teams was dynamic and collaborative – liaising with landlords, tenants, owner-occupiers, housing associations and local authority departments, such as Public Health. Housing Action Areas were area-based, with a defined objective, a budget and operational staff. Schemes were therefore relatively quick and easy to co-ordinate in order to make an impact. The scale was manageable. Each area included about 200–500 houses.

Special measures were adopted for consulting local residents. There was a strong commitment to encouraging them to stay, to counter the dislocation of clearance and its damaging effect on neighbourhood relations, continuity and social conditions.

In practice, because of the international financial crisis and the parlous state of the British economy, housing investment and public spending both began to shrink rapidly from 1974 and Housing Action Areas were declared only in a limited number of inner areas.

MIXED TENURES

The Housing Action Area programme was abandoned in the late 1970s under charges of over-municipalisation on the one hand and over-subsidisation of gentrification on the other. The needs of existing council estates began to take priority. But they had great symbolic value, marking the return to gradualism, disillusionment with single, clean-sweep solutions, and a willingness to accept a range of housing options. Above all, they emphasised the attractions of old neighbourhoods of street property.

Over time, areas went up-market and many existing residents moved out, but a mix of owner-occupation, council acquisition, housing association conversions, and private renting resulted. The process was, at least superficially, far less destructive, and was more attractive and popular than the wholesale clearance and estate-building that had previously been the order of the day.

Housing associations grew rapidly throughout the 1970s, playing a crucial role in inner-city renewal, diversifying the rented sector, complementing and in some ways compensating for local authorities, and offering a flexible, small-scale, non-governmental alternative to the large public bureaucracies that ran council housing. Whether they would retain these qualities as they stepped increasingly into the limelight was a question for the 1980s, and even more so for the 1990s.

OWNER-OCCUPATION

Post-war affluence fuelled the desire to own. The style of council building and the terminal decay of private-rented housing only served to reinforce the trend.

Conservative policies of targeting appeared both more just in times of shortage and quicker in producing bigger numbers of cheaper houses. The shift away from quality was never fully reversed and made council housing increasingly a welfare tenure. This, over time, undermined Labour's commitment to council housing. It served greatly to enhance the desirability of owner-occupation. Cheap owner-occupation seemed the obvious answer to unwanted, inefficient private landlords and overburdened local authorities. Even Labour eventually declared owner-occupation to be the 'natural' tenure (DOE 1977a).

Therefore after the austerity years, when builders had to have certificates before being allowed to build and resources were strictly directed towards the emergency council programme, all restrictions on private building were relaxed under the Conservatives in the 1950s to boost private building for owner-occupation. From 1952, the number of private completions leapt up year by year till 1968. From 1960, private building, largely for owner-occupation, generally dominated production and local authority building never recovered its early post-war pre-eminence. Table 21.2 illustrates the trend.

The main incentives to owner-occupation were so significant as to make the outcome almost inevitable. They affected virtually all social groups. The bars to other tenures were the strongest push factor. It was extremely hard to gain access to private-rented housing because controlled rents and security placed a premium on such tenancies, giving existing occupants a very strong incentive to stay put. Housing associations made a small, if important, dent in the total supply. Private renting was therefore a difficult and declining option. The growing council sector was in theory more promising and, particularly straight after the war, the hope was that it would provide for all – 'general needs'. However, the combination of shortage, slum clearance, cumbersome access mechanisms and discrimination

Table 21.2 Number of dwellings produced by local authorities and private builders[a] from 1919–89 in England and Wales (thousands)[b]

Years	Local authorities	Housing associations	Private builders	Total
1919–24	177	–	222	399
1925–29	326	–	673	999
1930–34	286	–	804	1,090
1935–39	347	–	1,270	1,617
1940–44	76	–	76	152
1945–49	432	–	126	558
1950–54	913	–	229	1,142
1955–59	689	–	623	1,312
1960–64	546	–	879	1,425
1965–69	761	–	983	1,744
1970–74	537	–	830	1,367
1975–79	641	–	667	1,308
1980–84	230	86	676	992
1985–89	102	175	884	1,161
Total	6,063 (40%)[c]	261 (since 1980)	8,942 (59%)	15,266 (100%)

Sources: Halsey 1989: 384, DOE 1991c.
Notes: [a] Housing associations are classed as private builders for the purposes of this table until 1980. They built fewer than 250,000 units between 1919 and 1980.
[b] All figures are rounded up to nearest 1,000 for clarity.
[c] Sales account for the difference between proportion of building and proportion of tenure.

continued to debar large sections of the population.

The growing value of tax relief made owner-occupation one of the most attractive investments possible. The impact was dramatic. From 1963, when an owner-occupied house was sold, households could move and trade up without the deterrent of a tax on their previous asset. Trading up freed up cheaper property for would-be owners, enticed by the same tax benefits.

Increasing incentives to owner-occupiers pushed up demand and therefore prices. As affluence spread with full employment, more and more people were willing to sacrifice more and more to purchase a home. Barriers to entry were lowered as building societies began to lend for the deposit as well as the conventional mortgage; they increasingly took account of women's earnings, thus bringing much larger numbers of lower-income households into the mortgage market; and they gradually relaxed restrictions on older property as improvement grants and the ending of slum clearance had an impact. Easy-term mortgages had become common.

Some local authorities had begun to sell their stock to sitting tenants in the 1960s, a move that was lastingly popular and culminated in the introduction of the compulsory right to buy for all council tenants.

RACE AND OWNER-OCCUPATION

Racial minorities were doubly disadvantaged in the scramble for access to housing. The move into owner-occupation was an obvious self-help solution to their problems, disadvantaged though they also were in borrowing money (C. Holmes, in North Islington Housing Rights Project 1976).

Minorities overwhelmingly purchased existing, older housing in cities (Henderson and Karn 1987). Therefore their concentration was conspicuous and their contribution to urban renewal important (DOE 1977b). They sometimes shared this role, somewhat unexpectedly, with affluent gentrifiers, although there were sharp and highly local geographical divisions.

LASTING POPULARITY

In 1974, there was a major property crash after the most intense period of mounting prices, speculative buying and rapid expansion. Housing association and council completions outstripped private completions for three years between 1975 and 1977. Over time, however, prices recovered and owner-occupation continued to expand, helped by improvement grants and tax relief. The cycles of growth and price rises had a serious effect on the wider economy, particularly inflation, but they did not deter owner-occupiers. The housing system was increasingly geared to meeting that demand. In the 1970s, the number of owner-occupiers rose by 2 million (Halsey 1989, Hills and Mullings 1990).

OVERVIEW OF THE 1970s

By 1979 several outcomes were inevitable. Council housing, by the nature of the conditions under which it was built and managed, was declining in popularity and was no longer, at least in crucial marginal areas, a strong vote-winner. The people who needed it most were generally in Labour strongholds in inner cities and often did not vote. Housing associations had a popular, new-found, if still small-scale, role. Owner-occupation, particularly cheap owner-occupation, was almost universally popular, even among those who could not buy. Surveys of householders showed that in younger age groups almost everyone expected to become an owner (Hills 1991a). Private renting, long paralysed by lack of incentives, continued to be rocked by scandals and sell-outs. Its decline appeared irreversible.

The year 1979 heralded the election of the most radical government in Britain since the Labour victory of 1945, but housing policy had already changed direction from the late 1960s onwards, when the crude number of dwellings decisively outstripped households for the first time (Halsey 1989, Donnison 1987).

The 1970s had witnessed four major new trends:

Small scale

The first was a new emphasis on community-based and small-scale developments. Community-based housing associations, housing co-operatives, advice centres, law centres, and other local service groups offering nursery, playground and youth activities, often with local committees and locally recruited staff, all attracted growing interest and grew up largely in inner areas. Youthful gentrification helped provide trained local staff and organisers.

Council problems

This move underlined the problems of council estates – particularly environmental decay, disrepair, management and social problems. Older estates increasingly stood out as areas of decline and newer estates as stark and alien worlds. The problems that only ten years earlier had been found overwhelmingly in the private sector now moved increasingly into the public sector.

Drop in output

By 1976, with better housing conditions generally and strong external financial pressures, the amount of public money spent on housing production began to slip as Table 21.3 shows.

Inner cities

Government wanted to change the inner-city drift. The swing away from demolition towards conservation and renovation had a major impact. About 1 million dwellings were renovated during the 1970s – 10 per cent

Table 21.3 Real net public expenditure (net capital spending) on housing in Great Britain, 1973–9 (1987/8 prices)

Year	1973/4	1974/5	1975/6	1976/7	1977/8	1978/9
Expenditure (£bn)	5.9	8.6	6.9	7.7	6.1	5.3

Source: Hills and Mullings 1990.

by housing associations, the rest by owners – while the number of council demolitions dropped from a peak of 340,000 between 1965 and 1969 to 209,000 between 1975 and 1979 (Halsey 1989: 385).

Housing Action Areas, particularly in London, were multi-racial initiatives. Many minority households gained access to council accommodation for the first time in the late seventies under renewal programmes which caused them to lose their private homes. The Homeless Persons' Act of 1977 marked the sharpest break in access patterns (North Islington Housing Rights Project 1976).

By 1979 there had been an irreversible shift in the attitude of the public, and the Conservatives won the 1979 election, partly on the exciting promise of introducing the Right to Buy for all council tenants. Councils had been on the ascendancy for almost 100 years. Now, for the first time since they were set up, that trend would be pushed into reverse.

Chapter 22

Legislative changes of the Thatcher years

When the Conservative Party won three decisive electoral victories – in 1979, 1983 and 1987 – it did not waste the opportunity to undermine the near-monopoly role of local authorities as providers of rented housing and to extend the opportunities for privatisation and home ownership. It opened up to the winds of competition many sectors of public or publicly sponsored activity, including housing associations.

Choice was part of democracy. Therefore people should be able to choose their form of tenure, their type of landlord, their pattern of spending. In Britain choice was more restricted than in continental countries because, by 1980, 72 per cent of all tenants rented from a local authority landlord – a near monopoly in the rented market – involving over 30 per cent of the total population. An ending of this monopoly was the underlying goal of six main Acts aiming at privatisation, choice and diversification.

RIGHT TO BUY

The 1980 Housing Act introduced the Right to Buy for virtually all secure council tenants (after three years' tenancy) at discounts of up to 50 per cent depending on the length of tenancy. The sale terms were made even more generous in 1986, with discounts of up to 70 per cent for flats. Under the Act, 1.2 million council dwellings were sold to sitting tenants between 1980 and 1989. The 1980 Act also imposed the obligation on local authorities to consult their tenants about changes in the housing services, and it gave security of tenure for the first time.

The Right to Buy was vaunted as 'the sale of the century', the greatest redistribution of wealth ever. However, sales were heavily concentrated among better-off tenants on better estates (Forrest and Murie 1991). The sale of flats was very slow, though it accelerated after more generous discounts were introduced in 1986. By 1990, about 10 per cent of sales were of flats (DOE 1990). None the less, on the poorest and most stigmatised estates virtually no sales took place and overall the Right to Buy increased the social polarisation of the poorest areas by highlighting their unpopularity to more affluent tenants.

The Right to Buy was extremely popular with most tenants, traditionally solid Labour supporters, even where they could not buy their property and even where they lived in property they would not choose to buy. It broke through a very large psychological barrier, opening up the prospect of ever-wider, lower-income home-ownership (see Saunders 1990 for a full discussion of this). It shifted the ideological ground and by the time of the 1983 election, Labour was guardedly supporting the Right to Buy, having traditionally been muted in its support for owner-occupation and having initially attacked the Right to Buy as unjust.

The long history of growth in home-ownership across the classes took a massive new turn as a result of the measure, and a new social division emerged between secure, satisfied owners, sitting on the bounty of a government-sponsored asset; and a large minority of tenants who generally aspired to ownership but could not achieve it. This cleavage, deepening as low-income owner-occupation spread, represented a much wider social reorganisation than ownership itself conferred, leading to growing segregation and polarisation in unpopular areas, where minorities, low-income and unemployed households, one-parent families and higher than average numbers of children and young people were concentrated (Power 1987a, Saunders 1990).

URBAN DEVELOPMENT CORPORATIONS

The 1980 Housing Act marked the beginning of a 'policy boom' in housing. The government saw local authorities not only as monopoly land-lords but as overgrown bureaucracies, inefficiently hoarding urban development land and feather-bedding their unproductive workforce at the expense of tenants and ratepayers. They therefore began a frontal assault on local authority powers and privileges. The Local Government Planning and Land Act of 1980 paved the way for the disposal of development land by public bodies, requiring local authorities to set up a register of unused land. It established two Urban Development Corporations to tackle the vast but now defunct dock areas of London's East End and Liverpool. They took over planning powers from the local authorities with a brief to attract private investment and transform the docks into attractive, renovated, commercial and residential areas. This was a totally new approach to inner cities, overriding local authorities' traditional role, focusing on enterprise and private investment through strong government action, and was akin to the role of government in the establishment of New Towns. Housing was to be primarily private and the emphasis was clearly 'up-market'. The approach was condemned by Labour councils as anti-democratic and irrelevant to existing residents (*Weekend Guardian*, 8–9

April 1989). But over the 1980s, eleven urban development corporations were set up, which introduced a new dynamic into decayed inner areas.

DIRECT LABOUR ORGANISATIONS

Controls on local authority repairs organisations were introduced by the 1980 Planning and Land Act, with obligatory competitive tendering for major contracts and the transformation of giant and often inefficient council repairs departments into pseudo-private building firms. This Act also forced local authority rents to rise steeply as government subsidy was tied to assumed rent increases.

HOUSING BENEFIT

The 1982 Social Security and Housing Benefit Act introduced a somewhat clumsily tied together, universal housing benefit system for all tenants (Kemp 1986). But housing benefit became less and less generous for those in low-paid work (Hills 1991b). Arrears rose steeply throughout the 1980s under the impact of both higher rents and steeper tapers. Housing benefit largely removed the issue of rent levels for the poorest. At the same time, it led indirectly to an expanding income to local authorities for spending on management and maintenance, paid for by central government, through housing benefit. A crisis arose over the abolition of housing benefit for young, single people in 1987, dramatically pushing up the incidence of begging in London (Hardwick 1991).

1988 HOUSING ACT – TENANTS' CHOICE

The 1988 Act was supposed to be the 'jewel in the Conservatives' crown', the Act that decisively took away local authority power and gave it to the consumer. In theory, tenants who wanted to leave the local authority could set up or find an alternative landlord. The Act was supposed to provide tenants who could not buy with choice and control. In practice, the Act relied largely on landlords taking the initiative. The alternative landlord had to be 'approved' by the Housing Corporation.

Tenants' Choice was constructed as a way of rejecting local authorities; it was inevitably extremely unpopular with them but in addition it created widespread suspicion among tenants. All early attempts to woo tenants away from public housing met with hostility, fear and opposition. Most Tenants' Choice groups disintegrated in the face of legal and financial complexities and no transfers happened in the first three years. There were few alternative landlords ready to take on the uncertainties of asking tenants to vote for higher and freer rents. Housing associations proved reluctant to jeopardise their sometimes fraught relations with local authorities. Quality Street, a unique and imaginative new private landlord venture, was rejected by well-organised council tenants as offering more uncertainty than quality. In the large, urban local authorities, where Tenants' Choice was more likely because of badly run-down estates and disaffected tenants, local authority hostility and tenants' fears over rent levels, security and displacement prevented change. None the less, it shifted attitudes among both tenants and local authorities.

Walterton and Elgin pioneers

Only one tenants' organisation had come near to success by 1992. Walterton and Elgin Community Homes registered their company with the Housing Corporation under Tenants' Choice, as the first approved land-lord in 1989. Their aim was to foil Conservative-controlled Westminster City Council's thinly disguised aim to privatise the two estates, along with much of its other stock. The government was caught between public-spirited, socially committed tenants wanting to gain control of their community under the government's umbrella and a hard-line, right-wing authority wanting to shed its social obligations in favour of private afflu-ence. Walterton and Elgin tenants received the strong backing of the Housing Corporation. But the hostility and highly organised opposition of Westminster City Council caused great embarrassment to the Conservative government and three years' delay, involving over half a million pounds in legal and other costs. The irony was not lost on the tenants.

The Westminster experience showed that Tenants' Choice required active local authority support. There were numerous ways in which the local authority could obstruct the process. Central government had less power in practice than it imagined to counter the powerful position that local authorities had acquired over the century, even a Conservative local authority. None the less, Walterton and Elgin tenants finally took over the ownership of their homes in 1991.

HOUSING ACTION TRUSTS

While Tenants' Choice withered on the vine, other attempts at break-up encountered even fiercer opposition. Under the 1988 Act, government-sponsored housing action trusts were to spear-head the break-up of vast bureaucratic council housing departments by expropriating the most decayed city council estates and improving them before passing them on to alternative landlords. Draconian measures were considered necessary, by-passing altogether local authority or tenant involvement and co-operation, along the lines of urban development corporations. The full implications of the measure were poorly thought through, and at the last minute, amid bitter, large-scale protests by tenants and under intense pressure in the

House of Lords, the government agreed to allow tenants in Housing Action Trust (HAT) areas to vote on whether the trusts should go ahead.

Six areas were proposed. Only two areas reached voting stage and in both, the government's proposals were resoundingly defeated. As a result, a new form of local authority-proposed, voluntary Housing Action Trust emerged in London, Hull and Liverpool, as both government and local authorities sought compromise mechanisms to spend the allocated HATs money. Two large areas, with 6,000 tenants, voted in favour of voluntary partnership HATs in 1991. Housing Action Trusts, based on co-operation between central and local government and the tenants, won the support of tenants through public guarantees on security, rent levels, investment and, crucially, the right to return to the local authority (Chapman Hendy Associates 1992).

Tenants had rallied their ranks forcefully when they saw their direct interests threatened. The government had explicitly aimed to withdraw public protection from vulnerable groups and individuals on the grounds of providing more individual incentives through many of the changes introduced in the 1980s. Its proposals to privatise marginal estates were almost bound to be voted down by those same vulnerable groups whose very status and security were undermined by the proposed changes.

The government's relations with local housing authorities were seriously marred by the 1980 and 1988 Acts. For the first time in 100 years local authorities saw their ascendancy put into reverse. Their housing stocks began to decline rapidly. Central government had given tenants the right to vote out their landlord or to buy themselves out of local authority ownership. But local authorities could obstruct change, the one power left after their primary role as providers of housing was restricted. Central government could not easily do without local elected bodies. Government was too complex and top-heavy to be able to dispense with them. Nor would its own constituency want local authorities to disappear. A majority of local authorities were Conservative-controlled in suburban, rural and small town areas, with fewer and more manageable problems than the Labour-controlled big cities. Housing Action Trusts and Tenants' Choice, focusing on Labour strongholds, had forced government to face the limits of its power, to recognise the significance of the local authority role, and to counter the real fears of tenants in poor areas. At the same time, the Act had forced into the open the problems resulting from local authorities' near-monopoly of low-income rented housing.

HOUSING ASSOCIATIONS

Housing associations were expected to expand rapidly under Tenants' Choice, but in practice very little happened. None the less, housing associations were deeply affected by the 1988 Act, as the main alternative providers to local authorities. They now had to raise private loans for a proportion of new investment at commercial rates of interest, to be fully repaid by rents, although most of their funding still came through capital grants. Rents for new, assured tenants rose steeply. Smaller housing associations found it hard to raise private finance. Since 1988, housing associations increasingly amalgamated into larger bodies in an attempt to raise investment loans. Housing association programmes expanded far more slowly than hoped, below the levels of growth achieved in the 1970s. Suddenly they became more vulnerable, more dominated by central government, and more exposed to private pressures. But gradually housing associations expanded, replacing local authorities as new social housing providers and finding themselves involved in a huge range of initiatives (Best 1992).

VOLUNTARY TRANSFERS

The 1988 Act brought to prominence a measure that had almost passed unnoticed when first introduced. Under the 1985 Housing Act, local authorities could transfer their housing stock to housing associations. Affected tenants had to agree to the transfers and major publicity campaigns were run in the areas proposed, both for and against local authority 'opt-outs'. By early 1991, voluntary transfers had taken place in sixteen local authorities. Nearly 100,000 units were transferred to specially established, new housing associations and more were in the pipeline. This was a far more significant change in practice than either Tenants' Choice or HATs. Voluntary transfers generally took place in smaller, more rural and more affluent district councils, owning between 3,000 and 9,000 properties. But by mid-1991, several large outer London boroughs, with more than 10,000 housing units, were exploring the idea.

Voluntary transfers were contested affairs but over half were successful, thanks to two critical elements: firstly, local authorities themselves were the instigators of change; secondly, both the guarantees and the incentives for tenants to vote in their favour had to be significant (Gardiner *et al.* 1991). Large-scale voluntary transfers increased the housing association sector by 20 per cent and looked like becoming a significant, if somewhat arbitrary, tool in the transfer of power away from local authorities. Other changes stemming from the 1985 and 1989 Act are discussed later under 'Tenant Participation' (see Chapter 23).

A changing public role

LOCAL AUTHORITIES CHANGE

Decentralisation

Right up to 1989, the large, urban local authorities reacted defensively to the Thatcher government's proposals and actions, even though they had begun to recognise the need for radical management change more than a decade earlier (Cole *et al.* 1988a, 1988b, 1989).

But the threat of privatisation galvanised local authorities into radical action. Throughout the 1980s, virtually every major local authority undertook some form of decentralisation of its housing service.

Behind the central–local crossfire, the reality of local authority housing practice was undergoing a radical reorganisation that was beginning to bear fruit at the end of the 1980s in spite of all the pressures of polarisation.

Glasgow District Council, for example, one of the largest landlords in Europe with 142,000 properties housing over 60 per cent of all Glasgow households, adopted a policy of inviting tenants on its giant peripheral estates of 10–15,000 units to take over the ownership and management of parts of the estates. Glasgow accepted the need to transfer at least 25 per cent of its stock to other landlords (Glasgow District Council 1986). Glasgow innovated in training, provision of furnished accommodation, tenant involvement, and 'executive' renting, while other local authorities joined the scramble to change.

In 1982, Walsall City Council in the Midlands entirely emptied its central housing department, leaving only the director, his secretary, and empty phone lines at headquarters. All other housing staff were pushed out to thirty-two neighbourhood offices, each covering about 1,500 properties. Islington followed suit in 1984 (Mainwaring 1988, Hambleton and Hoggett 1987).

Tower Hamlets in London underwent the most radical decentralisation of all, dividing up the large and extremely poor borough into seven autonomous neighbourhoods, each with its own chief executive, housing department, budget and all direct services.

These moves did not solve local authority management problems. By 1990, local authorities were still extremely large landlords, with the average city authority owning over 55,000 properties and the inner London boroughs 34,000 each (DOE 1991c). Many local authorities found that carving up and handing down budgets and decision-making was much more difficult in practice than originally envisaged (Power 1991).

The scale of councils, their history of centralised administration, their complex structures and their growing welfare role meant that they still faced major problems in a climate of financial stringency. Many local authority services were being cut and the biggest cities faced a crisis in core services that menaced urban stability (Audit Commission 1986, 1987; London Weekend Television documentary, September 1991).

None the less, voters were customers of increasingly threatened public services and local authority landlords irrevocably shifted their emphasis from production to management, and from central control to service delivery, in an attempt to preserve local support. There was growing evidence that local housing management was effective and sustainable. By 1991 it was no longer assumed, even in Labour Party circles, that local authorities would continue as dominant landlords (Gould 1991). The idea of creating once-removed housing companies under local authority sponsorship was emerging, indicating a broad if unstated consensus over the future direction of council housing – a gradual distancing from direct political ownership and control (Raynsford 1992, Merrett and Cranston 1992).

WORST ESTATES

A major actor in housing management change in local authorities in the 1980s had been the Priority Estates Project (Audit Commission 1986).

Priority Estates Project

Running parallel with the Conservative legislation that undermined the dominance of council landlords was an unprecedented initiative, inherited from Labour in 1979, to tackle the problems of difficult-to-let council estates, based on the findings of the Department of the Environment Investigation into the causes of rapid decline in the 1970s (Burbidge et al. 1981).

The Priority Estates Project (PEP)[1] was set up in 1979 as an experiment. A primary aim was to involve tenants directly in decisions affecting their estates as the only basis on which improvements were likely to work. A second aim was to localise the housing service as a way of escaping the bureaucratic inertia of the town hall systems and of responding directly to the needs of the estates. The intensive, integrated, tenant-oriented, local management effort had immediate impact, which local authorities and central government were keen to endorse (Power 1982, 1984).

The concept was extremely simple, appealing to all parties, government, landlords and tenants. It involved a reorientation of resources rather than major new spending (Power 1987a). It could largely be done within the existing framework of local authority ownership. Tenants on these estates were unlikely to exercise the right to buy. Ironically, this appealed to the government in spite of its radical rhetoric, partly because the worst estates were regarded as tinder-boxes. The Brixton and Toxteth riots of 1981 reinforced this fear. Interestingly, PEP projects were called *Partnerships* throughout the 1980s. About 30,000 units, involving 100,000 people, formed the basis of the PEP initiative.

The estate initiatives built on local authorities' home-grown attempts to decentralise and to shift their emphasis to better services, as the building programme slowed.

The success of PEP hinged on two main factors – the conditions on the worst estates and the popularity of local housing services.

The PEP model of management involved the following steps:

- 'ring-fence' the estate as an organisational and financial entity;
- set up a local office on the estate to run all day-to-day services;
- put locally based staff in charge of rents, lettings and repairs;
- consult tenants fully and provide an open door to their representatives;
- help tenants form representative organisations;
- support tenants in controlling anti-social behaviour, preventing the 'dumping' of disruptive households, and enforcing tenancy conditions;
- provide discrete and targeted budgets for repairs and improvements in line with available funding;
- monitor progress to show the costs and benefits of local management;
- provide training for staff and tenants' representatives;
- bring in other services to help improve social, economic and environmental conditions.

(See Power, April 1987, Vol. 1.)

Areas like the Broadwater Farm estate, Haringey, scene of one of the worst British urban riots in 1985, and the Orchard Park estate in Hull, one of the poorest peripheral estates in Britain, were among the Priority Estates Projects. The impact of Priority Estates Projects on tenant satisfaction and tenant involvement in some of the most run-down and stigmatised council housing areas was widely documented in the early 1990s and fuelled the belief that management mattered and that tenants could play a decisive role in their communities (Gifford 1986, Niner 1991, Hope and Foster 1991).

Self-help, local involvement and decentralised control, fitted the Conservative philosophy of 'economy, efficiency and effectiveness' by enhancing performance (Power, April 1987, Vol. 2). If the government were to narrow the scope of welfare, it had to continue to support those areas which offered least to the private sector. Over the 1980s, parallel with increasing privatisation, the government became more and more involved in targeted estate initiatives.

Estate Action

Following on the success of PEP, an ambitious and broad-based programme of estate renewal was launched in 1985 which, in 1987, became Estate Action. Initially it focused on the fifty most problematic urban authorities and, in its first year, allocated to individual estates £50 million of capital resources for improvements in security, the environment, the construction of local offices, and in some cases the buildings themselves, on condition that the local authorities set up local management offices and consulted their tenants over improvements. Estate Action money came primarily from the government's existing Housing Investment Programme, retargeted at problematic estates (Pinto 1992).

At the outset, one of the aims was to pave the way for privatisation. But such initiatives were rare because of the difficulty of 'privatising' welfare-dependent tenants. Estate Action concentrated on working with local authorities, in sharp contrast with other housing initiatives. The goal of privatisation was not central to the approach, though diversification through tenants' initiatives and right to buy was consistently encouraged. Conservative ministers were frequently photographed visiting Estate Action schemes in their unusual attempt to help beleaguered local authorities. By 1991, Estate Action had mushroomed, with capital spending on 350 schemes, making up 20 per cent of the local authorities' much-reduced capital allocations from central government. The total allocation for 1991 was £270 million, nearly six times the 1985 level (DOE 1991b).

Partnership again

By 1991, the government was making new overtures to local authorities as the steam ran out of Tenants' Choice. The partnership approach gained ascendancy. After the downfall of Mrs Thatcher in late 1990, Michael Heseltine, back in the DOE driving seat, relaunched Estate Action with a massively expanded programme of targeted help. A combination of physical, management and social measures was required to tackle the 'witches' brew of the most depressing conditions for tenants' (DOE 1991e). City Challenge was born of this change, combining partnership with local authorities with diversification and competition. The emphasis was on job training and enterprise, social initiatives, physical upgrading, management reform, and community involvement. However, the method of allocating funds continued to be a major source of controversy between central and local government (Pinto 1992).

TENANT PARTICIPATION

A key lesson of the Priority Estates Project was that without the tenants, money would be wasted, priorities wrongly assessed and improvements would quickly disintegrate. The bitter experience of the 1988 Housing Act led to a new approach to tenant participation, building on the PEP experience. Local authorities were massive landlords that could not be quickly dismantled, whatever the political imperatives. On the other hand, giving tenants the right to choose had the salutary effect of forcing local authorities to 'sell themselves' to their tenants, to take seriously the quality of the services they offered and the possibility of alternatives. They had lost their automatic dominance of the rented market and partnership became a new theme, possibly the most important consequence of government pressure.

In spite of deep hostility to alternative landlords, the idea of encouraging tenants to take over their estates and run them for themselves was attractive to all parties. Private landlords and housing associations had proved reluctant to take on the local authority mantle, so the argument for helping tenants to form their own landlord bodies appeared strong. They could do this either under the umbrella of the local authority, as tenant management co-operatives and estate management boards, or independently, as ownership co-operatives. Both had been proved possible in Glasgow, Liverpool and Islington, where tenant-based experiments had flourished since the 1970s (Power 1988).

The government review of co-operatives

The government had set up a review of co-operatives in 1987, partly with Tenants' Choice in mind, partly as a way out of local authority management problems. Its brief was to examine how tenant participation in management could advance tenants' choice, greater autonomy and better management. The Review Report, *Tenants in the Lead*, published in 1988, set up a new form of grant for tenant-based initiatives, both through the Housing Corporation (Section 83 grants) and the DOE (Section 16 grants). In 1990, £4 million was allocated to tenant participation initiatives, of which three-quarters of a million pounds went directly to tenants' organisations wanting to develop co-operatives, estate management boards (see p. 228), or other forms of tenant participation. The tenant grant regime was supposed to complement the 1988 Act – 'Tenants' Choice'. In practice, the new financial regime for housing associations made it very difficult for new ownership co-operatives to get off the ground. As a result, management co-operatives and estate management boards were the most important legacy.

The tenant bodies cost about £1 per unit per week to support – in the region of £30,000 per estate per year (Power 1988, DOE 1987). Tenant participation grants created a new wave of breakaway initiatives and the possibility of tenant training on a significant scale. A total of eighty tenants' groups had been direct recipients of tenant grants by 1991. Bodies like the Tenant Participation Advisory Service (TPAS), PEP and the National Federation of Housing Co-operatives also received grants from the government under the Co-operative Review.

Estate management boards

Estate management boards (EMBs) grew up in disadvantaged urban areas where tenants were desperate for change but too vulnerable to opt for total transfer (Zipfel 1989). They emerged from the Priority Estates Project, where tenants on large, run-down estates had elected a board to run the estate, with the local authority participating as full partner and member.

Local authorities had long been able to delegate their management to approved registered bodies such as co-operatives. Local authorities had experimented with tenant-member co-operatives for twenty years (Power 1988). EMBs had a similar legal form to a tenant management co-operative and they attracted wide support from central government, local authorities, and tenants. The experimental development of EMBs fed directly into the Review of Co-operatives, and the Section 16 grants helped a number of new EMBs off the ground.

Each estate management board registered as a company or a friendly society. All tenants in the area could become members of the organisation. Tenant support had to be demonstrated. The local authority became a corporate member. The agreement between the local authority and the tenants had to be formally approved by the Secretary of State (Zipfel 1989). Twenty-seven run-down estates had voted to form estate management boards by 1990, with overwhelming turn-outs and votes in favour. Another ninety-eight feasibility programmes were in progress, many of which would lead to estate management boards or co-operatives.

Local politics had become suddenly highly dynamic under the new tenants' voting powers. Through estate management boards, local authorities became active partners in change and break-up, an extraordinary and unexpected development resulting from the strong pressures for change among tenants. A number of important local authorities, such as Birmingham, Southwark, Hackney, and Liverpool, began to draw up significant proposals for a full range of breakaway initiatives.

1989 LOCAL GOVERNMENT AND HOUSING ACT

The final major Act of the 1980s defined and restricted council landlords in a radically new way. The aim of the Act was to separate landlord responsibilities from the more general responsibilities of local authorities by ring-fencing the Housing Revenue Account. It gave government a powerful lever on rent and spending levels that would almost certainly push rents up. But it would give housing managers self-contained budgets with which to run their services. It would make problems more directly visible by preventing subsidies from the general account to housing services, such as was common in big cities; and it would give tenants and housing managers a better deal by preventing subsidy to general spending from rent income. This practice was common in district councils.

Performance standards

To help drive the new controls and force better standards of management, the government introduced performance monitoring of local authority landlords for the first time ever.

The 1989 Act was heavily interventionist. But it was timely in at last imposing standards of housing management – a role that government should arguably have played from the outset (Stewart 1988). As part of the

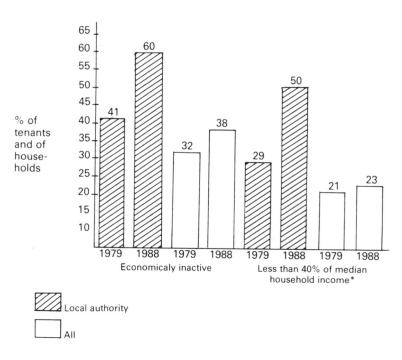

Figure 23.1 Concentrations of poverty in local authority housing in Britain in 1979 and 1988

Source: Office of Population Censuses and Surveys 1979, 1988

Note: *Median household income was £97 per week in 1979 and £196 per week in 1988.

'ring-fencing', local authority housing departments were required to prepare performance monitoring reports for the government, the general public, and specifically for their tenants.

POLARISATION AND REVIEW OF HOMELESSNESS

The long drift of council housing towards a 'welfare tenure' was greatly intensified by the policies of the 1980s. The proportion of economically inactive households in council housing rose from 41 per cent in 1979 to 60 per cent in 1988; the proportion of very poor council tenants with less than 40 per cent of the median household income rose from 29 per cent to 50 per cent. These steep rises contrast sharply with the overall increases. At the same time, the proportions of those with the lowest incomes dropped among owner-occupiers but rose steeply among council and housing association tenants. Figure 23.1 and Table 23.1 illustrate these trends. The shift could be explained from a number of angles. Firstly, nearly one-fifth of the council stock was sold to tenants who could afford to buy, with at least one earner. Only acceptable stock was bought, often concentrated on more popular estates.

The remaining stock was generally less popular and housed generally poorer people, also concentrated on unpopular estates (Forrest and Murie 1991). Meanwhile, additional owner-occupation drew a large section of households who otherwise might have competed for rented property. Demand for council housing was therefore increasingly biased, by the great expansion in owner-occupation, towards low-income households who could not buy.

Family break-up creating, at least temporarily, two households from one greatly increased demand for council housing in the 1980s – those alone with dependent children were guaranteed rehousing through the homeless legislation. A very high proportion of households admitted to council housing as homeless were lone parents – as high as 80 per cent in London. Local authorities in Britain accepted 56,750 homeless households in 1979 and 170,000 in 1990 (*Social Trends* 1992). The statutory obligation to

Table 23.1 Lowest fifth of population in Britain by income as distributed between tenures (1979, 1985), based on General Household Survey data

	Owning outright	Owning with mortgage	Local authority tenants	Housing association tenants	Other	All
1979	20	24	43	1.6	11	100
1985	16	13	57	3.9	10	100

Source: Hills and Mullings 1990, Table 5.16.

house 'priority homeless' left local authorities with little room for manoeuvre. These households, because of the high stress of their situation and often lack of immediate income, did not have other solutions in the short term. A high proportion of homeless acceptances were of ethnic minority origin.

By 1990, nearly half of all Afro-Caribbean households were in council housing, compared with a quarter for the population as a whole.

Unemployment rose steeply in the 1979–84 period, affecting 3 million people. This made a very large segment of the population dependent on renting. The loss of private-rented units pushed many newcomers, young people and transient households to look to the council, but they were often excluded from council housing as not being in the highest priority category. This created a crisis of street homelessness as people, who previously might have lodged in a private hostel, bed and breakfast or rooming house, found themselves with nowhere to go, unable to pay the premiums and deposits often required in the private sector.

Policies bringing about the closure of long-stay, large institutions – from children's homes to mental hospitals – often came to fruition in the 1980s and meant that many vulnerable households requiring support and care looked to local authorities. If they were housed, they often created new demands and pressures, increasing the proportion of needy. If they were not housed, they increasingly ended up on the streets (Greve 1990a).

Through a whole range of changes, the large, local authority housing stock and the much smaller housing association stock became the refuge for people desperate for somewhere to live.

The Homeless Persons' Act (1977) increasingly became, in the 1980s, a way of underpinning the whole council allocation system. Poor or vulnerable people could not buy, nor in most cases, gain access to private renting – the supply had dried up. Council housing was the one vehicle, and it was still a very large tenure with over 5 million dwellings in 1989 in the whole of Britain. Access for the most needy was guaranteed by legislation, as long as people fitted the legal definition. Councils had little choice but to go 'down-market'. Those that were excluded, such as young, single people, often resorted to the streets.

This sequence explains why privatisation was much less promising a solution than at first hoped. Buildings were sellable in most cases, but buildings housing very poor people on a secure basis were generally not. In any case, the government could no longer afford to lose the housing that kept most people off the streets.

By 1988, it was approaching a crisis of policy between the need to house the very poor – a growing demand – and the desire to push rented housing away from local authorities, a policy as unattractive to private investors as it was to local authorities themselves.

Thus, when the Secretary of State for the Environment decided to review the homelessness legislation in 1988, it was almost certain in advance that the government, for all its harsh anti-welfare rhetoric, would keep this valuable law in place, in spite of the fact that it was making council housing increasingly into a one-class tenure. By 1991, the Secretary of State for the Environment was calling on local authorities to be partners in the attempt to rescue the worst estates, now housing a large majority of households dependent on welfare, and offering special incentives to local authorities and housing associations to help house people sleeping rough on the streets. Welfare housing had come into its own (DOE 1991b, 1992a).

PRIVATE HOUSING IN THE 1980s

Owner-occupation

Nearly 1.5 million new owner-occupied homes were built between 1980 and 1989 in addition to mass sales of local authority dwellings to tenants (Hills and Mullings 1990). Shared ownership, with the occupier buying only part of the property and renting the rest through a housing association became a popular new form for those who could afford less than a whole house. But popular as owner-occupation was, its ascendancy was threatened by over-extension as well as economic and demographic change.

In the late 1980s, high interest rates, restrictions on tax relief and a collapse in the property market slowed the galloping spread of owner-occupation. House prices tumbled from 1989 onwards and repossessions of owner-occupied houses due to defaults on mortgages reached alarming new levels. By 1991, 10 per cent of all mortgages were in arrears of more than two months (Wibb and Wilcox 1991) and over a quarter of a million owed more than six months' payments (Best 1992). Government confidence was rocked by the spectacle of distraught families becoming homeless through building society repossession. A battery of emergency measures was assembled, including proposals to turn mortgages into rents, to use repossessed, empty houses for homeless families, and to open up private renting with housing associations acting as agents.

Private renting

The number of private-rented units continued to plummet throughout the 1980s and by the end of the decade, only 7 per cent of households rented from private landlords compared with 12 per cent in 1979. The Business Expansion Scheme, a generous tax break for small, new investors, was applied to rented housing in 1988 but its incentives were too short-term and too geared to the advantages of selling into owner-occupation to provide a lasting framework for expanding rented housing. Re-sale was

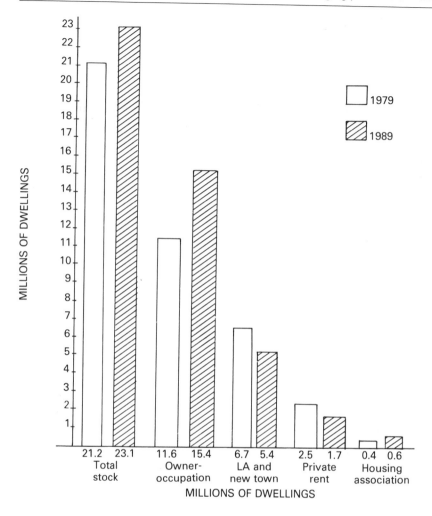

Figure 23.2 The changes within housing tenures in Britain during the Thatcher years, 1979–89

Source: DOE 1990

allowed after five years. The new private-rented units were far outweighed by losses in other parts of the market. None the less, by 1991 a new cross-party consensus in support of private renting was emerging (*Roof*, September 1991), and the collapse of the property market resulted in a growth in short-term lets, including repossessed, formerly owner-occupied, houses.

The Thatcher years in retrospect

By the time Mrs Thatcher resigned in 1990 every aspect of housing had been shifted. Local authorities as landlords were dwindling, their systems overhauled and their focus on management and maintenance enhanced. Housing associations were growing – their stock had doubled between 1974 and 1989 (Hills and Mullings 1990). They were being forced to act increasingly like private businesses, while also taking on a bigger social role. But they did not take over inner-city council stock as the government had hoped. Tenants had been given remarkable power to broker their limited fortunes and were experiencing a real stake in their housing. The Citizen's Charter, introduced by John Major in the build-up to the 1992 Election, and proposed compulsory competitive tendering for housing management were the most recent ideas to continue the shake-up of state-run housing.

Figure 23.2 underlines the major changes that we have described – the growth in owner-occupation, the decline of private renting, the emergence of housing associations as a growing, if small, force, while in the 1980s local authority renting steadily declined.

NOTE

1   The author was commissioned by the Department of the Environment to help set up the Priority Estates Project in 1979. Its work is described in detail in Power (April 1987; 1987a).

Chapter 24

Conclusions

OVERVIEW OF CHANGE

Over the century, housing change and political intervention had gone hand in hand. The resulting changes in condition and in tenure had been radical. Table 24.1 presents an overview of how the different tenures developed since 1900.

Table 24.1 Housing stock and tenure in Britain, 1901–89

	Total stock (million)	Owner-occupiers (%)	Local authority or new town (%)	Private rented (%)	Housing associations (%)
			(a) England and Wales		
1901	6.7	NA	NA	NA	NA
1914	7.9	10	0.3[a]	90	0.6[b]
1939	11.5	33	10	57	
1953	12.7	35	19	47	
1961	14.5	44	24	31	
1971	17.0	52	28	19	0.9
1979	18.8	57	29	12[c]	2.0[c]
1989	20.5	68	21	7	2.8
			(b) United Kingdom		
1979	21.2	55	32	12[c]	1.9[c]
1984	22.1	61	28	9	2.4
1989	23.1	67	24	7	2.8

Sources: 1901 stock figure from Halsey 1989, Table 10.1; 1914 figures from DOE 1977a, Technical Volume, Table I.23; 1939–71 figures from Holmans 1987, Table V.1; 1979–89 figures from DOE 1990.
Notes: [a] About 20,000 dwellings; [b] about 50,000 dwellings; [c] breakdown between private renting and housing associations estimated for Wales.

POPULARITY OF LOCAL AUTHORITIES

Underlying the developments was the pivotal role of local authorities. They emerged in the 1920s as major providers because of wartime pressures, their success in tackling urban problems, and their links with government. They proved largely effective bulk-builders and they were generally popular with tenants throughout their history. They often experienced conflict over who to house, and the 'need' versus 'merit' debate came to a head in the 1980s with growing homelessness pressures. Council housing became ubiquitous because of renting shortages in villages, towns and cities alike. While the scale of operation and of estates often made for difficulties, this was not always or necessarily the case (Holmans 1992). Renewed shortages at the turn of the 1990s led to calls by the Conservative-dominated Association of District Councils for renewed council building (Anson 1991).

Table 24.2 sums up the main housing events of the century.

SOCIAL HOUSING REFORMED

A say for tenants

The most significant change in housing away from the course of the last 100 years was the new direction of local authority landlords in the 1980s. They were emphasising to tenants their right to exact good quality services and giving tenants a say in how these were delivered, through the introduction of service agreements (London Boroughs of Camden, Islington; and York). Tenants were asking for more control and better services and were exercising their voice powerfully. But they did not want local authority protection and support to go, even where they voted for estate management boards, co-operatives or a transfer of stock.

Management devolution

The management patterns of local authority services were likely to go on changing rapidly and radically, as the global shift towards informal, decentralised and fluid structures continued (Mingioni 1989). Contract management, devolved autonomous units, and voluntary and tenant-based initiatives were all likely to expand. Councils were allocating budgets to managers with a brief to make the service work, recruiting trained housing staff and putting repairs out to contract. They were spending more on repairing the stock. Thus the local authorities would no longer remain large, inflexible, bureaucracies. They would lose much of their centralist power and move towards more front-line services, playing a back-up, enabling role at the centre.

Table 24.2 Sequence of main housing events in Britain, 1862–1992

1862	Peabody Trust established – start of philanthropic housing movement.
1885	Local authorities housing services established.
1912	First garden city founded.
1915	First World War; rent strike, tight rent controls.
1919	Homes fit for Heroes' campaign; councils build cottage estates on garden city lines.
1924	Grants to builders for low-cost home ownership.
1930	Slum clearance; outer cottage estates plus inner flats; flat subsidy and overcrowding subsidy introduced.
1940s	War; shortage; rent controls; Labour government.
1945	Aneurin Bevan, visionary Minister of Housing; mixed developments; quality council housing; expensive sites subsidy.
1951	Conservative government; a 'people's house'; cheap, utilitarian, mass council housing.
1956	Renewal of slum clearance; mass housing estates; high rise subsidy.
1957	Attempt to free rents; 'Rachmanism' scandals in private renting.
1964	Labour government; big drive for industrialised building; grants to renovate; Housing Corporation set up; housing associations take on new life; improvement grants.
1968	Ronan Point disaster; high rise abandoned.
1969	General Improvement Areas; shift from clearance.
1974	Housing association grant; Housing Action Areas.
1976	Race Relations Act.
1977	Homeless Persons Act guaranteeing access to statutorily homeless.
1979	Priority Estates Project; Conservative victory.
1980	Right to buy; Urban Development Corporations; competitive tendering.
1982	Social security and housing benefit reforms.
1985	Consolidating Act allowing voluntary transfers and estate management boards; Estate Action set up.
1988	Tenants' Choice; Housing Action Trusts; Co-operative Review; Tenant Participation Grants; private finance for housing associations.
1989	Ring-fencing of Housing Revenue Account.
1991	New Partnerships – City Challenge.
1992	Compulsory competitive tendering for housing management.

Housing associations

Local authorities were collaborating with housing associations in a wide range of homelessness initiatives, no longer able to attempt a monopoly role. Nor did housing associations look likely to supplant local authorities, although their role had expanded and their pace of development had certainly overtaken local authorities. They were catching up with the notion of tenant control, partly under the influence of local authority reform.

Cost of welfare

Private finance was not a magic solution and whichever way money was raised, the community as a whole had to shoulder the cost of welfare or face growing social disarray in run-down city areas. As youths rioted in cottage estates in the North, in Wales and in the Midlands at the end of a hot summer in 1991, new questions arose over the deeper problems of social housing, the poverty of its tenants, the alienation of its youth, and the clashes with police, with minority shopkeepers and with the wider society. Local authorities could not necessarily respond adequately to the underlying tensions within their increasingly unstable marginal estates. Meanwhile, housing associations faced growing strains as they increasingly shared in the welfare role.

OUTCOMES

It could be argued, as the government has done, that its policies worked to shift local authorities out of their complacency, out of a low-cost, low-service mentality into a more modern, more responsive and more efficient approach.

It could also be argued that, in many ways, the role of local authorities had been validated. They provided a necessary umbrella for needy people who could not compete evenly; they were popular service providers, for all their limitations, with those people who relied on them; they could operate more effectively than previously, if constrained by some of the disciplines of competition and cost control; they were used by the wider society to play an essential social role in the increasingly complex environment of the 1990s.

Part IV

Denmark

There is little leaning on shovels ... there is a social ethic ... a realisation that if they don't work, none of them will have anything.

F. Scott, *Scandinavia*

Denmark is a small country with a population of just over 5 million on the edge of the European Community, with only a narrow sound between her and Sweden. She shares a border with Germany, with which she has had tense and contentious relations.

Denmark shares much of her history with the other Scandinavian countries. The three main languages of Scandinavia are closely related, and Danes, Norwegians, and Swedes can, with a lot of effort, understand each other, both in written and spoken form. So can Icelanders, although Iceland has preserved a more archaic form because of its isolation. The Finns generally speak their own language but often use Swedish in co-operative Nordic initiatives. The frontiers between Scandinavian countries are open and there are mutual rights of residence, work and welfare. There is a common and distinctive culture, history, standard of living and way of life that marks Scandinavians out from the rest of Europe. The Nordic Council, through which they deal with their common interests, provides a loose federation for the 22 million people who talk of themselves as Scandinavians, as well as Norwegians, Swedes, Finns, Icelanders, and Danes. Denmark itself includes Greenland and the Faroe Islands, although they have limited 'Home Rule' within the kingdom of Denmark.

In spite of this, there are some tensions and rivalries, certainly between the Danes and the Swedes. There is frequent condemnation of the pollution of Swedish industry, whose waste is blown towards Denmark, and of the Swedish nuclear fuel industry, an energy option rejected by the Danes. There is even a myth that the Swedes invented concrete housing! Because of Denmark's size and proximity to Sweden, she cannot always shield herself from national Swedish policies and sometimes feels dominated by a relative industrial giant.

Denmark is the only Scandinavian member of the European

Community, a link that increasingly outweighs her ties with Scandinavia. She is included in this study because of her unique housing system, with elected tenants' representatives forming the majority on the boards of all social housing organisations in the country.

Her history as a great seafaring nation, once the leader of the largest established kingdom in Europe, was replaced, after she became the small country she is today, by her success in 'carefully nourished sufficiency'. Denmark's wealth was based on a co-operative agricultural organisation, popular education, brilliant design – 'beauty in things of everyday utility' – practical inventiveness and vigorous trading relations with Britain, continental Europe, America, and the developing world. Hans Christian Andersen and Soeren Kierkegaard may be the most famous Danes, but her modern, tasteful style and her co-operative, democratic organisation give her a celebrated image abroad, far outstripping her size or economic importance.

Denmark lies at the same latitude as Scotland and is afflicted with strong, cold winds from the North Sea, Finland and the Baltic, a fact that has encouraged Denmark to develop modern windmills for electricity. These windmills are now a major export success and a growing asset at home. A new hyper-market on the outskirts of Copenhagen is lit, heated and air-conditioned by adjacent wind-powered generators. The windmills are futuristic rather than quixotic and reinforce the outsider's impression of a highly innovative and adaptable society. Windmill parks to provide a more general electricity supply are being developed, an idea that has been exported to California, using Danish equipment and know-how.

Denmark suffered under the two major recessions caused by the oil crises of 1974 and 1980, and went into an escalating balance of payments deficit that was concealed for a while by her historic position as an avid international borrower with assiduous attention to repayments. The conservative Danish government of the early 1980s took draconian steps to force savings, cut spending and inflation, and wean investment away from extravagant consumption into export-oriented enterprise. There was a strong push towards saving fuel. Danish housing became much more energy-efficient, and by 1991 Denmark was no longer importing oil.

Europe is vital to Denmark's future. Denmark, in spite of her agricultural and rural base, small population and peripheral status in the European Community, has a strongly anti-isolationist attitude and highly internationalist outlook, possibly resulting from a seafaring tradition. Denmark is close to Eastern Europe and is historically associated with the Baltic. She is alive to upheaval there. A high proportion of foreign workers in Denmark come from Eastern Europe.

Danish products are aggressively marketed abroad, not least her housing products. Danish industrialised building components are world-famous, along with Copenhagen china, interior fittings, the Lur sign on the international butter mountain, and of course Lego building toys.

How Denmark came to such prominence and then decline in international relations, how she developed her economic, social and industrial institutions, and how she organised her housing in possibly the most innovative ways in Europe, are interlocking questions that we explore in this section.

Part IV.1 Inner-city tenement blocks

Part IV.2 Haraldskær – Denmark's national tenant training centre

Chapter 25

Background

FACTS ABOUT DANISH HOUSING

Denmark's population of just over 5 million has been growing very slowly over the last generation and is expected to decline slightly after the turn of the century. Denmark's birth-rate since 1981 has been below the replacement level, giving her a small proportion of young people and many small households. In 1986, 58 per cent of *all* households comprised single people (Boligministeriet[1] 1988f). Denmark is not densely populated, although over four-fifths of all Danes live in villages and towns of over 1,500 inhabitants and the Greater Copenhagen region houses over one quarter of the population. Denmark had until recently the highest income and highest standard of living of the five countries in this study, although its level of unemployment rose steeply to 11 per cent in January 1992. The biggest change in Denmark's population was the growth in one-parent families and decline in two-parent families over the last twenty years. Table 25.1 summarises some important aspects.

Denmark has almost one dwelling per household, in spite of the very large number of small households. Seventy per cent of the population lives in single-family houses. Denmark has built at a fast rate since the war. Over 60 per cent of the stock has been built since 1960. The average size of dwellings has fallen steeply in both public and private sectors because of the growth in small households. Three-quarters of post-war building has been for private owners. Denmark has a large stock of old flats – over three-quarters of a million – from before the Second World War, most of which are private. It has a further 400,000 flats owned by non-profit housing companies, mostly built since the war. Table 25.2 summarises the Danish housing stock.

One quarter of Danish private consumption goes on housing. Denmark has one of the highest housing standards in the world (Boligministeriet 1988f). Table 25.3 illustrates the distribution of housing between different tenures.

Table 25.1 Demographic facts about the Danish population, 1970, 1985, 1990

	1970	1985	1990
Population	4,900,000	5,100,000	5,140,000
Density – no. of people per sq. km	–	–	119
Population in urban areas	–	–	87%
Households	1,800,000	2,200,000	2,400,000
Size of households (persons)	2.7	2.4	2.2
No. of children per woman	2.6 (1965)	1.5	–
Crude birth-rate (per 1,000 pop.)	–	–	10.8
Age of population (%):			
Under 15	–	–	17
Under 20	31	27	–
20–34	22	23	–
35–64	35	37	–
65+	12	15	15.4
Household structure (%):			
Single – no children	24	31	–
Couples – no children	26	27	–
Single with children	4 (73,000)	5 (110,200)	–
Couples with children	38 (679,000)	30 (627,000)	–
Others without children	9	7	–
Non-Danish ethnic groups (%)	–	–	2

Sources: Boligministeriet 1987b, Commission des Communautés Européennes 1987, *The Economist Pocket Europe* 1992.

A LOOK AT HISTORY

Early history

Denmark's early history is little known beyond Scandinavia, yet it greatly influenced the way her urban settlements grew and the way she tackled her housing and social problems. Danish history reads like a magnificent tapestry, evoking legendary figures and epic events. The Vikings, the original Scandinavians, set the scene with powerful ocean-going boats, distant colonies, trade and raid. The Vikings had a primitive system of democratic

Table 25.2 Facts about housing stock in Denmark, 1960, 1970, 1985

	1960	1970	1985
Total no. of dwellings	1,300,000	–	2,200,000
Dwellings per family	0.7	–	1.0
Average size of dwelling (new completions):			
Private	–	150m^2	100m^2
Public	–	90m^2	75m^2 (1980)
New units built as houses	–	70%	73%

Sources: Boligministeriet 1987b, 1988f.

government that included elected chieftains, participatory law-making meetings of farmer-citizens, and some egalitarian ideas. But the wind- and sea-swept Scandinavian landscape and harsh winters limited the scope for home-grown wealth and encouraged plunder, as well as trade, by sea. Britain was long tied to Denmark through both, including a short period in the tenth century when most of England was part of the Kingdom of Denmark. By the thirteenth century, Denmark virtually controlled the Baltic, a position she maintained until Sweden broke away 300 years later.

Denmark was gatekeeper on one of the most important trading routes in the known world, collecting tolls from all ships, thus making herself wealthy and powerful.

She followed an unusual system of government that was more democratic than most. Her monarchs were elected from among royal blood and could be removed if they failed to follow the law-making courts of the people: 'If you do not wish to do as we say, so will we go against you and kill you, for we will not tolerate lawlessness or disorder of you' (Scott 1975: 43).

Table 25.3 Distribution between housing tenures in Denmark, 1991

Tenure	%	No. of units
Owner-occupation	58.1	1,381,000
Non-profit housing companies	17.9	426,000
Private renting	16.3	386,000
Public housing (local authority)	3.0	70,000
Private co-operatives	4.7	112,000
Total no. of units in 1950, 1,100,000	100.0	2,375,000

Source: Danish Statistical Office, April 1992.

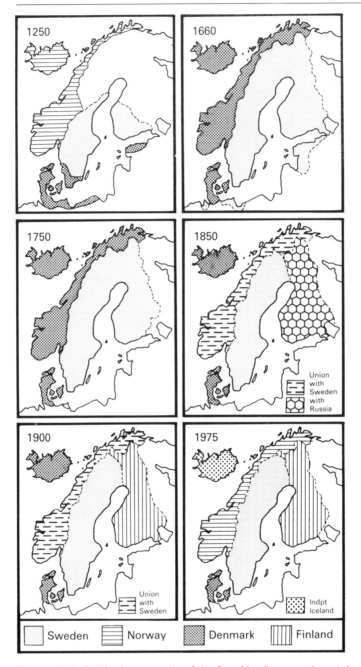

Figure 25.1 Political geography of the four Nordic countries at the times of the cross sections

Source: Mead 1981: 4

Sweden was the first part of the Danish kingdom to break free. Much longer-lasting was Danish control of Norway, which continued for over 400 years till the defeat of Napoleon in 1814; Danish control of Iceland only finally ended in 1944.

Denmark's controlling position on a major world trade route led her to try and remain neutral in major European wars, so as to trade with all sides. At the end of the eighteenth century, when Napoleon was threatening British interests, this neutrality enraged her British trading partners. In 1807, Britain laid waste Copenhagen, stole the Danish fleet and left the Danes so bitter that they joined with Napoleon. When Napoleon himself was defeated, Denmark lost everything. The Danish state was declared bankrupt in 1814 and Norway was lost to Denmark, becoming an independent country, having formed part of Denmark since 1380. This humiliation was followed fifty years later by the invasion of Jutland by the Prussian army, resulting in Denmark losing one-third of her remaining land and population as Schleswig and Holstein were taken from Denmark (see Figure 25.1).

Growth of towns

Copenhagen became an important walled city early in its history. After the capital was largely destroyed by fire in 1795 and the British bombardment in 1807, it was rebuilt within its ramparts in a stern, classical style of which much still survives.

By the mid-nineteenth century, Copenhagen had 120,000 inhabitants, about the same number as Dublin. The next largest town, Odense, had 8,700 inhabitants. Elsinore, Aarhus, Aalborg and Randers, all had between 5,000 and 8,000 – almost villages by modern standards. Apart from these centres, Denmark was a land of farms and villages, very much as she had been for 800 years. There was very little industry, although a modern railway network was in place by 1830, based on successful trade in farming produce.

Copenhagen and other towns started to grow in the 1850s. The size of Copenhagen was contained by the city ramparts, with four gates that closed every night, till 1870. Her population doubled up along 'labyrinths of passages and courtyards'. As the century wore on, agriculture was mechanised and people began to move off the land more rapidly: small craft-based industries grew up, not just in Copenhagen but in other towns, which expanded fast but not explosively.

Because Danish urban growth and industrialisation were so gradual and because her farming was so prosperous, Denmark experienced many fewer social problems, pressures and upheavals than France, Germany or Britain in the late nineteenth century. Therefore, the imperative to take drastic reforming measures was absent. So was the conflict that normally precedes radical change. Rather there was a nurturing of innovative, organisational experiments to maximise Denmark's superficially limited economic potential and her historic skills in trade and internal organisation – 'a late and sophisticated adaptation to the industrial age' (Jones 1986: 44).

Education and farming

Two developments in the nineteenth century greatly influenced the shape of modern Denmark, agricultural co-operatives and folk high-schools.

Folk high-schools

Denmark from the sixteenth century was strongly Protestant, individualistic and nationalistic in character. Great Danish thinkers like Kierkegaard argued for religious and political freedom and for the personal development of the individual.

Nicolai Grundtvig, a Danish theologian, educationalist and visionary, set up the first folk high-school in 1844 in order to emancipate rural peasants from ignorance and straight-jacketed education. The idea was based on a rejection of strict classical education, rote learning of grammar and narrow privilege. It was meant to develop the whole mind, to build a sense of Danish culture, to relate ideas to practical life, to appeal to rural peasants. The folk high-schools encouraged arts and crafts, a sense of taste and design, and a love of books, all of which have strongly influenced modern Denmark at every level. The agricultural developments were partly made possible by educational reform and the folk high-school movement. Many other aspects of modern Danish life, such as the politically independent trade unions and the co-operative housing movement, also derived part of their tradition from folk high-schools.

Agricultural credit and co-operatives

Peasant reforms in the late eighteenth century and early nineteenth century enabled a majority of peasants to buy back their land from large owners. While 60 per cent of peasants had regained ownership of their farms, the state had intervened to provide smallholdings for landless farm labourers. Agriculture developed rapidly. In 1849, the first credit society among farmers was formed, a kind of co-operative lending body to help poorer farmers and smallholders buy their land. Credit organisations formed the basis of a powerful, agricultural co-operative movement that later greatly influenced housing and other developments.

In 1874, Hans Broge of Aarhus in Jutland proposed co-operative dairies in every village. Farmers and smallholders would share in the buildings, equipment, marketing and proceeds of their dairies in proportion to their share in production. But all members would have equal standing in decision-making with one vote for each member. This helped smallholders to be as active as farmers, leading to economies and innovations. It brought increased wealth for everyone. The co-operatives were able to modernise and mechanise, building on the rural education of the folk high-schools. The state provided advisers to regulate standards and help the poor-quality dairies.

The co-operative idea spread rapidly – eighty were formed between 1874 and 1884. Five years later there were nearly 700 and by 1895 there were 830. The Viking horns (*Lur* in Danish) were chosen as the emblem for co-operatively produced butter, a sign of Danish quality evoking the earlier co-operative tradition, and Lurpak became a household name internationally. Britain took 90 per cent of Danish exported butter on the old sea route up till present times. Over 90 per cent of all Danish milk and dairy produce still comes from the co-operative sector.

Pigs and bacon followed suit, only a few years behind the dairies. Co-operatively produced bacon quickly became a rural factory product, with pigs specially bred to please English breakfast taste and processed in co-operative factories. Poultry and eggs then followed. Co-operative stores were also set up, based on the Rochdale Pioneers' model; but unlike Britain, where urban squalor and exploitation drove the movement, Denmark's strong co-operative movement was almost entirely rural and created an organisational base for later urban and industrial development.

The co-operative movement epitomised the Scandinavian approach that has puzzled many visitors – 'a capitalist system of production, paired with a socialist system of distribution'. The combination of efficiency, quality control and profit orientation along with participation, profit-sharing, education and social development – which other countries find hard to emulate – flourishes in Denmark on a long history of trade and exchange, immense difficulties in survival, and a rural economy that has been nurtured through enlightened education.

Multi-party structure

The Danish two-chamber system of government was established by constitutional law in 1848 and the parliament was set up in 1901. In 1953, this was reformed into a one-chamber parliament, the Folketingt. The Danish multi-party system, with power distributed among many main parties, led to a consensus approach to government, with policies built on compromises and a little for everyone rather than violent swings in favour of widely polarised interest groups (Alestalo and Kuhnle 1984: 48). Universal female suffrage was introduced early, in 1909, though only in the 1953 reforms was female access to the throne established.

After the First World War

Denmark remained neutral in the First World War. When it was over, as part of the German peace settlement, a referendum was held in the still-troubled border areas of southern Jutland, which had been lost to Germany in 1864; part of Schleswig was returned to Denmark. Her modern border became fixed, though not without further challenge. Iceland at the same time gained partial independence. Denmark continued to industrialise and her urban population grew rapidly, but it was a development based on small-scale craft industry, with skilled artisan labour and apprenticeships rather than mass production (Alestalo and Kuhnle 1984: 30); large industrial areas outside Copenhagen were unknown. Agricultural employment dropped steadily.

Strong welfare

As early as 1848, the Danish Constitution had proclaimed state responsibility to help those in need. In 1891, a law to provide for the destitute and for old age was passed. The earliest health insurance associations, descending from the old craft guilds, were set up in the 1890s.

In the 1930s, the Danish birth-rate plummeted, leading to exaggerated fears of population decline. This led to a new wave of social provision. Progressive health and child care measures were introduced to help women.

In 1933, during the world recession, unemployment in Denmark rose to a staggering 44 per cent (Jones 1986). In the same year social insurance was reformed, covering unemployment, accident, welfare and general need – more progressive provision than in any other country.

In the 1930s, the Nazis in Germany began to agitate for the return of southern Schleswig to Germany, and the Danish government began to question its neutral, non-military stance.

The Second World War

In 1940, the Germans demanded Danish capitulation under threat of bombing Copenhagen. Denmark lived through a five-year occupation during which her population found many means of passive non-co-operation. One form of Danish resistance was to secrete Jews to safety in Sweden. Very few were found by the Germans. In 1943, the Danish government was taken over by Germany, but was freed again by German capitulation in 1945.

Post-war Denmark

Economy

Denmark emerged from the war with little direct damage and with her national pride intact. She went on to develop rapidly as a modern industrial state with international help through Marshall Aid. By the 1960s, Denmark had begun to overtake Britain as a wealthy industrial nation. During the great growth period of the late 1950s and early 1960s, Denmark developed her foreign markets with aggressive trading. Short-cutting much of the pain and squalor of large-scale urbanisation, she none the less became an ultra-modern model of affluent living.

Social underpinning

A strong social underpinning was based on the philosophy that all should benefit equally in the new wealth regardless of personal setbacks. This open-ended commitment was to become an immense financial and administrative burden in the 1970s, but it coloured Denmark's unusual combination of enterprise, industriousness and welfare. Women were working on a wide scale by 1960. Forty per cent of under-2-year-olds and 56 per cent of 2- to 7-year-olds were cared for in nurseries. The folk high-school tradition lived on with far-reaching adult education programmes and a very progressive secondary school system.

Family change

Denmark's family and social patterns changed more rapidly than in other countries. By the 1970s she had the highest divorce rate in Europe. Young people left home earlier. Her birth-rate was very low. Her strongly secular and individualistic culture – 'the most de-christianised country in the world' (Jones 1986: 212) – created a desire for freedom from traditional and unadaptable constraints.

Foreign workers

In spite of Denmark's international links it remained a fairly homogeneous Nordic society until about twenty years ago. In the early 1970s, a little later than Britain, France or Germany, Denmark began to attract large numbers of foreign workers and refugees. By 1984, 1 per cent of the Danish population comprised first-generation immigrants, other than EC nationals and Scandinavians. The numbers have since grown, as families joined the immigrant workers. Some, including many East Europeans, were racially indistinguishable from Danes. But three groups in particular stood out: the Turks and immigrants of Arabic origin, of whom there were estimated to be about 21,500; and the immigrants of Asian origin (mainly Pakistani and Sri Lankan), of whom there were about 11,500. They were particularly noticeable, primarily because of the previously homogeneous ethnic make-up of Danish society. Their concentration within its major urban centres, Copenhagen and Aarhus in particular, made them more visible and seem-ingly more numerous than they really were. Their much larger families, general youthfulness, and higher birth-rate meant that school populations in certain areas were particularly affected. One school just outside Copen-hagen had no Danish children at all and twenty-nine language groups represented.

No incoming group had strong historic links with Denmark, unlike much of the post-war immigration to Britain and France.

Table 25.4 shows the different ethnic groups in Denmark in 1984.

Change in the 1980s

The economic crisis in the 1980s led the Danish consensus over welfare, government regulation and market orientation to break down. Interna-tional borrowing and export had provided the basis of Denmark's industrial success. But rapid inflation followed by economic slump left Denmark with the highest ratio of foreign debt to gross national product of any European country (*The Economist Supplement* 1988).

Social problems began to mount as cuts bit hard against more margina-lised groups, and real incomes fell. They took on ugly racial overtones as Denmark's homogeneity and high standard of living appeared simultane-ously threatened. Denmark has seen a real fall in standard of living,

Table 25.4 Ethnic proportions of total foreign residents in Denmark, 1984

	Total foreigners
Scandinavia	22,334
European Community	23,756
North America	4,668
Other categories	53,304
which include:	
Turkey	17,827
Yugoslavia	7,397
Pakistan	6,659
Arab countries[a]	3,628
South and Central America	1,942
Chile	776
South-East Asia[b]	4,680
Total	104,062

Source: Boligministeriet statistics 1989.
Notes: [a] Algeria, Morocco, Egypt, Jordan; [b] Philippines, Thailand, Vietnam.

although income per head in the late 1980s was still very high by European standards (Boligministeriet 1988f).

Danish chauvinism ran high and local political campaigns took on distinct racial bias. One mayor in an outer Copenhagen borough was re-elected in November 1989 with an increased majority on an explicitly anti-immigrant platform.

Denmark's unique experience is hard for outsiders to explain. One view of Denmark is that she concentrated her abundant energies on internal development when once she had lost her international role in the nineteenth century, producing a highly skilled, highly cultivated and highly efficient society. Another view is that her reliance on the sea, as a kingdom of 500 islands, made her into a successful trading nation, maintaining her international role long after her political leadership had diminished. A third view is that the Viking past, with its international ambitions, its commmunitarian base and its pervasive organisation, lived on to take new forms in twentieth-century Europe.

NOTE

1   Boligministeriet – Danish Ministry of Housing.

Chapter 26

Development of social housing

Danish urban housing movements developed slowly. In 1850, only one in twenty Danes lived in cities. The rest were still on the land or in small rural settlements. The standard Danish house was timber-framed with clay or brick infill and thatch roof. Many such traditional houses survive in town, village and countryside and there was a ready transference of frame building from ancient to modern building systems, often relying on the same concept, either in timber, steel, or concrete.

FIRST SOCIAL HOUSING INITIATIVE

Copenhagen was Denmark's only major city. The poor, hemmed in by the ancient ramparts, often shared a single room between two families.

A serious cholera epidemic in 1853 goaded the Medical Association of Copenhagen into action. Doctors collected money to help relieve distress and some of this was invested in 'The Doctors' Union Dwellings', a spacious, airy development of 500 flats, built along wide avenues in two-storey blocks, just outside the city walls in 1857 (Salicath 1987 – see Plate 26.1). This earliest 'social' housing project included shops, a library, a community centre, a nursery and gardens – features that were common to most of the initial social housing experiments in Europe but which only in Denmark became a common, and then a required, feature (Fuerst 1974). This early scheme was still occupied in 1970, when it came under threat of urban renewal. It was still very popular with tenants and strongly defended against demolition because of the pleasant, high-quality units and the tree-planted environment.

GOVERNMENT SUPPORT FOR SELF-HELP

The Doctors' Experiment was not copied but urban congestion quickly led to new efforts.

In 1865, a group of shipbuilding workers took the first self-help initiative, setting up a building association in Copenhagen to improve housing

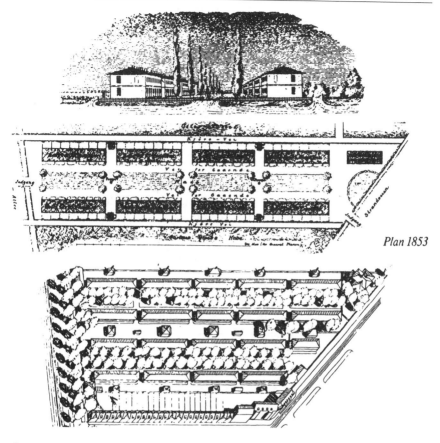

Plan 1853

Plate 26.1 The first social housing development in Denmark, 1853

Source: Salicath 1987, vol. I

conditions for the poor. This workers' movement created pressure on the government to act. In 1887, the government introduced the first state loans to local authorities or building associations to help provide cheap rented housing. The earliest co-operatives were formed through these loans but they could often only house better-off workers because of repayment costs.

The early initiatives were largely unregulated and shortage of capital led to speculation. Investors often realised their assets, cashing in on the boom in house prices by selling off their property. Thus, most of the earliest associations quickly folded and the impact of these initial social efforts was small, amounting to fewer than 1,000 units by the turn of the century.

PRIVATE LANDLORDS

Private landlords profited from the boom in demand for low-cost housing as people moved to towns and cities. In Copenhagen they built massive tenement blocks, packed with tiny flats off long internal corridors, for the lowest-paid workers. Poor quality as the workers' tenements were by comparison with the leafy Doctors' Union dwellings, many of them survived both world wars and are still in use today. Their wooden stairwells and closed inner courtyards, the small flats and unmodernised facilities, make the old central areas of Copenhagen stand out as a monument to Denmark's late-nineteenth-century urban development. For their time, they were considered very advanced and gave private landlords a higher status than they enjoyed in Victorian Britain. Private landlords were able to meet demand because Denmark was still a largely rural society and her cities, apart from inner Copenhagen, were relatively uncongested. The dense, high blocks, often of five or six storeys, housed large numbers of workers relatively cheaply. About 900,000 private flats were built in Denmark before the First World War, almost entirely by private landlords. Many of these are still occupied today (Boligministeriet 1988f).

The lucrative investment in low-cost housing led to an over-supply of cheap rented flats in Copenhagen by early this century. There followed a serious property crash, leaving 10,000 empty rented units and thousands of unemployed construction workers.

EARLY HOUSING CO-OPERATIVES

In 1912, after the property crash, the President of the Union of Building Carpenters, J. Jensen, founded the first, fully-fledged co-operative workers' housing association (Arbejdernes Andels Boligforening). It became one of the biggest co-operatives in Denmark. A co-operative building association was formed a year later, the Arbejdernes Kooperative Byggeselskab. Both these organisations grew slowly but became powerful models, copied all over Denmark, owning many thousands of units each by the 1980s. The property crash had underlined the instability of the private housing market and the vulnerability of individual workers within it. It led to the growth of a widespread co-operative housing movement that involved saving, building, renting and managing housing.

The early housing co-operatives built on the agricultural tradition of one member, one vote, with each member-tenant investing capital into the organisation, originally about 6 per cent of the cost of a dwelling. The individual stake was cut to 2 per cent in 1982 but still applies to all Danish social housing. Proceeds were reinvested in the organisation to produce more housing.

The idea had strong appeal among the skilled artisans and rural migrants to Copenhagen in search of housing. The unusually high level of education among recently arrived, low-income, urban households – thanks largely to folk high-schools – helped create the atmosphere of initiative and self-organisation that provided the base for Danish co-operative housing. Co-operatives were built on entrepreneurial and self-help principles that enabled them to grow alongside private landlords. The government was keen to work with such organisations, as it had been with the agricultural co-operatives.

BETWEEN THE WARS

After the First World War there was a chronic housing shortage due to the drying-up of private investment in housing and continuing urbanisation. The government gave generous grants and subsidised loans to housing associations and co-operatives for the construction of non-profit housing. The money was channelled through local authorities. In 1919, the National Federation of Non-Profit Housing Companies[1] was set up to work for the advance of member organisations, combining the resources of the early builders' co-operatives and the non-profit housing companies. This body evolved into the most powerful promoter of social housing.

The Danish housing companies were influenced by the British garden city movement, though inner-city building also continued, modelled on the private tenements. Gardens, balconies and allotments were very important aspects of all social housing developments and the vast majority of Danish social housing was built with balconies.

The principle of each incoming tenant investing savings in the company that was producing the housing was maintained. Danish social housing companies were strengthened by the threefold support: individual members' savings; government and local authority support; and private loans.

Meanwhile, private housing was being built in the form of 'villas' around the growing cities and towns. Affluent Danes aspired to housing close to their rural roots, and single-family, detached houses were the norm for the growing owner-occupied sector.

The great depression hit Denmark very hard and, between 1927 and 1933, there was very little building. Acute shortages developed and tight rent control was imposed in 1933, while unemployment affected nearly *half* the population.

In 1933 and then in 1938, the government introduced strict rules for the development of co-operatives and social housing companies. The registration of rules became obligatory. Rents covered costs, with government help taking the form of subsidies on the initial costs. Housing societies had to reinvest surplus rents, which rose over time, in modernisation and further building. This 'self-financing' principle of each organisation became an important fundamental of Danish social housing, building on the self-reliant and independent origins of the movement.

An essential element of the developing social housing societies was the relationship between the housing organisation and residents. The co-operative origins of the movement influenced the shape of many companies, making the participation of residents appear natural and logical. There was a less sharp distinction between landlord and tenant, owner and renter, investor and beneficiary, than in other systems and countries, not least because from the beginning the tenants invested savings in the housing organisation in order to gain access.

The growth of builders' co-operatives, alongside housing co-operatives, exercised a strong influence on developments. Workers were in fact putting up housing *for themselves.* There developed a steady, long-term relationship between the co-operative companies that built social housing, the companies that owned and ran social housing, and the investors who in the first place were member-tenants. From the earliest developments, the non-profit housing associations and Labour organisations, principally the unions, developed a strong working relationship. Government and local authorities were the backers, regulators and facilitators, rather than organisers or promoters. This made a big difference to the organisation of social or state-subsidised housing.

In 1938, Denmark introduced special help for families with children. This included rent allowances for large families, a very avant-garde measure compared with other countries, and subsidised loans to build child-care facilities and community centres as part of new housing developments. From the outset, housing companies had provided some community facilities. From then on, government help to social housing organisations was tied to the inclusion of community facilities. By the time war broke out there were about 20,000 social housing units in Denmark, about 2 per cent of the stock. The dominant private-rented sector was still much needed and a total rent freeze was introduced in 1939 to stop steep rises in rents. This move was basically the kiss of death to further significant growth in private renting and opened the door to 'mass' social housing.

Table 26.1 Estimated breakdown of housing tenure in Denmark at the outbreak of the Second World War

Private landlords	800,000
Owner-occupiers	300,000
Social housing companies	20,000
Total	1,120,000

Source: These figures are estimates only, based on post-war stock and building figures from the Boligministieret 1987b.

The stock of Danish housing at the outbreak of the Second World War was divided between three main tenures as shown in Table 26.1.

THE SECOND WORLD WAR

During the German occupation of Denmark in the Second World War, very little building took place, leading to serious shortages in spite of the lack of significant war damage (Fuerst 1974). The shortage led to a wartime Housing Commission which recommended longer-term planning for housing, with ten-year time-scales, to create more stability and to help develop modern building techniques by encouraging stable investment. This more planned approach became possible because of Denmark's multi-party political system. Stability was also encouraged by the trade unions which, in spite of close connections with the Social Democratic movement, never had direct links to the party-political structure. This encouraged a consensual approach.

Two wartime moves helped accelerate the development of mass social housing. In 1941, the National Federation of Non-profit Housing Companies combined with trade unions to set up the Arbejder Bo ('Workers' Home'), a national non-profit building and development agency which created new associations all over Denmark, often in liaison with local authorities. The company actually built 60,000 housing units directly in the post-war period. Alongside this nation-wide move, a special Copenhagen Social Housing Company (Koebenhavns Almindelige Boligselskaberne – KAB) was set up, which helped produce, and by 1990 was managing, 30,000 units, one of the largest Danish housing companies. But most Danish housing companies were small, local bodies, in many cases linked to the co-operative origins of the movement. Big companies often created new local associations which owned the homes. The larger sponsoring company then offered to provide management and other services to local associations. This was the Danish alternative to over-large housing structures.

NOTE

1   National Federation of Non-Profit Housing Companies – sometimes referred to as the National Federation or National Housing Federation.

Chapter 27

After the Second World War

The war provoked a big shift of population towards the towns. Denmark industrialised very rapidly. In 1945, empty property was requisitioned to help solve the chronic housing shortage. A year later a generous building subsidy was introduced for anyone prepared to build – private landlord, social housing company or individual owner. The non-profit sector was set to grow rapidly, as it had in place a national structure and local organisations ready to build.

By 1950, one quarter of Denmark's population lived in greater Copenhagen – over 1.25 million people. Eight hundred thousand people were packed into inner Copenhagen. Modern Denmark was largely forged around it and rapid building got under way to house the new masses in sprawling suburbs of private housing and dense developments of social housing.

STANDARDISED BUILDING PARTS

The government increasingly regulated and standardised building elements throughout the country and through all sectors in an attempt to speed up and reduce the cost of construction. At the same time, there was a strong emphasis on quality. In 1947, the Danish Building Research Institute had been set up to help the drive towards efficiency.

The government regulation of social housing followed the pattern of the dairy industry. It enhanced quality, building on the organisational independence of the self-help bodies. It did not take them over or mastermind them. It combined regulation with efficiency by maintaining the share principle and by forcing companies to recover costs through high rents. This in turn created impetus towards quality. Table 27.1 summarises developments to 1947.

The stock of housing had grown in the ten years from 1939 to 1950, with some losses of private-rented housing and growth in owner-occupation and social housing. One hundred and fifty thousand new additions had been achieved (Boligministeriet 1989).

Table 27.1 Development of Danish social housing organisations to 1947

1853	First social housing project
1865	First builders' co-operatives
1887	First government loans
1912	First workers' co-operative housing association
1919	National Federation of Non-Profit Housing Companies
1933	Government introduced registration of non-profit companies – principle of self-financing
1937	Rent allowances first introduced; communal facilities required in social housing
1939	Rent freeze
1941	Wartime Housing Commission – longer-term, planned approach
1941	Arbejder Bo – development body to set up housing companies all over Denmark
1945	Requisitioning of property
1946	Government standardisation of all building parts throughout Denmark; government subsidies to all types of house construction
1946	Non-profit Building Supply Company – to meet government requirements for standard parts
1947	Danish Building Research Institute established

POST-WAR DEVELOPMENT OF 'MASS' HOUSING

The trade unions, which had invested in the national building agency in the war, now were instrumental in forming a national building supply body to provide building materials and equipment to the co-operative and non-profit builders. All supplies matched government-imposed standards by 1960. The aim was to reduce costs, control quality, and get the fastest possible progress for the social housing companies. The National Federation of Non-profit Housing Companies became majority shareholders in this purchasing company.

INDUSTRIALISED HOUSING

Non-profit housing had a strong producer element, as well as being consumer or tenant-oriented. Government funds were tied to using industrialised building techniques, building components and internal equipment. About 90 per cent of all post-war housing production was industrialised and, as systems developed, the time to build a housing unit dropped to two months.

The early development of prefabricated and industrialised building techniques and components was fuelled partly by the innovative builders' co-operatives within the social housing movement, partly by strong government support and regulation, and partly by the modern style of Danish industry generally.

Another influence was the building apprenticeships system, a direct legacy of the folk high-school tradition, leading to a highly skilled workforce producing precision-made units. The quality of building parts and finishes, the speed of building and the price of housing production, all reflected a highly organised, modern and competitive building industry comparable to Sweden's and very different from Britain's (Dickens *et al.* 1985). One outcome was that virtually the entire post-war social housing stock was industrially built, about 80 per cent in medium-rise flats. Another outcome was the sophisticated standards of equipment and communal facilities. A third was the visible quality of social housing dwellings. Industrial methods in Denmark were part and parcel of an integrated organisational approach that revolutionised building systems (Kjeldsen 1976).

FLAT-BUILDING

The shift to industrialised building encouraged the concentration on flat-building in the social sector. This was partly because of the need to house lower-income workers in urban areas where they were arriving off the land in ever-growing numbers. There was a strong commitment to conserving farming land, preventing urban sprawl and protecting the green environment, although this did not take legal shape till the 1960s.

Industrialised building itself created a rationale for flats. It became relatively cheap and easy to double or treble the number of units by stacking uniform apartments on top of each other. 'Room units' could be purchased off the shelf, rather like their famous Danish play equivalent, Lego. Social housing became dominated for a time by concrete blocks of modern flats.

THE MOVE TO LARGE SCALE

A natural consequence of the support for industrialised housing combined with rapid urbanisation was a growing government emphasis on large developments. From 1960, the Danish government actively encouraged developments of 2,000 units. These large new developments were to embody a 'neighbourhood ideal' with a mix of social groups, ages and households, wide-ranging community facilities and a planned, modern environment. Conditions were laid down through regulation.

Social housing companies were firmly established all over Denmark and were leaders in this ambitious and modernistic approach. The rhetoric of 'machine living' and social structuring caught on. At the same time, the National Federation of Non-Profit Housing Companies had become a powerful organiser, promoter and lobbyist, with its own expansionist aims, both in building and in the provision of advice services covering finance, law, standards, equipment, and so on. The National Federation encouraged local authorities to promote and extend social housing through new or existing societies.

HIGH RISE

High-rise building was fashionable in the early post-war period. For a while 'everyone was hooked on high rise', though it was denounced by some architects very early on (Dybbroe 1989). Forty thousand were built in the 1950s. However, their general unpopularity meant that most flats were only three to six storeys high. About 40 per cent of all social housing was built in blocks of flats. Given the emphasis on industrialised, large-scale flat-building, which dominated all social housing for thirty years after the war, it is perhaps surprising that high-rise developments did not play a more dominant role. However, from the outset there was strong resistance to the style and the Danes were very successful in applying the same indus- trialised techniques to low- and medium-rise housing. The early 5-storey private tenements avoided the kind of industrial-revolution slum conditions experienced in bigger industrial cities and countries; they created a pattern of dense, medium-rise developments that could be successfully copied. The ultra-modern, high-rise designs offered few advantages, were more costly than low-rise flats, and were quickly denounced.

From the mid-1960s, the government encouraged the application of industrial building techniques to smaller-scale, low-rise social and private housing, using a 'cluster' layout around more traditional street patterns (Svensson 1988 – see Plate 27.1).

SOCIAL HOUSING COMPANIES AND EXPORT OF MASS PRODUCTION

Social housing was dominated by the building boom, by modern techniques and by the idea that all parts of a building could be mass-produced. Expan- sion fuelled innovation, and companies developed and marketed ever more 'off-the-shelf' housing units. However, although industrialised housing became identified with large social housing estates, the method was quickly applied to smaller and more low-rise developments.

In 1968, the National Federation of Non-Profit Housing Companies sponsored a new building corporation, Danalea, to develop prefabricated houses and flats. Danish housing units were widely promoted abroad and many a British local authority became enamoured of their clever design,

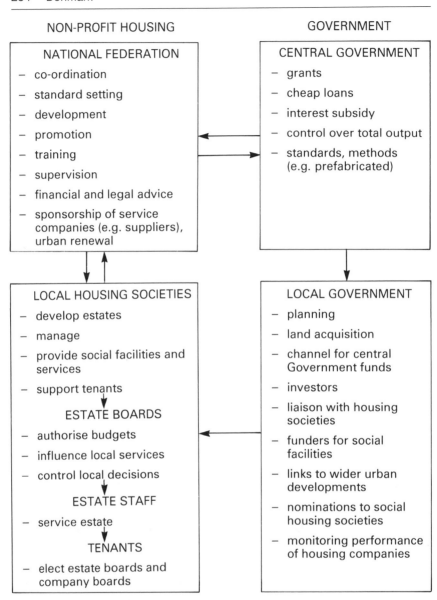

Figure 27.1 Links between government and the non-profit-making bodies in Denmark

their reputation for quality, and their trading skills. Third World countries also invested heavily in Danish prefabricated units. The Larsen and Nielsen system was just one of the industrialised prefabricated systems that gained international renown.

THE ROLE OF LOCAL AUTHORITIES AND CENTRAL GOVERNMENT

Local authorities sponsored social housing, used it to meet local needs, and often had close dealings with housing companies. Three hundred and thirty of the 650 non-profit companies were sponsored directly by local authorities after the war. Smaller local authorities were still forming companies long after the boom period was over.

Local authorities were the main channel for government funds and acted as planning authorities and partners in development with the social housing bodies. They played a major role in organising land deals. They were therefore the brokers of most housing companies' fortunes. Local authorities also had to put in some of their own funds for the development of social housing. Local authorities had responsibility for urban renewal but carried it out through independent agencies, usually in the non-profit sector. They had jurisdiction over non-profit housing companies, receiving government assistance and operating in their area, ensuring that access was arranged fairly, nominating tenants to 25 per cent of units. They were represented on the boards of housing companies in their area, making up less than half of all members. These different roles gave local authorities a significant influence over the activities of social housing organisations. They in turn were subject to and strictly regulated by central government and exercised control over housing activities within a complex national legal framework. Ultimately, of course, the national government could overrule local authorities within the agreed framework.

NATIONAL ORGANISATION

The government and the National Federation of Non-Profit Housing Companies played an important dual role in the development of housing. Figure 27.1 shows the links between government and the non-profit bodies, nationally and locally.

VOLUME OF PRODUCTION

In the twenty-five years from 1950 when post-war building began in earnest, social housing made up 28 per cent of all production. In the 1950s, due to acute shortages and tight controls, nearly 40 per cent of all housing was built by social landlords. The years with the greatest numbers of social units were the late 1960s and early 1970s before the oil crisis. But in the same period, private building also boomed and therefore the share of all building by social landlords dropped. Building momentum was then at its peak and the whole idea of mass urban housing was enormously attractive, meeting Denmark's need for a rapid urban response to the post-war industrial boom.

Table 27.2 Housing production in Denmark, 1950–75

	Number built	%
Social housing companies	251,500	28.6
Private building (for owner-occupation and renting)	627,500	71.4
Total	879,000	100.0

Source: Boligministeriet 1987b.

Table 27.3 Ownership of housing stock in Denmark in 1950 and 1975

	1950		1975	
	Number	%	Number	%
Social companies	68,000	5.3	319,500	14.8
Local authorities	40,000	3.2	60,000	2.8
Private landlords	750,000	59.0	670,000	31.2
Owner-occupiers	413,000	32.5	1,100,000	51.2
Total	1,271,000	100.0	2,149,500	100.0

Source: Boligministeriet 1987b.

Table 27.2 shows the volume of building between 1950 and 1975; Table 27.3 illustrates changes in the ownership of the stock over the same period.

The role of the social housing companies had been transformed and the volume of owner-occupation had risen steeply. As Denmark's urban population expanded, virtually all new building was in these two sectors. Private landlords lost their dominant position and increasingly sold out into owner-occupation. Local authorities, after a short emergency period during and after the war, adopted an organising, planning and enabling role, building only a very small amount, mainly in Copenhagen.

Chapter 28

Private house-building

While social housing companies became the major providers of new rented housing in the post-war period, nearly three-quarters of all new housing was provided through private building.

Owner-occupation had already become popular in urban areas before the war. After the war, government incentives to all kinds of housing investors encouraged the spread of single-family, owner-occupied houses. Whereas in pre-war Denmark nearly 100 per cent of urban housing was in flats, by 1970 over 70 per cent of it was in single-family houses for owner-occupation. The 'villa' style dominated, and rows of detached family houses, often with only two metres in between, were built in sprawling suburbs. Owner-occupied housing was normally timber or brick-built and much more traditional in appearance than social housing, even though industrially made components and fittings were widely used.

With tax relief on all interest charges on loans for owner-occupation, there were strong incentives to save, build and become an owner. By 1960, three-quarters of all new housing was for owner-occupation.

Growth in owner-occupation continued throughout the 1970s, and 410,000 new owner-occupied units were added to the stock between 1970 and 1985.

As a result, by 1975 over 50 per cent of households lived in owner-occupied dwellings, a majority of them single-family houses. In particular, families with children were drawn to owner-occupation. It offered more control, more space and more independence. Larger households, usually with children, were concentrated there.

However, there were changes along the way, culminating in the major reform of housing subsidies in the early-1980s, which greatly slowed the growth in owner-occupation. Before we look at those reforms, we will mention briefly the developments in private renting and wider urban policy.

PRIVATE RENTING

After the war, government subsidies were available to private landlords to build new housing, but strict rent controls, in place till 1958, provided a strong disincentive to invest. In addition, private landlords were not allowed to sell their rented property into owner-occupation, cutting off another form of return. The government therefore introduced special loans for private landlord building in 1950.

By 1966, the overall number of housing units was beginning to match households. Many post-war controls were relaxed, including the prohibition on the sale of rented units for owner-occupation. A flood of sales led to about 200,000 units being sold off – about 20 per cent of all private rented units.

By 1975, there was growing concern over the exodus from inner-city areas and the continued decline of the older private-rented stock, with the conversion of many units into owner-occupation since the war. The pressures this created on poorer tenants led to a movement into outer social housing estates as the only other avenue for displaced, low-income families. In 1976, the government therefore reintroduced tight controls over the sale of private-rented flats into owner-occupation. The restriction applied to all units built prior to 1969. Unless the sale had already been agreed before 1969, or unless the tenant vacated the property voluntarily, it could not be sold. Private tenants were given the right to form ownership co-operatives to take over their blocks, subject to all tenants agreeing, if the landlord should want to sell up. This led to the formation of a number of private co-operatives from the late 1970s.

The controls on sales reintroduced in 1976 stemmed the rate of conversion from private renting to owner-occupation, but it did not reduce the landlords' desire to do so. Therefore ways round the law were found. Whereas private renting made up nearly 40 per cent of the stock in 1970, its share had dropped to 20 per cent in 1988.

SLUM CLEARANCE

Demolitions also caused some change, although Denmark did not embark on a big slum clearance programme. Firstly, its urban stock was on the whole relatively recent, dating from the late nineteenth century onwards. Secondly, the private-rented blocks were in high demand and were popular due to their cheapness and inner location. Thirdly, social housing, while popular with better paid workers because of its high quality, was too dear for poor, insecure newcomers. Denmark was still urbanising rapidly throughout the post-war era and needed its older urban stock, while greatly expanding its new stock. Therefore, only inner-city housing that was literally unrescuable tended to be demolished.

The National Federation set up its own non-profit urban renewal company in 1969 in order to carry out urban renewal schemes in inner areas of Copenhagen and other cities and towns. The demolition and rebuilding of the Norre Bro area of Copenhagen, a very poor central neighbourhood, by several housing companies was criticised by many Copenhagen residents as too heavy-handed, displacing traditional communities and expanding several big social companies into large and bureaucratic landlords.

But most inner areas were not given such radical treatment. Generally, the process of inner-city renewal was piecemeal. Therefore, in all, little of the old private-rented stock was actually lost but much of it was converted to other tenures.

GOVERNMENT INTERVENTION

The changes in housing patterns were facilitated by changes in the way housing and urban policy moved. By the mid-1960s, two things had happened. First of all, the supply of new dwellings had greatly expanded so that space and amenity standards had risen and household size had fallen far beyond other countries.

Secondly, the cost of housing to the individual households had risen dramatically. New urban building land was increasingly scarce and expensive and agricultural land was being lost at a very rapid rate. Cost rents for new social housing, which were set on the basis that each development had to meet its own costs, were too high for low-income households and were a deterrent to applicants. Cost rents on older property were far lower and artificially depressed. This coincidence led to government changes in financial support to help make rented housing more viable. It also led to changes in the legal framework of urban development.

DECONTROL OF RENTS

In 1966, rents on older property which were very low were freed so that increased income could fund renovation. At the same time, direct subsidies to building for owner-occupation were cut. Rents for new social housing units were more heavily subsidised because of excessive costs. The prices of standardised building parts were also controlled, but inflation in land and house prices continued.

BUILDING FUND

In 1967, a national building fund was set up into which housing companies paid a share of their higher rents from older property on which little or nothing was owed. This building fund was established jointly by the

Table 28.1 Changes in Denmark's pattern of land use, 1951–82

| Use | Number of hectares | | | |
	1951	1982	Rate of change	Total land (%)
Towns over 200 inhabitants	74,000	189,400	+115,400	4.4
Farms and rural buildings and gardens	185,700	221,000	+35,300	5.1
Summer cottages	5,000	42,100	+37,100	1.0
Agricultural land	3,158,500	2,896,600	−261,900	

Source: Boligministeriet 1988f: 33.

government and the National Federation of Non-Profit Housing Companies as an independent investment company. Its funds could be used for new developments, renovation, or the rescue of companies or estates in crisis. It was increasingly used to meet growing deficits and building problems in the newer stock. In addition, each housing company was required to set up a building fund.

LAND ZONING

In 1969, tighter land use controls were introduced with strict zoning across the whole country.

Denmark was divided up into urban and rural zones. The allotment movement had remained strong and many recent city dwellers had built summer cabins on their plot of allotment land. This growth in summer cabins was to be restricted to designated areas.

The Danish government was also anxious to preserve the farming base. Farms could no longer be turned into second homes. The government, through the strict zoning measures, protected the coastline from ribbon development and tried to stop urban sprawl.

At the same time, it recognised the strong desire to enjoy the country-side and it therefore designated certain areas for summer cottages. Some housing companies set up summer holiday centres. The summer cottage areas attracted a burst in another innovative style of prefabricated housing – timber-frame, chalet-style cabins. Danish prefabricated chalet-cabins have been another major export product!

The effect of urban development on land use can be clearly seen in Table 28.1.

REORGANISATION OF LOCAL GOVERNMENT

The other major change was the reorganisation of local government in

1970 from 1,400 councils, based on ancient settlements, into 277 local authorities. As a result, the average municipal size went up from 4,000 inhabitants to 18,000. In addition, there were twelve county or regional councils. Only Copenhagen and Frederiksberg were kept out of the reorganisation and continued with their own city governments. Copenhagen had grown rapidly till there were well over a million inhabitants in the metropolitan region (Boligministeriet 1988f).

IMPACT

The effects of these measures were not always as intended. The relaxation of controls in the private rented sector led to a boom in sales to owner-occupation. The higher rents for older property drove many more households to try and buy. There were many other more complex influences, such as the tax system, the state of the economy, inflation and mortgage availability, but all enhanced the seeming desirability of owner-occupation. The greater subsidies to new social housing contained rapid inflation a little but rents continued to rise steeply. The reduction in incentives to owner-occupation were less strong an influence than the advantages that remained for those able to buy. Thus in the late 1960s and early 1970s, owner-occupation continued to expand, private renting to decline, and social housing to move further in the direction of low-cost, mass building in outer estates.

NEW TOWNS AND PERIPHERAL ESTATES

The stock of social housing units grew from 200,000 in 1965 to 400,000 by 1985. Much of this growth was in large, peripheral estates that were not attractive to families and too far out for young single or elderly households – two growing groups.

The larger administrative units and the stronger central planning of urban and rural zones encouraged new town developments, both to try and contain urban sprawl and to foster 'rational' building. The major housing companies played a big role in building a series of 'new towns', particularly around Copenhagen but also in other major cities. Although originally conceived as very large social housing developments, in practice the peripheral estates rarely grew to their target numbers. Plans were often over-ambitious and reflected an earlier ethos of mass building that was falling from favour. People and employers did not generally want to relocate away from existing cities and the level of crude need had fallen. Danish towns, by international standards, were small and relatively uncongested. Breaking out of the settlement pattern was neither urgent nor popular. The large, peripheral 'new town' estates heralded the end of 'boom times' for social housing.

The peripheral areas of the 1970s proved to be some of the hardest to let in the early 1980s. The problems took two main forms: social isolation in unsettled, new communities; financial difficulties due to high costs, low demand and growing poverty among new tenants leaving private-rented housing.

AGAINST MASS HOUSING

By 1974 when the oil crisis hit Denmark with special ferocity because of her lack of energy, the Danish public had begun to react strongly against mass housing. Industrialised methods had gathered speed and multiplied their impact on the urban landscape and environment. There were immediate difficulties in letting some flats, particularly large family dwellings, on the largest, most modern and most peripheral estates such as Taastrupgaard and Ishoejplanen outside Copenhagen. From 1974 onwards, the overall production of social housing dropped from a high of 15,000 units a year to a low of 5,000 units in 1979 (Boligministeriet 1987b). Large-scale 'mass' estates were no longer being planned, although many that were in the pipeline continued to completion during the 1970s.

CONSERVATION

In 1974, following the oil price hike, the government introduced special energy conservation grants. Owner-occupiers could deduct half the cost of energy-saving measures from tax; landlords could deduct one-third. At the same time, the average size of social housing flats was cut from 100 square metres to 85.

The Danish preference for detached houses became an ideal that families on moderate incomes could often no longer aspire to. The need was for more economical group or 'cluster' houses, which were much more energy-efficient, both in construction and in heating – saving 25 per cent of the cost. Cluster housing usually included communal facilities that were the norm on social housing estates – shared laundries, playrooms and, in some cases, eating and social rooms. The Danish Building Research Institute helped innovate in lower density 'cluster' housing, attracting international acclaim for its attractive layout, impressive facilities and low energy use. It was often organised as a form of private co-operative housing with individuals owning their own private space but being members of a co-operative body for the shared areas.

The changes in the 1970s – away from mass housing, away from detached houses, in favour of more energy-efficient and more community-based solutions – led to a burst of interest and activity in inner-city conservation and renewal. There were about 400,000 pre-1919 properties in urgent need of modernisation, mostly owned by private landlords, the majority of them in Copenhagen but some in other cities too. The shift in emphasis towards conservation included both the unpopular estates and the private tenements. The social housing companies had a major role to play in both. Before looking at how the changes were brought about we will outline briefly the way non-profit housing was run.

Plate 28.1 Energy efficient cluster housing

Chapter 29

Social housing organisation and tenants' democracy

TENANTS' DEMOCRACY

A complex organisational jigsaw had emerged in the process of building and managing mass housing estates. With government funding, channelled through local authorities, non-profit and co-operative housing organisations had grown and spread, becoming semi-public bodies with a strong participatory element. The democratic and voluntary co-operative tradition had remained strong. In an attempt to retain social cohesion and commitment, the tenants' role was gradually extended.

A radical new approach was adopted through national legislation. From 1958, tenants had been allowed to elect representatives to housing company boards and to be consulted over rents. In 1970, elected estate boards were introduced, which had to be consulted over budgets and decisions affecting their estate. An estate was a group of dwellings built together as a planned entity by a housing company.

From 1984, estate boards were strengthened and tenants could form a majority on the company board. These estate boards became responsible for decisions and priorities affecting that estate. Each estate had its own income and expenditure account. The books had to balance between rent income, repayments and management and maintenance costs. The central management organisation, which also had to balance its books, would be paid for services from each estate budget and each estate would have its own sinking fund for planned maintenance. Arrears and empty units would directly deplete the local budget.

The aims were: to ensure tenant responsibility and involvement in hard decisions, along with a tenant voice; to protect societies from escalating costs and possible bankruptcy due to problem areas slipping out of control; to promote efficient management; to highlight problems quickly by 'ring-fencing' each discrete area of housing. The idea of giving tenants a direct say was to reinforce their stake in the new communities and to inject direct feedback into the companies.

The estate-level organisation was in part provoked by the growing

Table 29.1 The functions of different parts of a social housing company

ULTIMATE AUTHORITY
- sponsors the company
- nominates or elects representatives to the board
- oversees general direction
- intervenes in serious crises
- has authority to dissolve company

Government
Funders
Members

COMPANY BOARD
- legal and financial responsibility
- approves accounts and budget
- fixes policy for allocations
- supports estate boards

Tenants
Sponsors
Government
representatives

CENTRAL OFFICE
- administers company affairs
- hires staff
- organises and controls finances and balances income and expenditure budget
- implements policy on rents, allocations, etc.
- organises and runs development programme
- supervises estate-level operations
- provides support to estate boards
- services and responds to main board

Company
director plus
employees

ESTATE BOARD
- holds annual meeting to elect board and approve accounts and budget
- approves house and estate rules and charges
- approves works
- checks standards
- draws up and renews planned maintenance programme
- oversees local facilities and clubs
- has input into caretaking appointments and standards

Elected
tenants'
representatives
Staff
Company
representatives

LOCAL STAFF
- custodial caretakers/technical staff/tenant adviser/local housing administrator – work at estate level
- run local office
- order and carry out small repairs
- maintain heating systems, lifts, gardens and communal areas
- liaise with tenants and help sort out problems
- supervise contract work on estates

Company
employees

problems of managing a large and complex stock, and in part by the need to break down the units of organisation into a manageable size so that both tenants and managers themselves could handle problems effectively as they arose and became quickly visible. Difficult areas could be identified, isolated and targeted.

The right of tenants to have a controlling and decisive say in the non-profit housing companies was called 'Tenants' Democracy'. It involved a carefully balanced power-sharing system rather than a direct transfer of authority. The boards themselves were carefully regulated by government and the sponsors of the company were the 'ultimate authority'. Table 29.1 shows how companies operated at different levels.

THE COMPANIES

There were 650 non-profit housing companies in Denmark. They were generally small, with an average of about 700 units each. Almost all companies were locally based and closely linked with and influenced by the local authorities responsible for the area they operated in. Both these elements made tenants' democracy more possible. The local authority was always represented on the main board, with up to one less than half of the board members. Employees, unions and financial sponsors such as banks were also usually represented on the board if they had contributed to the company's finances. Nearly half of all non-profit companies were founded as co-operatives governed by members alone.

Management services

A number of the larger associations offered management services to small local associations. There was no general agreement about whether larger or smaller associations were more efficient or more responsive. However, costs rose with size in several cases due to the greater organisational complexity and also possibly to the more advanced services that became possible (National Federation of Non-Profit Housing Companies 1989). In spite of that, the management services contracts appeared to be a very useful way of helping small associations survive (Arbejdernes Kooperative Byggeforening 1991).

Local management

Systems and services in Denmark generally worked to high standards. Estate services were no exception. All estates, no matter how small, had locally based staff.

Custodial caretakers had a significant and respected role, carrying the main responsibility for keeping the estate and its services operational. Caretakers combined responsibility for repairs, grounds maintenance, cleaning, building problems, the maintenance of heating and other services, and tenant liaison.

A majority of maintenance was carried out through 10–16 year renewal programmes. By law each estate had to keep a repairs sinking fund for this programme and each estate board annually approved the programme prepared by the company. This prevented the kind of decay so common on British and Irish council estates, and kept even the worst estates in reasonable condition. The cost of management and maintenance was £1,250 per unit per year in 1989 (National Federation of Non-Profit Housing Companies 1989). Allowing for Danish wage levels, this was not much higher than British spending, a cost that was borne through higher rents. Most of the spending was ploughed into direct services, which were clearly visible on the ground – maintenance, caretaking, smart local offices and tenant services.

Character of estates

Two-thirds of social housing was built between 1960 and 1980 – a quarter of a million units out of a total of 426,000 social housing units. Almost all social housing was built in estates of over fifty units. The style of construction was conspicuous, combining non-traditional design and modern materials such as concrete or corrugated sheets.

In spite of the modern industrial style, estates were usually well landscaped, with plenty of greenery; were maintained to a very high standard by comparison with other countries, and were extremely clean and well cared for. In all, about 80,000 dwellings or one-fifth of the stock was built on large problematic estates, about fifty having 1,000 or more flats. Four-fifths of the stock was in smaller enclaves that were well managed, popular, and attracted a wide range of tenants. The beautiful fitted kitchens and bathrooms, the polished wooden floors, and the standards of decoration, all enhanced the attraction of non-profit flats.

Danish social housing provided extensive social facilities. By law, 3 per cent of space in social housing developments can be devoted to tenant activities – common rooms, classes, gyms, playrooms, activity and games rooms, tenants' offices, cafés. All estates, even small ones, had remarkable social facilities by international standards. This made 'tenants' democracy' easier to make into a reality.

RENTS AND HOUSING ALLOWANCES

In order to understand the way estate budgets and estate boards operate it is important to have an idea, in outline at least, of how rents were set and were evolving as building continued in the 1970s.

Rents and costs

There was no direct rent pooling in Denmark, though rents rose with inflation and older property eventually generated surplus funds. These surpluses were paid into the National Building Fund to help pay for new construction, modernisation and remedial work. The majority of the increased rents on older property went on repairs, so that Danish property was maintained to high standards. There was therefore not enough saving to pay for remedying all the defects of industrialised housing or for maintaining it as fully as the older, more conventional stock.

Rents on modern estates were very high, sometimes £100 a week or more. There were some estates where the costs of loan repayments, management and maintenance, simply did not match with rents people were prepared to pay. This acted as a severe brake on lettings. In these extreme cases (usually caused by a combination of intensely unpopular estate conditions and relatively high capital costs due to recent construction), additional loans were made available to keep rents at an affordable level.

Housing allowances

Almost all elderly tenants in social housing (about 60,000 in all) received rent rebates. Up to 70 per cent of their rent was paid. In addition, about 120,000 non-elderly households received some help, mainly families and in particular one-parent families, making the total numbers receiving help close to half of all social housing tenants. Non-pensioner households received rebates up to 25 per cent of rent. If they were dependent on income support they received that separately and paid their share of the rent out of it (Salicath 1987: 155). A family with two children earning £300 a week and paying £70 a week rent would get a rebate covering 14 per cent of rent, which would still leave the family paying 20 per cent of gross income – nearly 40 per cent of net income – on rent. Housing allowances were graded to help with children and other dependent members of the household (Salicath 1987: 155). All households eligible for assistance were required to pay at least 15 per cent of gross income in rent. In the case of higher income households, 25 per cent of income had to be paid. Where rents were higher than the limit of income for low-income households, a housing allowance would cover 75 per cent of the additional cost. The tenant would always pay for a proportion of the increased rent. Therefore, even the poorest households paid about 30 per cent of their housing costs.

This system made tenants aware of housing costs and responsive to differences in rent levels, which depended on a number of factors including floor size. Rents were charged by the square metre, as in other continental countries. This made large social housing units particularly unpopular, even though rents per square metre fell slightly with size. Estate-level budgets and rent-setting were real, because even the poorest tenants paid something. Tenants had a strong incentive to exercise some influence over financial decisions, precisely because they paid a lot towards costs, even when on housing benefit, and because rents had to cover the full cost of the housing service.

Housing allowances were part-funded by local authorities. Authorities thus also had a direct incentive to watch housing costs and to keep down the number of households dependent on allowances. Social housing rents could deter local authorities from nominating their full quota of lowest-income households because of the burden of contributions towards rent allowance payments. By the same token, the lowest-income households would often go for cheaper housing in the private-rented sector, if it was available, to reduce their housing costs. Both the local authority and low-income tenants had a powerful reason for wanting cheap private-rented accommodation to stay in existence on a substantial scale. The way access to social housing was organised became an increasingly important issue as pressures on private renting grew.

REHOUSING – A SOCIAL ROLE

There was no income limit or other bar on access to social housing. None the less, non-profit housing companies were set up to help those in need and therefore focused their efforts on housing lower-income groups. They particularly aimed to house families with children, giving them priority in their waiting lists. They had a parallel aim of building communities, providing for a broad range of incomes and integrating different types of households, young and elderly, single and family. For example, in 1992 there was a government quota of 2,500 new units for young people's housing.

The housing needs of poorer groups, such as one-parent families and minorities, meant that companies were asked to play a growing 'welfare' role. The training officer at the National Federation argued that the very term 'social housing' had become stigmatising because it implied housing for people in need and therefore somehow incapable of solving their own problems. The companies faced increasing problems of polarisation and residualisation with growing concentrations of poorer groups and an exodus of better-off groups (see Figure 31.1, p. 295).

Housing companies opposed a strictly low-income rehousing role, strongly defending their integrated housing policies and fighting hard to preserve their function as providers for a wide range of people. In order to do this they maintained a consistent attempt to provide good quality services, charging high rents compared with other countries.

Social mix

Because there was no direct income limit for people wanting to be housed, about 30 per cent of social housing was occupied by people who would not currently gain access on the basis of need. People were encouraged to stay as their incomes rose, as they helped economic viability and social stability. Thus social housing was not just an immediate solution to housing need but also a way of helping people gain greater economic and social stability. Moving out to make room for those in greater need was *not* considered a positive solution to housing problems.

Waiting lists

Most housing companies held a waiting list. Applicants could ask for particular estates but had to queue. Each company developed its own system. For example, in Arbejdernes Kooperative Byggeforening (AKB) tenants could be registered for up to five estates, though many only applied to one. Applicants could register with as many companies as they wanted, though in practice most people only registered with their local company. In major towns and cities there would be several local companies. Applicants had to pay a deposit of about 100 kroner (£10) to register on a waiting list. This was a way of keeping the waiting list for people who were sure they wanted rehousing. People were made offers of rehousing according to date order, based on when they registered, and people were told where they were on the list.

Property was normally allocated to the next household in the queue requiring the size of unit available, although social need could override the waiting list and companies were obliged to accept a proportion of nominated households. People applying for transfers within the company had first priority for empty units. This represented the explicit commitment of the housing companies to keep tenants satisfied within the company rather than drive them elsewhere; tenants could thus upgrade their conditions if they were originally forced to accept the worst property. Better property tended to go to transfers and worse property that had been vacated to incoming tenants.

Tenants' deposit

Although non-profit companies provide rented housing primarily for households who could not afford to buy, applicants were expected to contribute a share in the capital cost of the dwelling – normally 2 per cent of the value. This stemmed from the Danish co-operative tradition. Tenants were expected to leave their flats clean and ready for letting, and any costs to restore lettable conditions came out of the tenants' share.

Tenants on low incomes were offered easy-term loans for the obligatory 2 per cent deposit or 'stake'. They got the deposit back when they moved out, but the company could hold the deposit to clear any arrears and also to make good any damage to the property. The special loans to low-income households and their repayment were not the responsibility of the housing company but normally of the local authority (see below).

Local authority nominations

The local authority nominations could override the waiting list in a maximum of 25 per cent of vacancies for 'social need' or 'hardship' cases and in a further 30 per cent where there was urban renewal activity. Urban renewal nominations applied mainly to Copenhagen and other big cities. Local authorities often did not take up all their nominations and tended only to use them in extreme circumstances to prevent homelessness. As in Britain, local authorities were obliged to pay for the homeless in hotels if nowhere else was available, but local authorities in Denmark avoided nominating poor households if at all possible because of the high social rents. Private landlords provided cheaper housing for these groups.

The companies were supposed to respond to specially needy cases, even if they were not nominated. Local authorities had responsibility for ensuring that housing companies rehoused in accordance with these rules, with a balance between date order waiting lists, priority to families with children, and overriding nominations for a limited proportion of lettings.

For nominated households, the local authority had to underwrite the rent and make the 2 per cent deposit available through loans to the family. The incentives for social housing companies to house the needy, particularly those with lettings difficulties, were therefore quite significant.

Tenants' role

Tenants did not play a direct or specific role in allocations. The boards determined policy in line with government requirements, and local authority needs and nomination rights. Company staff ran the system. Estate boards did not have any direct jurisdiction over access.

None the less, particularly in the co-operatives and in companies with active tenant boards, pressures were strong, particularly to avoid over-concentrations of very needy or deprived households. On estates where there were major concentrations of ethnic minorities, tensions over allocations could be severe, particularly if the local politicians also played an active role. Conflict over allocations and over the social role of non-profit housing was growing in Copenhagen especially.

Thus, allocations were determined by a combination of housing and social need, ability to pay, eligibility for local authority assistance, political pressure, and willingness to move into the available empty property – invariably the less desirable because of the transfer system. The companies clung to social stability as a goal but their direct social role in helping the most needy tended to dominate. There was serious concern in housing companies and the National Federation that the requirement of tenant involvement and the legal responsibilities of estate boards and company boards might become more difficult if social housing continued to house more marginal people. Many pressures, particularly the continued loss of cheap private renting and the intrinsic unpopularity of some modern estates, were moving companies in that direction.

SUPPORT FOR TENANTS' DEMOCRACY

There were only very limited and exceptional circumstances where the rule of 'tenants' democracy' did not apply. Tenants' democracy had three main organisational supports: training, practical support, and professional housing staff.

Training

Training for the tenant board members was essential if they were to cope with the legal and financial systems. Within the local estate budgets there was an allowance for tenant-based activities which should include training. The National Federation of Non-Profit Housing Companies had set up a residential training centre in a large farmhouse in Jutland in 1970, accommodating sixty people at a time in single rooms and was in permanent use, teaching tenants how to run meetings, how to speak in public, how to prepare budgets, how to draw up planned maintenance programmes, and how to work with tenants from other cultures, etc. It also taught housing staff how to work with tenants as partners and how to improve management. Caretakers and repairs workers were also trained in housing management, tenant relations, service standards, equipment innovations, etc. The centre was self-financing. Each estate board and company allocated money for training.

Practical support

Small-scale, self-governing social and voluntary organisations appeared to flourish in Denmark. However, on some of the more depressed estates, the estate boards were not strong. Where there was a high turnover of tenants and an influx of people from different countries or from stressful backgrounds, reliance on self-organisation posed special difficulties. In spite of this, special interest clubs often did spring up. These clubs were helped by the special funding laws.

Some larger, city-based companies employed tenant-advisers. Companies also laid strong emphasis on local authority involvement, as they helped with social initiatives and facilitated educational initiatives.

About 90 per cent of all estates had functioning estate boards (Salicath 1987).

Professional housing staff

A third requirement was a balance between the power of the tenant boards and the professional housing staff. 'Do-it-yourself housing management' did not arise among tenants in Denmark. Staff were employed to manage rented housing. The law on tenants' democracy stated that estate boards did not have power to hire and fire staff. This had to be done through the administration which, in turn, was answerable to the elected board. Nor did estate boards control allocations. The relationship between boards and paid staff was a microcosm of the relationship between parliament and the permanent civil service. Tenants' powers were constrained by law and regulation as well as by the balance between staff and board. There were obvious reasons for this limitation – fear of bias, prejudice and favouritism; lack of experience and expertise; potential instability; staff opposition; the need for decision-making authority by the manager.

TENANTS' INFLUENCE

The role of tenants in social housing was far more developed in Denmark than in other EC countries in the study. According to Jespar Nygard, tenant activist and chairman of the AKB company, there were three main incentives for tenants to become involved in their company and estate affairs.

1   Most people feel some sense of identity with their local area and care about conditions there.
2   Many tenants express a desire to influence events, although they often find this difficult to put into practice.
3   People are naturally curious and like to know why and how things are done in their name.

Therefore identity, influence and information seem to offer many tenants a reason for at least registering an interest in their local area and, if they see a way, for getting directly involved. In addition, emergent leaders often enjoy the experience of a share of power and recognition. Figure 29.1 shows how tenants are involved in housing companies and how company staff respond to tenants.

At the estate level, tenants exert influence and have powers of representation. At company board level, tenants normally have a decisive majority and can exercise control within the limits of the law. The law is enforced through the ultimate authority, usually, though not always, the local authority. Company policy is implemented by the staff, who are answerable to the company board, rather than at estate level. This carefully evolved system of checks and balances is illustrated in Figure 29.1.

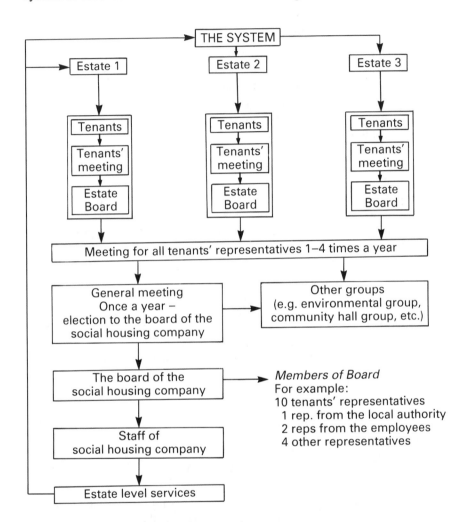

Figure 29.1 Model of the tenants' representation in a social housing company

Note: This diagram is based on the AKB non-profit housing company's structure.

Chapter 30

Conditions in the 1980s

At the beginning of the 1980s, the Danish economy was in deep crisis and, for the first time since before the war, unemployment was becoming a serious problem. Denmark's export trade was far slacker than its import trade – in spite of vigorous attempts to sell its off-the-shelf housing units abroad! Mass social housing estates in and around Denmark's main cities, Copenhagen, Aarhus, Odense, and Aalborg, were proving hard to let, hard to manage and costly. Boosting owner-occupation had diverted savings into more and more expensive housing, rather than into more productive investment. Popular, private flats with low rents for the standard of accommodation were still rapidly disappearing, and many of the poorest households, previously occupying the oldest and cheapest inner-city units, were being squeezed out into the low-demand mass housing estates. In the 1970s, small but not insignificant numbers of immigrant workers had come into Denmark and, by 1984, 2 per cent of the population comprised foreigners. This growing minority population, previously concentrated in the inner-city tenements, began to spread out and become more conspicuous. Racial tensions rose as Denmark's homogeneity was threatened. Housing management was stretched to its limits. The government acted in four ways: changes in support for owner-occupation; the development of private co-operatives; changes in private renting in urban renewal areas; and changes in social housing.

CHANGES IN SUPPORT FOR OWNER-OCCUPATION

From 1984, tax relief on mortgage loan interest was only given at the lowest rate of tax. Most house-owning Danes pay tax at higher levels, so this had an immediate effect. Repayments were also rephased so that part of the capital repayments were fixed throughout the life of the mortgage, making it more expensive at the beginning for new buyers. In addition, a special tax on loan finance was introduced. By 1985, 55 per cent of households owned their own home, but the rate of growth in owner-occupation slowed down for the first time for thirty-five years. In 1986, the government radically cut the incentives to owner-occupation, raising the threshold for first-time buyers and adding new taxes. Even higher quality detached houses were to receive lower subsidies. These moves were extremely unpopular with middle-class Danes as real incomes fell along with house prices.

The government clawed back privileges to the better-off by cutting spending on ever better quality private housing, though it continued to help inescapably poor households through social housing.

The package of economic measures, including the cuts in owner-occupier support, caused inflation to fall to a very low level. This reduced popular expectations of the gains to be made from owner-occupation. The effect of the cuts in tax concessions and the plummeting of inflation caused house prices to fall in real terms, paralleled by a fall in real incomes through other anti-inflationary measures. Therefore, a fear of buying, an inability to meet repayments, and difficulties in selling cut demand for owner-occupation drastically. The measures succeeded beyond anticipation. A knock-on effect was a rise in demand for social housing and increased attempt to prevent the loss of private-rented units.

There was a steep fall in the volume of private house production from 15,000 units in 1985 to 4,000 in 1988 and 1989. Owner-occupation may continue to grow slowly and to house the majority of Danes, but it is unlikely to recover the overwhelmingly dominant position it seemed set to enjoy at the turn of the 1980s.

The increase in row and cluster housing was an important by-product of steeper costs. There was a great expansion in condominium housing (small blocks of owner-occupied, purpose-built flats) and private co-operatives, which aimed to bridge the gap between owning and renting.

DEVELOPMENT OF PRIVATE CO-OPERATIVES

A second major housing initiative of the 1980s was the encouragement of private co-operatives. These were different from 'social housing' co-operatives in that they were largely privately financed and each member bought a 20 per cent stake in the property. Their stake could not exceed this, but they could sell their 'share' at market value when they moved on, although the resale value of the capital stake was still a relatively uncertain quantity in the early 1990s. The government helped promote this 'mixed' tenure from 1981, through favourable index-linked finance, at a time when support for owner-occupation was being cut and private renting was declining. The aim was to help people who could afford some, but not the full, housing costs of owner-occupation. It was an attempt to satisfy demand for quality individual homes in a more economical and environmentally sensitive way than previously. Private co-operatives aimed to rebuild the informal economy, social networks and mutual support on the basis of individual effort.

The private co-operatives were generally either purpose-built 'cluster' houses or converted old tenements. They often formed part of mixed tenure schemes including owner-occupied and social housing too. Although popular, they sometimes experienced difficulties in managing common facilities under this structure.

In the wake of the economic crisis they were seen as a possible in-between tenure, creating new housing, reducing demands on the state for services, and being affordable to people now excluded from more traditional and expensive owner-occupation. They appealed to people able to afford more than the straight rent.

Co-operative units made up over a quarter of all house production at the end of the 1980s. The government had a target of 3,000 new private co-operative units a year in 1992. Groups were forming faster than help was available. Most co-operatives were small, involving 30–60 households. In all, there were 112,000 private co-operative units in 1990 making up nearly 5 per cent of the total housing stock. They were supported by the National Federation of Non-Profit Housing Companies, even though they formed an 'in-between' tenure with the possibility of individual gain.

The advantages of private co-operatives were:

- They were cheaper to the government and the individual than owner-occupation.
- They involved less direct subsidies from the government than social housing.
- They offered some of the control of owner-occupation.
- They offered more economical housing than detached or large-scale units.
- They offered communal facilities and co-operative ideals for those who liked them – a popular concept in Denmark.
- They were an energy- and building-efficient form that was also attractive.
- They involved small groups working together.

There were some disadvantages:

- Very low-income households could not afford a 20 per cent share.
- Little was known about the sale value of a 20 per cent share if many such units were built. People were attracted in through lack of alternatives. They could end up stuck there.
- Their real cost in lost tax because of the financing of index-linked loans was much higher than the government subsidies implied.
- Many social housing companies did not like private co-operatives as they attracted away from social housing people with 'psychological reserves' (Boligministeriet 1989). They also disliked the shift away from a purely non-profit objective.

- Existing owner-occupiers did not like them either as they attracted away potential buyers, further depressing the market.
- Private co-operatives were seen as unfairly favouring those who could get in.

However, private co-operatives appeared set to grow further, as an innovative and small-scale compromise between individual and social ownership and control. Their greatest appeal was their lower direct cost to purchasers and government.

CHANGES IN PRIVATE RENTING IN URBAN RENEWAL AREAS

In the old private-rented areas each flat was usually small and self-contained within large blocks. The blocks were often solidly built – rather like the Peabody flats or the early London County Council buildings. Therefore, they survived remarkably well and continued in use without significant reinvestment. It was this that made them cheap, relatively manageable and popular with groups needing small, economical, centrally located housing.

Copenhagen's large square tenements with multiple flats, common, wooden stairwells and shared plumbing, gas and electric supplies, did not always lend themselves to conversion. They often involved a large number of tenants in one block – maybe 400 – and required significant structural repair, e.g. to main joists on roofs. If there was timber decay, then the problems could be insuperable and demolition might be the only solution.

Forty-three per cent of the poorest households still lived in private-rented housing in the late 1980s (Boligministeriet 1987b, 1989). Its tenants comprised disproportionate numbers of elderly households, young people leaving home, immigrant workers and refugees. The problems of ageing stock, increasing structural decay, obsolescent facilities, low rents and inner-city congestion, made private renting unattractive to economically secure families compared with a house in the suburbs. On the other hand, the growth in single person households, particularly young, newly formed households, and increasing energy and transport costs, made inner-city locations more attractive and gave private landlords a ready market.

Until recently, minimal slum clearance helped retain the private rented sector, with few direct subsidies. Private landlords could claim depreciation against tax and some tax relief on repairs and management costs. They were also able to claim energy conservation grants. These factors, coupled with low running costs, encouraged them to stay in business. The law in 1976 preventing the sale of private-rented accommodation to owner-occupiers stabilised the sector after a period of rapid decline in the 1960s and early 1970s, when the gains from owner-occupation became far greater than those attached to renting.

Tenants could form a private co-operative to take over ownership if the owner wanted to sell up. They preserved a form of renting while giving private tenants a greater stake and allowing reluctant landlords to withdraw. They offered part individual ownership, part collective ownership. Tenants in urban renewal areas had the right after renovation to form a private co-operative or a non-profit co-operative society to take over the block of flats they lived in, as long as the majority supported the proposal.

Special funding for urban renewal

From 1983, the government increasingly encouraged urban renewal through special funding. Local authorities and housing societies, particularly in Copenhagen, were active in clearing and replacing some of the most run-down areas. Renovation and piecemeal renewal gained in popularity against demolition, partly because of cost, partly because of resident opposition to demolition.

The Vesterbro area of Copenhagen, right behind the central railway station, was the most ambitious renovation scheme. Poor conditions, opposition to renewal, and conflict with residents led to an incremental approach. The area was the biggest and most renowned slum area of Copenhagen, only five minutes' walk from the elegant town hall square. Conditions there were dismal by any modern standards. The shops were almost all run by Asians and Turks, creating a sense of both a ghetto and an alien world to Danes. The inner courtyards of the tenement buildings, surrounded on four sides by high, packed flats, were deserted of people but full of bikes and rubbish bins. There were few young children to be seen anywhere. The streets were dirty, unlike almost any other housing or commercial areas in Denmark. There were tourist hotels around the edges of the area, creating a major pressure for its renewal.

The few people on the streets were mainly drug-addicts of North European origin, or Chinese, Indian, Pakistani, Turkish and Moroccan men. There were few women to be seen, although the area was renowned for prostitution. There was graffiti on some walls. A plastic bag, full of excreta, had been thrown at an entrance door to one of the stairwells and stuck there, hanging off the window. It seemed a bitter gesture to a minority household. The atmosphere of the area was tense and hostile. On one corner outside a cafe, there was a large gathering of young addicts on the street. Strangers were accosted for money. Police had been trying to move the 'junkies' on without success. There had been street clashes. The Danish National Urban Renewal Company had been trying to establish plans for the renovation of the area, with selective demolition of houses suffering from dry rot. The block structure made it difficult to remove a single house, as the tenement flats were built around an integral courtyard structure (Danish National Urban Renewal Company 1989).

Vesterbro was the worst example of slum private-rented housing (National Federation of Non-Profit Housing Companies 1989). But it was easy to see that the area was needed in its present form by many low-income groups as a refuge or a first stop or a marginal area. There was strong defensive action against take-over, particularly by the young 'drop-outs'. Renewal plans were caught on the horns of a dilemma – where would the residents go if it was renewed; how could they stay in a smart and expensive central district if it was renovated? As a result, progress was extremely slow.

Some of the unpopular 1960s and 1970s estates with vacant units in the early 1980s took disproportionate numbers of urban renewal rehousing cases. But vacancies through this route were less and less available. High costs, resident opposition and a shortage of cheap housing, made large-scale renewal less and less likely.

The survival of a large, old, private-rented sector looked likely for some while, though the conversion to private co-operatives was an avenue of change. Low-cost, private renting looked more and more attractive to the government, the occupants and the population as a whole, because there were too few alternatives for many individuals and groups. Therefore, there was new talk of improved subsidies to private landlords.

CHANGES IN SOCIAL HOUSING

Danish governments, even the Conservative government of the early 1980s, encouraged social housing when private production lagged seriously. The government relied on social housing to meet urgent needs. An expanded social house-building programme in the early 1980s helped to keep the building industry operational during the recession, alleviating some unemployment and generating economic activity. The squeeze on owner-occupation created new demand for social housing among more marginal households, while the loss of private-rented units and urban renewal created demand among very low-income groups. The style of social housing had changed dramatically over the post-war period. More recent estates were limited to between 40 and 150 dwellings. Cluster-style houses with brick finishes were strongly favoured. Industrialised building methods were adapted to fit new needs. Renovation was popular and traditional styles were enjoying something of a come-back.

Under the impact of the recession, the total number of new housing units dropped dramatically, but social housing was protected from the worst effects of the recession as Table 30.1 shows.

Table 30.1 The trend in housing developments in Denmark, 1970–84

	1970	1973	1978	1982	1984
Total units built by social housing companies (%)	30	24	18	54*	36
Units built as detached houses (%)	48	61	65	18	32
Units built in clusters, terraces or semi-detached (%)	10	7	18	46	41
Built in blocks of flats (%)	42	32	18	37	27
Total number of units built	50,600	55,600	34,200	20,800	26,800

Source: Boligministeriet 1987b.
Note: *Large rise due to recession and building slump – government intervened.

Financial reform

Generally, the Conservative governments of the 1980s introduced more stringent financial regimes for social housing. Some specific measures made the system more responsive. In 1982, the government introduced index-linked finance for social housing in order to reduce the cost of new-build and renovation. This had the effect of reducing rent levels for new or renovated housing in order to increase demand and make it attractive to tenants in work, as well as affordable to those dependent on benefit.

Special housing

Other initiatives in the 1980s included an innovative and ambitious programme of housing for the elderly, introduced in 1987. This housing was to be integrated into wider neighbourhoods to avoid any hint of segregation and was to help elderly people stay in their own adapted homes as long as possible.

The elderly (over 65) population reached 750,000 in 1990 and is expected to reach 1,000,000 by 2020. Therefore great importance was attached to modifying the sheltered housing concept to allow maximum flexibility and to reduce costs. Local authorities had responsibility for allocating funds and nominating elderly applicants, but developments were generally carried out by social or private companies. Altogether in 1991 there were 109,000 'elderly units' – over a quarter of all social housing units.

There was also a special allocation of funds for young people's housing and about 4,000 units were being provided each year by the end of the 1980s.

Table 30.2 summarises the main housing developments from 1958 to 1990.

Table 30.2 Main housing developments in Denmark, 1958–90

1958	Tenants' Democracy introduced
1958	Rent controls relaxed
1966	Reduction in incentives to owner-occupation
1967	National Building Fund set up
1968	Danalea founded to develop prefabricated housing for home and export
1969	Urban Renewal Agency set up; strict land zoning introduced
1970	Local government reorganised
1970	Introduction of estate boards
1976	Control on sale of private-rented flats
1981	Government support for private co-operatives
1982	Indexed financing for social housing
1984	Strong reduction in support for owner-occupation
1984	Tenants gain majority on non-profit company boards
1985	Rescue programme for unpopular estates
1987	Housing for elderly initiative

Chapter 31

The rescue of mass housing areas

In 1985, the government identified eighty-one estates, involving 25,000 social housing units, which were in serious difficulty, based on a Commission it had set up in 1984. This represented 5 per cent of the social housing stock, but up to three times this number were in some difficulty and needed special treatment (Vestergaard 1988). The most distressed areas were to have their loans rescheduled through index-linked finance in order to release money for major repair and renovation.

The problems of unpopular estates and the cost of rescuing them had generated widespread disillusionment with 'mass' social housing. But the rescue programme on the difficult estates attracted a lot of resources. This was primarily because of the evidence of growing polarisation and the isolation of the industrially built estates, an outcome that was anathema to the social and political make-up of Denmark (Vestergaard 1988). It was also because of the greatly increased need for the estates.

LOW DEMAND

Lettings and management problems quickly emerged in areas that were intrinsically unpopular. In the period of low housing demand in the early 1980s, far more households with acute social problems gained access to social housing than previously. This was largely because the volume of empty units greatly exceeded the size of the quota available to local authorities. At the same time, major urban renewal schemes in Copenhagen were becoming more important. This meant that groups previously excluded gained access to non-profit housing through nominations, particularly recently arrived immigrant communities. Households displaced by urban renewal were entitled to have their new social rents subsidised to the old private level for five years after moving to compensate for the generally lower private rents in old property. This rule made urban renewal nominations popular with housing companies with lettings difficulties, since rent income was assured. It also made social housing more affordable and therefore acceptable to very low-income households. The combination of vacancies, urban renewal, and increasing need, tipped the balance of social housing sharply in favour of poorer groups.

SOCIAL CHANGE

Figure 31.1 shows growing under-representation of the top three socio-economic groups in social housing; a big decline in the proportion of middle-income groups and skilled workers; an increase in the proportion of unskilled workers; and a major leap in the over-representation of unemployed households. In the Danish context, this shift was extremely alarming. As soon as demand picked up in the mid-1980s, housing societies were only too keen to limit access for minorities and very low-income or unstable households and give priority to better-off households at their expense in order to redress what they saw as a damaging decline in the social mix on some estates (Vestergaard 1988, Boligministeriet 1987b).

Increasing Polarisation

Table 31.1 emphasises the changes in rehousing patterns between tenures over fifteen years. The changes had a dramatic effect on the role, status and stability of social housing. Only one-third of households comprised married couples with children, whereas this used to be the norm (Salicath 1987). One-parent families, low-income households, unemployed households, and single people were all seen as particularly vulnerable groups and greatly increased in numbers.

The working-adult age-band was increasingly under-represented. Established households tended to move out. The new incoming households tended to have more children than average or to be young, single or elderly.

In 1970, social housing tenants had more members employed than private renting and nearly as many as owner-occupiers. By 1985, 40 per cent of social and private tenants had no working member. Owner-occupiers were in a much more favourable position, with half of all households having two working members and a further quarter having one working member.

No figures were available for the rehousing of ethnic minority households. However, in 1970, it was almost unheard of for minority households to live in social housing. By 1985 there were large concentrations of minorities on some of the most difficult estates – up to 40 per cent in certain areas of Copenhagen and Aarhus. A combination of low demand, high vacancies and urban renewal had changed both the population and role of social housing from being part of general housing provision to being increasingly an option for households who did not have a secure position in the Denmark of the 1980s.

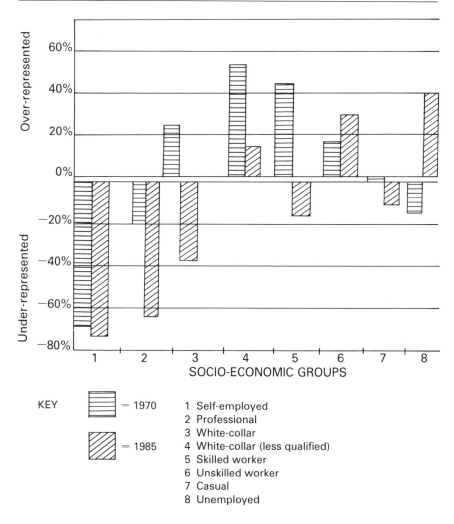

Figure 31.1 Under- and over-representation of different socio-economic groups in Danish social housing in 1970 and 1985

Source: Boligministeriet 1987b: 86

GOVERNMENT COMMISSION ON PROBLEM ESTATES

The government quickly mounted a rescue operation in two stages. The government commission in 1984 had looked at the causes of the problem and proposed solutions. This collected detailed evidence from seventeen of the most distressed estates, identifying key elements in the decline:

Table 31.1 Under-representation and over-representation of households with different characteristics in Danish social housing, 1970 and 1985

Type of household	1970	1985
Single	+10	+45
Families with children:	+24	−8
Couples with children	+22	−35
Single parent with children	+126	+161
Household without working member	−9	+42
Households with income below		
150,000 Danish kroner p.a.	+28	+45

Age groups	1970	1985
0–21	−6	+27
22–24	+13	+13
25–34	+27	0
35–49	+16	−15
50–64	−14	+4
65+	−27	+9

Changing proportion of households with no workers, with 1 or with 2 workers in different tenures

Tenure	Number of workers	1970 (%)	1985 (%)
Social housing	0	20	39
	1	42	35
	2	30	23
Private renting	0	28	40
	1	44	38
	2	24	19
Owner-occupation	0	28	17
	1	38	24
	2	34	50
All	0	22	27
	1	40	30
	2	30	7

Source: Boligministeriet 1987b.
Note: These measures of over- and under-representation are in comparison with the population as a whole.

- 'the poorest areas had sunk very fast because of the economic independence' (ring-fenced budgets);
- some large-scale developments were 'visually repulsive';
- stairways were 'cold, utilitarian, unhomely';
- hallways were 'spartan and devoid of imagination, desolate and a source of conflict';
- play areas were 'poor, few, not for all age groups, unsheltered' (from cold Danish winds);
- the 1960s 'love of repetition, size, simplicity, functionalism' was unsuccessful.

Some of the most serious problems to emerge were:

- a large increase in female-headed households and disproportionate numbers of children;
- overcrowding in smaller flats in contrast to lettings problems for large flats, often because of costs;
- isolation because of lack of cars;
- tensions between groups, leading to withdrawal and apathy;
- tensions between staff and tenants;
- the collapse of some estate boards;
- resentment from local politicians;
- increasing vandalism and general physical decline;
- racial tensions arising from the recent access by minorities, particularly Turkish households with visibly different lifestyles and appearance, to social housing;
- a decline in standards of management and supervision as problems accelerated;
- acute stigmatisation of the estates;
- wider problems, for example in local schools.

(Vestergaard 1988)

The housing companies were trapped in a dilemma. Much of their stock was built in a style and on a scale that had become unpopular. Rents were higher on new estates than for older flats, which were becoming more attractive to lower income groups as problems mounted on the new estates. Rents therefore deterred marginally better-off households, as well as poorer households. Companies then had to rely on local authority nominations. Rent incomes dropped. Nominations carried 'guaranteed rents', but very poor households still had to pay a proportion of their rents. Arrears mounted. Should companies risk arrears and spiralling decline in status and demand or should they keep some properties empty while they tried to attract better-off households? Traditionally, Danish housing companies had given top priority to financial viability and had screened households for ability to pay. They were now faced with a totally unforeseen and new situation. The immediate reaction was a 'siege mentality' and a 'fire-brigade strategy'. Some companies tried to block lettings to more than a small percentage of minorities; some, threatened with bankruptcy, maximised their nominations. The worst estates faced a crisis of financial and social survival.

SOCIAL INTEGRATION

There was a consensus about the primacy of social and lettings problems: 'The number of non-resourceful tenants has increased and is definitely one of the reasons for the problematic situation... The selection of applicants is the only solution. Social stability is more important than ability to pay...' (Kristensen 1989a).

Tenant board representatives on the worst estates were often the strongest lobbyists for social integration. They knew that the economic survival of their estate depended on it. They were also frequently demoralised by the passivity or alienation of more depressed households. While having no choice but to accept a significant level of nominated households, they were seriously concerned at their ability to keep the estate functioning. The housing companies shared this concern and fostered the aim of social mixing with the backing of government. One of the key features of the rescue was to attract more self-sufficient tenants to the most decayed areas. There was often also a political and racial dynamic, with Turkish households and other minorities seen as a direct cause of decline and growing stigma.

FINANCIAL RESPONSE

From 1978, limited funds had been available for repairing building defects and upgrading environments. But this had not stemmed the growing problems. Making the estates attractive was seen as pivotal to drawing in more secure tenants. The housing companies cried out for a major cash rescue.

In 1985 the government applied the index-linked finance system, established for new-building, to outstanding debts so that a margin of spare cash could be immediately created that could be used to finance a further indexed loan for renovation and improvements. The additional loan would be repaid on an extended index-linked basis so that only minimal rent rises would be necessary at the outset. The effect of the new arrangement was that 60 per cent of the existing rent could be divided between:

1   payment of the outstanding existing debt;
2   payment of the new indexed loan for financing the improvements.

The other 40 per cent continued to go on management and maintenance.

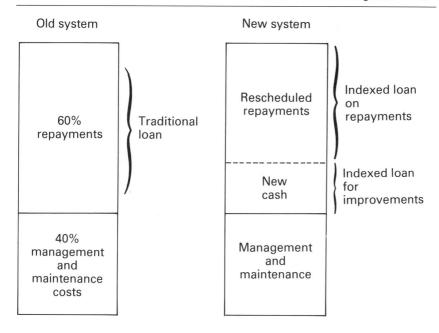

Figure 31.2 Funding arrangements for special programmes on target estates in Denmark, 1989

Source: Arbejdernes Kooperative Byggeforening 1989, based on the experience of Tastrupgaard

The refinancing allowed special funds for building defects that related to industrialised construction, such as sprawling concrete and damaged flat roofs. Conversion of unwanted large flats into smaller units was another priority. Social initiatives were also supported. Altogether, 2 billion Danish kroner were allocated in 1985 to finance the conversion of traditional mortgage loans into index-linked loans. This special measure to help companies facing extreme difficulties on certain estates was available only for a limited period and was restricted to the recent estates with major structural problems.

As a result, major rent rises were avoided – rents actually fell on some estates where previous loans had forced uneconomically high rents – and cash became available to rectify some of the problems. The extra money was targeted at the external problems of the buildings, communal areas and facilities, and the environment. In most cases, the dwellings themselves were popular and attractive, with a very high standard of amenity and maintenance. They presented few internal physical problems, only excessive rent levels.

Today, the National Building Fund and the Building Defects Fund, to which all social housing bodies are obliged to contribute, channel support to problem housing. Figure 31.2 illustrates the way rescheduled loans created additional resources.

In addition to the refinancing, which prevented rent rises, rents were actually reduced for a few of the largest and most difficult-to-let flats on the worst estates. Rooms were blocked off or bricked up to reduce the floor space and therefore the rents. Occasionally, an additional flat could be produced.

FOUR-DIMENSIONAL APPROACH

The aim of the new resources for unpopular estates was to encourage the housing companies to combine with local authorities and the government in addressing four interlocking targets:

- physical upgrading;
- financial solvency;
- management responsiveness;
- social development and integration – including employment and training initiatives.

Implementation

- Improvements were popular with tenants – the use of bright colours, attractive entrances and lamps, changed appearances radically.
- The management of the housing companies had been localised and expanded at the estate level.
- Support for the tenant boards had been enhanced by the rescue programme. Tenant activities had increased, though often only with considerable outside support.
- Turnover of tenants had plummeted, partly, according to experts, because of the knock-on effects of the 'crisis' in owner-occupation; empty dwellings had all but disappeared, and waiting lists had expanded significantly. This had transformed the financial viability of the companies and had helped to make estate boards workable.
- Access for the poorest groups, and particularly minorities, became severely restricted as general demand rose. Some housing experts believe this was a very important factor in enhancing the viability of areas that had been in steep decline (Vestergaard 1988).

The four-sided approach recognised that problems were complex and interlocking. There was no simple cause and effect. A number of key issues quickly became clear:

- physical solutions might be a prerequisite, but of themselves they did not overcome the problems of these estates;

- estate budgeting protected the mass of social housing from a spiralling effect and quickly isolated and highlighted the problems, but it did not solve the cash problems of the estates in greatest difficulty;
- financial solvency and high-quality management were paramount, but could not be achieved in ghetto housing areas without special support;
- retaining or restoring a social mix in each housing estate was a most fundamental condition of success;
- front-line staff could not solve complex social problems without additional resources;
- tenants' democracy would not survive great social instability;
- poor areas needed a supportive local network which was broken by constant turnover;
- non-profit housing companies might not be strong enough as social support agencies where tenants themselves faced wider social breakdown.

(Kristensen 1989a, 1989b)

OUTCOMES

To outside eyes, the Danish estates in the Rescue Programme looked spruce. The level of tenants' activity was impressive. The management problems seemed less serious than in other countries. Social tensions had not displayed themselves in major disorders, even though there was constant reference to vandalism and crime. Physical conditions and state of repair were high by international standards.

The government commission report (1987) concluded that social housing would continue to have problems requiring constant attention because it attempted to house people in real difficulty. There were no 'once and for all solutions'. The most stigmatised, isolated and ugly 'mass' estates would continue to pose serious difficulties. Over half the non-profit stock was in this form and the growing needs of marginal groups would most likely be met in unpopular estates. For these reasons, in Denmark there was as much emphasis on management – organisational, financial and social – to provide long-term support for the estates as there was on physical remedies.

A revival in demand for social housing from better-off groups offered the prospect of restoring the fortunes of unpopular estates, precisely because it ensured the place of social housing as a resource for mixed communities. This wider goal still attracted support at all levels of Danish society.

Chapter 32

Post-war housing achievements

Denmark was transformed from a well-developed, village- and craft-based agricultural society into a modern, international, industrially successful urban society in the post-war era. House-building reflected that growth and change, with a high output right up to 1985. Both social and private housing grew rapidly, as Table 32.1 shows. The proportion of total output contributed by social housing varied widely, as Figure 32.1 illustrates, being lowest when total production was highest. Tenure changed significantly, away from private and local authority renting in favour of non-profit, owner-occupied and co-operative ownership, as Table 32.2 shows.

The number of dwellings has almost doubled since the war. Nearly 1 million new owner-occupied houses have been built. As private landlords declined, social renting more than made up. By 1991, there were slightly

Table 32.1 Social and private house-building in Denmark between 1950 and 1989

Years	Social housing	Private housing	Total	Social housing as % of total
1950–54	42,000	72,000	114,000	37
1955–59	47,000	76,000	123,000	38
1960–64	40,500	128,500	169,000	24
1965–69	57,500	161,500	218,000	26
1970–74	64,500	191,500	255,000	25
1975–79	37,000	145,000	182,000	20
1980–84	43,500	82,500	125,000	35
1985–89	40,000	40,000[a]	80,000	50[b]
Total	372,000	897,000	1,266,000	29
Average per year	9,300	22,425	31,650	–

Source: Boligministeriet 1987b.
Notes: [a] This figure includes co-operatives. [b] Estimates based on Danish government figures 1989.

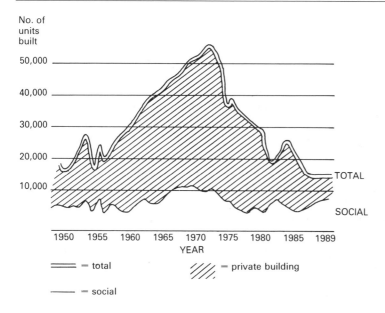

No. of
units
built

Figure 32.1 Housing units completed in Denmark per year, 1950–89 (total
output and that of the social housing sector)

Source: Boligministeriet 1987b

more social rented than private rented units (National Federation of Non-
Profit Housing Companies 1991).

The Danish Ministry of Housing believes that overall shortage is a thing
of the past.

Table 32.2 Danish housing stock by tenure in 1950, 1970 and 1991

	1950		1970[a]		1991	
	%	Numbers	%	Numbers	%	Numbers
Non-profit social housing	5.3	68,000	13.9	255,000	17.9	426,000
Local authorities	3.2	40,000	3.3	60,000	3.0	70,000
Private co-operatives	(not known)		(not known)		4.7	112,000
Private landlords	59.0	750,000	36.5	670,000	16.3	386,000
Owner-occupiers	32.5	413,000	46.3	850,000	58.1	1,381,000
Total	100.0	1,271,000	100.0	1,835,000	100.0	2,375,000

Sources: Boligministeriet 1988f, 1989; National Federation of Non-Profit Housing
Companies 1991; Danish Statistical Office 1992.
Note: [a]Estimates based on information from Boligministeriet 1988f.

CONCLUSIONS

Danish housing policy had evolved to incorporate many special features by the beginning of the 1990s.

Owner-occupation no longer enjoyed total primacy over other tenures but continued to represent a popular choice for families and economically secure households. Many owner-occupiers faced falling real values for their property, difficulties in selling and high costs. New households and first-time buyers faced new barriers to purchase through much-reduced tax concessions.

Private renting provided cheap, old, inner property which was desperately needed, but was being lost too rapidly due to controls, costs of renovation, and lack of special incentives. It was likely to receive growing support but increasingly would be converted or replaced, often in the form of private co-operatives or social housing.

Private co-operatives provided a mixed tenure solution that combined low cost, quality and communal provision. These were a major source of new, state-supported housing, although their future role was far from certain.

Non-profit housing companies offered intensive, tenant-controlled responsive management and high-quality maintenance for large groups of needy households. They would play a growing role in inner-city renewal and their unpopular 'concrete' stock would house more and more marginal households because the outer estates were often intrinsically unattractive and because poorer groups were squeezed out of older housing as it was renovated. A form of 'market-conscious management' helped contain the most serious problems.

All sectors of the housing market were closely regulated, offering high standards by international measures and offering many unusual features, such as resident control, communal facilities, an individual stake in ownership, strict energy conservation and high levels of repair and services.

But Denmark had not escaped common international problems:

- growing tensions resulting from the concentration of low-income minority households in run-down inner areas and increasingly in large social housing estates;
- serious physical, social and management problems on the industrially designed and built flatted estates;
- a growing distance and dislocation between the system of government and the needs of low-income and vulnerable households.

The stigma of the worst estates, the difficulties in keeping them financially viable, and the unstable conditions in the most marginal areas, contradicted the general stability and smooth operation of Danish society. The government would continue to support social housing in order to contain some of the growing problems among poorer groups on the basis of a strongly equalising tradition. But the squeeze on incomes, resulting from wider economic measures and a very low level of inflation, was creating political tensions that were expressed in increasingly racial terms. A steep rise in unemployment, which accompanied economic change, made for even greater strains.

Housing companies faced an uphill battle in keeping anything like a cross-section of Danish society in some areas. There would be growing pressures and problems as tenants' democracy lived through the problems of exclusion and under-representation by marginal groups.

The cost of constantly trying to put these problems right would be high but lower than the cost of facing trouble on the streets. So far Danes have been willing to pay this price and their problems have consequently remained more manageable than larger neighbouring countries with more complex social and political systems.

The Danish model of tenant involvement, co-operative organisation, communal social provision and citizen control can offer some useful and inspiring lessons for more heterogeneous societies lacking a strong co-operative tradition.

Giving people a stake, however small, in their housing; involving them in decision-making; building and retaining mixed communities; providing training to tenants' representatives; organising management at the front line, estate by estate; having local budgets to focus services and decision-making – these are all ideas widely accepted in the international housing world of the European Community.

Denmark, as a small, relatively homogeneous and industrious society, has executed these principles, applied the ideas organisationally, and adapted the co-operative tradition – originating in Rochdale, Lancashire, not Aarhus, Jutland – to the urban housing problems of Denmark. The successful application of creative ideas in Denmark seems to underline a peculiar national characteristic of Scandinavians; but the fact that these ideas have been implemented successfully makes the Danish housing experience particularly relevant to the European Community as a whole.

Part V

Ireland

In the particular we understand the universal.
James Joyce, *A Portrait of the Artist as a Young Man*

The decision to include the Republic of Ireland[1] in this study of European social housing was not obvious at the outset. (Note that the Republic of Ireland, formerly Eire, is referred to throughout as 'Ireland', with the six counties being referred to as 'Northern Ireland' or 'Ulster'.) Experts commented that Ireland was too small, too marginal and too dependent on Britain to be worthy of separate study. Its social housing is modelled on the British local authority ownership pattern.

However, there are a number of reasons why Ireland is of particular interest. Firstly, its links with Britain are so long, so tenacious and so abrasive as to make its experience interesting. The continuing problems and unresolved conflict in Northern Ireland make the whole situation in the Republic of singular importance to Britain. The almost uninterrupted outflow of Ireland's people from 1845 onwards to Britain and America has continued to colour Irish development, making its internal struggles international in significance.

Britain's role in Europe and the world has frequently been over-shadowed in the past 200 years by her controversial relations with Ireland. The problems of Ulster and its bitter sectarian divides have taken up a great amount of energy, while posing problems, largely unrelated to housing issues within the Republic. The failure to include Ulster's housing structure and problems in this study of Ireland was a serious but calculated omission, dictated by space and the complexities of the issues.

The ties between Ireland and Britain remain strong, while Irish ties with America are fuelled through Irish–American political and financial links with Irish nationalist movements, leading to complex trans-Atlantic relationships which make its experience important in the context of Europe to which it has firmly joined itself. Thus Ireland's unique political context makes her place in Europe also unique.

Ireland's peculiar housing structure compared with the other countries in the study also makes Ireland fascinating. Ireland has by far the highest rate of owner-occupation of any EC country, now running at 80 per cent and still rising. It also has some of the most run-down and marginalized public housing estates in Western Europe, reminiscent of the deeply blighted and semi-abandoned inner areas of Liverpool or outer Glasgow. Ireland's poverty is disguised by the well-housed majority, but shows up in the bleak, uniform and decayed social islands of local authority housing estates – the crowded, inner-city balcony blocks of Dublin and the acres of cheap mass housing estates, some industrially built, on the outer edges of Dublin and Cork.

Ireland's late entry into Europe belied a deep commitment. From the outset, Ireland saw Europe as a possible way forward out of its unhealthy dependence on Britain, its festering nationalism and its poverty. It had little to lose in terms of industrial competition and possibly much to gain as an investment opportunity. Ireland's strong commitment to Europe and its dependence on European policies has changed Ireland's self-perception from being an adjunct to Britain into an adjunct to Europe. Irish pre-occupation with Europe is beginning to overshadow obsession with the North and with Britain and is provoking ambitious proposals for high-speed ports in Waterford and Cork to link directly with France, in competition with the Channel Tunnel which is seen to give Britain yet another huge reason for bias against Ireland. Ireland's historic ties to the Continent, through religion and through anti-British sentiment, make her a strong European partner.

Within the European Community there are sharp contrasts between conditions in the powerful and wealthy pivotal nations of France and Germany and the small, peripheral nations, such as Denmark and Ireland. Understanding something about a much poorer, more rural and more economically dependent 'region' of the European Community, especially one with as divided and tragic a recent past as Ireland, may offer insights into development and co-operation that will shape our thinking. If Ireland's past has influenced Britain and America, its present and future may influence Europe.

Ireland comes last in this study for two reasons. Firstly, it would be hard to understand Irish housing developments without first understanding the British history which so dominated it. A second reason for discussing Ireland last is the stark difference in economic, social and housing conditions between Ireland and the other four countries in the study.

Chapter 33

Background

FACTS ABOUT IRELAND

Ireland is different from the other countries in the study in a number of important ways. Apart from Luxembourg, Ireland is the smallest country in the European Community with only 3.5 million people. After more than a century of decline, her population has been increasing faster than the other four countries in the study since 1960, though it is projected to decline slowly after 2000. The trend is largely explained by a higher than average birth-rate. However, the birth-rate has fallen rapidly from four children per woman in 1965 to 2.5 in 1985. Ireland is still a very youthful society with nearly half the population aged under 25 and one-third of all residents under the age of 16. Ireland's family pattern is moving rapidly towards the rest of Europe but it is still significantly different from Britain and Continental countries.

There are four central features which single Ireland out:

1   The population density, with only fifty-one inhabitants per square kilometre, is much lower than the other four countries.
2   Nearly half of Ireland's inhabitants live in rural areas or villages – a much higher proportion than in the other four – and one-third still work in agriculture. Rural areas are losing population, however, and this trend is almost certain to continue.
3   Unemployment, particularly long-term unemployment, is very high with almost one in six of the workforce out of work in spite of extraordinarily high emigration. Ireland has less than half the income per head of France, Germany or Denmark.
4   Immigration and ethnic minority issues, which dominate so much of the debate around social housing in the other countries, barely figure in Ireland. Constant emigration and high levels of unemployment make non-Irish immigration very unusual. Ireland is ethnically extremely homogeneous. Emigration, not immigration, dominates emotive political debates. Table 33.1 shows the continuous losses from the earliest official records. Prior to independence, emigration levels were even higher (Foster 1989: 352).

Table 33.1 Net outward migration from Ireland since 1926 (i.e. numbers leaving minus numbers returning)

1926–35	1936–45	1946–51	1952–56	1957–61	1962–66
−17,000	−19,000	−24,000	−39,000	−42,000	−16,000
1967–71	1972–76	1977–81	1982–86	1987–91	
−11,000	+14,000	−3,000	+15,000	−50,000	

Source: Department of the Environment (Ireland) 1986, 1991b.

IRELAND'S TOWNS AND CITIES

Ireland has only five towns with more than 30,000 people (see Table 33.2), although over half the population lives in settlements of over 1,500, the Irish definition of a town (DOE (Ireland) 1986). Dublin dominates Ireland. The capital houses one-third of the country's population with over 1 million inhabitants, although its inner-city core area has experienced a population decline of 8 per cent over the last fifteen years; so too have those of Cork and Limerick.

IRELAND'S CURRENT CONDITIONS

Ireland is the poorest country in the study. She enjoys better housing standards than her general economic position would indicate. Over 90 per cent of Ireland's stock is in single-family houses, many of them detached bungalows standing in their own grounds. This has resulted partly from a strong tradition of individual ownership and self-building in rural areas, partly from a ready supply of land and labour, and partly from savings and remittances stemming from rural traditions and emigration. She still has many fewer dwellings per population than other countries and only after 1970 did government investment in housing move towards continental levels, during which period she overtook Britain in her level of investment. In some periods in the early 1980s, Ireland's investment in housing was actually ahead of all the other countries as a percentage of national wealth.

Irish people spend less of their personal income on housing than people in the other countries – 11 per cent. House prices are lower. Rents are also

Table 33.2 Population of towns in Ireland of more than 30,000 inhabitants, 1986

Dublin Co-Borough*	502,749	Galway Co-Borough	47,104
Cork Co-Borough	133,271	Waterford Co-Borough	39,529
Limerick Co-Borough	56,279		

Source: Department of the Environment (Ireland) 1986.
Note: *The Greater Dublin population is 1.2 million.

Table 33.3 Basic facts about Ireland

Population[a]	3,500,000
Population (%):[b]	
Under 20	39
20–60	47
Over 60	14
Persons per household[c]	3.1
Population in rural areas (%)[d]	43
Economically active unemployed (%)[e]	16
Number of dwellings	1,057,000
Stock of dwellings per 1000 inhabitants:[f]	
1960	240
1986	300
GDP per head in US dollars[d]	$11,952
GDP per head in purchasing power parity*	41 (cf. France 70)

Sources [a] *Eurostat*, Demographic Statistics, 1989; [b] Commission des Communautés
Européennes 1987; [c] Irish Government Census 1986; [d] *Economist Pocket Europe* 1992;
[e] EC *Labour Force Survey* 1991; [f] Eurostat, *Social Indicators for the EC 1977–1986*.
Note: *Income in dollars adjusted for purchasing power parity shows that Ireland ranks
     tenth among EC member states in income per head and is much poorer than the
     other four countries in this study, which are very close together at the top of the
     wealth table, averaging 70.

low, particularly in local authority housing where tenants on average spend
only 5 per cent of their income on their housing. This reflects the overwhelming dependence on state support of 70 per cent of tenants and the
much lower incomes generally. Space standards and the level of amenities
are also somewhat lower than other countries, although the vast majority of
dwellings have all the basic services.

Table 33.3 presents some of the basic facts about Ireland and its
housing.

IRELAND – HISTORICAL GLIMPSES

> *History is not so much a matter of learning from the past as of stirring old
> grievances to keep them on the boil.*
>
> D. Walsh, *The Party Inside Fianna Fail*

Unravelling the wool of history is always presumptuous. Irish history, as it
has influenced the development of its cities and its housing, is more fraught

than most. The extraordinary power and force with which Britain sought to dominate and control Ireland from the seventeenth century onwards and the paralysing disasters of the last 200 years make Ireland's continuous struggle for nationhood hypnotising, yet misted over with confusion.

Celtic Ireland, on the furthest edge of Western Europe, is shrouded in mystery – the Gaelic language; a Celtic Catholicism; the monastic tradition; peasant farming, some secure and prosperous, some barely subsisting; island cultures; and a rich poetic and artistic legacy. It is on this rural and religious tradition that much nationalist fervour has fed, standing apart as it does from Britain's industrial, secular and imperial might.

From the seventeenth century, Irish land was taken over by the British aristocracy until by 1800, when the official Act of Union with Britain was passed, the ownership and control of Ireland was in the hands of largely Protestant British Lords, and the established Church of Ireland was a branch of the established Church of England. The Irish peasantry had become tenant farmers, holding feverishly to their traditional beliefs, regarded as 'ignorant, superstitious, even barbaric' by the dominating Anglo-Irish aristocracy. The nineteenth-century British Prime Minister, Disraeli, regarded Irish affairs as an onerous but inescapable liability: 'a starving population, an absentee aristocracy, an alien Church, and in addition the weakest executive in the world'.

By 1800, in contrast to the Irish hinterland, Dublin had grown into a powerful and elegant Georgian city, encompassing the original Viking port and the medieval city that had grown up around that 'foreign' settlement. Dublin had become Britain's second city. In the whole British Isles, only London was larger and Dublin had over 100,000 inhabitants. The landed gentry had built a beautiful city of spacious terraces and squares. The prospering cotton industry drew in landless workers from the Irish countryside. Constant dealings with Britain, coupled with Ireland's traditional links with France and Catholic Europe, made Dublin an important international centre. The other Viking ports of Cork, Waterford, Limerick, Galway, and Belfast, all on Ireland's main river estuaries, developed too as secondary centres. By the beginning of the nineteenth century, the population of Ireland was over 8 million.

Following the Act of Union with Britain in 1800 and the ending of a separate Irish parliament (Foster 1989), two things brought a sharp change in Ireland's fortunes. Much of Dublin's aristocracy abandoned Ireland for more prosperous England, while retaining ownership of the land. Ireland no longer ran its own affairs and the elite therefore lost its prestige and much of its stake in a separate Irish identity.

By contrast, Belfast developed rapidly in the nineteenth century as a massive industrial centre, undermining Dublin's cotton and other industry and enhancing its trading ties with Britain. Belfast, even more than Dublin, was Protestant-owned, dominated by and tied to Britain's booming economy, with its narrow sea straits over to Scotland.

Dublin declined steeply and rapidly from a wealthy capital to an impoverished outpost. Yet tenant evictions in the rest of Ireland ensured that she would continue to receive floods of rural immigrants seeking better fortunes.

Until the emancipation of Catholics in 1829, the vast mass of the Irish population was not only dispossessed of its land, but culturally suppressed, often at gunpoint. The deep alienation that this brought reached the extreme in the Irish Famine of 1845, caused by a blight of the potato crop, but leading to mass starvation through the British Corn Laws which aimed to protect British farmers from the competition of cheap imported food. As the Famine spread, over 1 million Irish people died. Hunger afflicted millions, and in the following ten years a million survivors of the Famine left Ireland for Britain to the east or America to the west. By the end of the nineteenth century, the population of Ireland had almost halved to just over 4 million, in spite of the very high birth-rate.

There followed a period of rapid change while great British Prime Ministers, such as Robert Peel and Gladstone, made and broke their international reputations on Ireland's fortunes which had become a scandal, a scab and a running sore in Britain's epoch of unrivalled world supremacy.

In 1846, the Corn Laws were repealed, partly under the impact of the Irish famine; in 1869, the Church of Ireland, tied tightly to land ownership and the 'Protestant Ascendancy', was disestablished; and in 1870 the first Land Act was passed giving tenants and farmers some right to security on the land in an attempt to stop mass evictions of famine-starved families for failure to pay rent. In 1891, the Second Land Purchase Act allowed much of the remaining rural population to become owners of their homes and plots of land, though this did not alter their basic poverty.

It took thirty years for tenant-farmers to regain any real hold and it therefore did not stop the Fenian Nationalist Uprisings and the violent 'Outrages' committed by the desperate rural Catholic population against 'foreign' landlords. The reforms, loosening Britain's grip, shifted Anglo-Irish relations inexorably towards final rupture. This was not Britain's aim. Rather the hope was to 'kill Home Rule with kindness', as Gladstone believed was possible.

Ireland's struggle for independence was led partly through the Irish Party in the House of Commons, brilliantly headed by Charles Parnell; partly through the Land League, fighting for Irish repossession of the land. But Home Rule for Ireland, Gladstone's brave compromise, first proposed to Parliament in 1886 and allowing Ireland limited self-determination, was too slow in coming to stem the problems. Political autonomy became umbilically tied to the Land Struggle, with Parnell desperately trying to lead the two and serving a prison sentence for supporting evicted tenants.

Gladstone lost his battle for Home Rule. He was outvoted first in the House of Commons, then six years later in the Lords, and fell from power over failure to solve the Irish question. In 1891, Parnell died in ignominy, aged only 45, having betrayed religious sensitivities through a personal scandal. The Irish problem continued to create bitter divides, the most bitter of which became the internal war between Ulster Protestants and Southern Catholics. It also dominated British politics. In 1911 the British Parliament finally passed the Parliament Act to limit the power of the House of Lords to obstruct new laws after their third rejection of a new Home Rule Bill. Home Rule finally became law as a result, on the eve of war in 1914. It was bitterly opposed by Protestant Unionists and failed to appease embittered Nationalists, many of whom were Republican sympathisers or supporters. The Ulster question was still unresolved and the Act of Home Rule was immediately suspended at the outbreak of war.

Political ferment was brought to the boil by conscription. In 1916, a small group of armed Irish nationalists led the 'Easter Uprising', declared the formation of a provisional government of the Republic of Ireland, but were quickly defeated with heavy casualties and ignominious executions of the leaders. Central Dublin was devastated by shooting, looting and fire. Sackville Street, 'one of the most magnificent streets in Europe' (Lyons 1973), was entirely destroyed. The aftermath caused the final struggle for Independence to gain almost universal support. But many Protestants continued their fevered allegiance to Britain.

In 1921, the establishment of the Irish Free State turned Irish eyes from its bloody and famine-struck past to an independent future. Its power-base was the rural Catholic population; its assets were virtually non-existent. And it had to live with the bitter breakaway of Ulster where a majority of the population was fiercely Protestant and where most Catholics, for historic reasons, were the 'navvies' of Belfast's industrial wealth.

Independent Ireland had very little industrial development; its farming was largely small-scale, poor and backward; its cities, while expanding in land area, acted as conduits for emigration and had very slow population growth or economic development compared with the rest of Europe. At the time of independence, Ireland had only 3 million people.

Independent Ireland had very little help with which to build its brave new future. Foreign investment flowed to Belfast through its continuing links to Britain, rather than to a proud and independent but economically decimated republic. The Irish overseas – 4 million settled in America in the fifty years after the Famine – fuelled Ireland's legendary hostility to Britain while sapping Ireland's energy and talent.

Many would argue that the Catholic Church played a decisive role in all these events, with one religious representative (nun, priest, monk) to every 350 inhabitants – most of them originating from the same rural communities that fuelled the struggle for national Irish sovereignty. It is impossible in this brief outline to unravel the role of religion in the division and underdevelopment of Ireland. The religious and cultural divides run very deep and their examination cannot be done justice here.

In the twentieth century, Ireland has remained conflict-ridden, poor, rurally based, Catholic and in many ways still dominated by Britain – at the very least by British involvement in and attempts to control the North. Her economic fortunes and her legislative structure have largely followed Britain's. Britain's determination to equalise conditions between Ulster and Britain led inexorably to Ireland emulating the British Welfare State in an attempt to equalise conditions between the Republic and the North.

EMIGRATION

Between 1845 and the end of the nineteenth century the population of Ireland was halved by famine and by emigration. Since 1871, over half of those born in Ireland emigrated. In spite of this, the population in most of the period increased slowly but steadily, mainly through a very high birth-rate. Uniquely, in the years between 1971 and 1981, there was a net inward migration, the only period since records were kept when this happened. Since the mid-1980s, emigration ran at a very high rate – between 30,000 and 40,000 a year, mainly in the 20- to 40-year-old age group – threatening to overtake the rate of natural increase in the popula-tion. For over a century and a half emigration took its toll on the young adults of working age, on the educated, and on the enterprising and the ambitious. Many famous Irish citizens have lamented the 150 years of 'bleeding'. Poverty played a big role in emigration.

POVERTY

Income levels in Ireland are far behind Britain's and even further behind those of Germany, France and Denmark. Irish incomes are just over one-third of Danish incomes and two-thirds the British level (EC 1990a). Fully one in ten live at below the Irish supplementary welfare allowance level. Half of these are eligible for benefit but are not getting it, either through not claiming or through getting less than their entitlement. The other half fall through the poverty net because they are in full-time employment – a disqualification even if wages are very low – or because they have no children or are single or are students – all additional disqualifications. The numbers dependent on welfare rose steeply in the 1980s.

One-third of all Irish households have no one in work – many of these are one-parent families who form a rapidly growing proportion of all families. There are no official figures for this (DOE[1] 1992) but a steep rise has been witnessed in local authority lettings (SUSS Centre 1987).

There is a large travelling population of 25,000 people in Ireland, still commonly referred to as 'tinkers'. Many of these households are very poor, eking out a marginal existence in the countryside or around Dublin.

ECONOMIC DEVELOPMENT

Ireland's economic development was handicapped by lack of indigenous energy sources: no coal, no oil, some natural gas and large acreages of peatbog that were difficult to extract and produced less energy (relative to the power consumed in extraction and conversion) than other sources. For centuries Ireland depended on burning peat in open hearths. Attempts to generate a modern power supply from it proved expensive and it now depends heavily on imported coal for heating and industrial use. No nuclear power programme or other alternative sources of energy have yet been developed.

Since the war, and especially since the advent of the EC, Ireland became a cheap, unrestricted and virtually tax-free base for off-shore manufacturing, particularly in chemicals and assembly for high technology industries. In this way it was very much like a Third World country, with a high level of unemployment and under-employment, a large supply of land, a small internal market, and most of its modern economic development dependent on the import of foreign capital. Ireland's manufactured products used largely imported foreign raw materials which were then exported for sale. Ireland thus provided a cheap processing base for firms from America, Europe and the Far East. In the new multi-national enterprises that grew up in the last forty years, a relatively low use was made of labour. Even so, 40 per cent of all of Ireland's working population is employed in these foreign industries. Much of the work employs women, often on a part-time basis. The new industries are much less dependent on large urban areas than earlier industrial developments. Therefore, cities like Galway and some other smaller towns in the west have experienced rapid growth.

Ireland still relies heavily on imported goods, including energy, many of which are expensive. This makes Ireland far poorer than her neighbours, fuelling emigration. Since Ireland educates a higher proportion of young people to an advanced level than most other European countries, a large proportion of these young people do not find sufficient opportunities in Ireland and leave.

Ireland benefited from the EC's common agricultural policy, regional policies, and its poverty programme. Irish dependence on agriculture led to favourable subsidies. Much of this money was absorbed in building the Welfare State and investing heavily in housing. Ireland received windfall infrastructure grants from the EC in preparation for 1992.

THE NORTH

The division between the Republic and the North of Ireland dominated much Irish thinking, and greatly influenced the form its institutions took and the priority it gave to social and political over economic considerations. There were two main elements that influenced policy.

Firstly, the North enjoyed a standard of public service and development that was as close to mainland Britain's as the British government could make it. This stood in stark contrast to the economic difficulties of the Republic where the ability of the government to provide advanced welfare services was limited by the poorly developed economy. British welfare services and subsidies from central government to Northern Ireland outstripped any other region of the United Kingdom and Ireland could not hope to match this. But as one commentator on the link with the North put it, 'How can we expect a million Ulstermen to want to join us as things are?'.

Secondly, the violence that pervaded the politics of Northern Ireland since 1968 spilt over to the Republic, both in the strain that it put on political relations with Britain and in the deterrent effect on tourism – one of Ireland's main earners and economic hopes for the future. Ireland attempted to capitalise on its green, agricultural landscape, its low density, its relatively unspoilt and harmonious environment, and its castles and great houses, but with limited success: in large measure because of the 'Troubles', although some would blame the rain! There are some signs, however, that tourism is growing again after the set-backs of the last twenty years.

As long as South and North remain juxtaposed, with Belfast almost as alien to Dublin as Paris, and frequently more so than Liverpool or London, Ireland may remain economically weak and politically trapped in a debilitating internal struggle that diverts it from vital issues of wider significance.

Ireland's housing history builds on a peasant tradition, coloured by nationalist fervour and a strong attachment to the land, with an urban tradition dominated by British industrial and administrative patterns. Ireland's troubled history from the Act of Union to the present day has placed an indelible mark on her housing developments.

NOTE

1   DOE – in this section (Chapters 33–40) this is the Department of the Environment (Ireland).

Chapter 34

The start of Irish housing policy

From 1800 to the First World War

Two factors dominated Irish housing history from the eighteenth century to the present: firstly, the agricultural base of the economic, social and political life of Ireland; secondly, the dominance of Dublin as the major centre of Irish cultural and urban development.

The vast majority of Ireland's dense rural population at the time of union with Britain in 1800 lived in single-room, mud and turf huts. Dublin, by contrast, represented Ireland's wealthy landowners and was arguably greater than Glasgow, Liverpool, Bristol or Newcastle until the end of the eighteenth century. Ireland's importance to Britain was symbolised by Dublin and led to the Act of Union itself in 1800. Dublin comprised the old medieval city port areas, occupied largely by traditional workers in the cotton, machine and shipping industries, and Georgian terraces around squares and avenues, built with gracious, spacious interiors, elegant façades, and enclosed gardens for the ruling classes.

As the nineteenth century progressed, three things affected Irish housing developments. The agricultural economy became deeply polarised between absentee landlords with large rural 'seats' and impoverished and famine-struck tenants who were evicted if they failed to pay their rent. A landless mass found its way to Dublin and the other major cities – in a majority of cases en route for the New World or the booming industrial centres of Britain. The impact on rural and urban housing conditions was dire. People could barely subsist, let alone afford rents. People doubled up, threw up shacks, were evicted again and again. This situation only changed after the Land Purchase Acts, which by the turn of this century had created widespread rural owner-occupation, a change that was to shift the structure of Irish housing in the twentieth century into ever-higher levels of support for low-cost owner-occupation.

The beautiful country mansions of the Anglo-Irish aristocracy became legendary in contrast to the rural hovels, often depicted as picturesque crofts. The aristocratic mansions themselves were frequently seriously neglected. The Famine, in decimating the population, undermined much of the rural base for the economy. The landed class itself declined both in power, wealth, and size. The loss of rent and cost of poor relief under famine conditions combined to make them far less prosperous. Wages in Dublin were significantly lower than elsewhere in Britain; a rising proportion of the population was unskilled and, in spite of intense overcrowding, the constant exodus to Britain undermined any incentive to build working-class housing.

Dublin's population rose from the beginning of the nineteenth century, when it numbered 170,000, to 318,000 by 1850. By 1901, it had reached 382,000, representing a much slower growth than other cities in Britain, including Belfast. Much of the growth from 1800 onwards was outside the old city boundaries. While Dublin's population doubled in the thirty years to 1873, its land area increased eightfold (Arnold Horner, in Bannon 1985). New peripheral settlements for the Protestant-dominated elite developed. This marked the beginning of an endemic conflict over Dublin's future development. The city itself was now largely Catholic and poor. It was administered under British rule by an inefficient city corporation, which often failed to enforce even basic sanitary laws or take even minimal action against transgressing landlords (Mary Daley in Bannon 1985). Very little new housing was built within the city and, as the elite left Dublin, so the spacious Georgian terraces were gradually broken up into tenements, let out room by room. Overcrowding in Dublin was worse than in any other city in Britain and the death-rate was much higher.

The scale of the problem constantly overtook any action since, in practice, Dublin Corporation was closing unfit dwellings at a faster rate in proportion to the population than any other British city. The corporation was widely accused of causing dereliction as a result, an ironic outcome, given that it was simultaneously accused of taking too little action.

The conditions in the city and the reputed incompetence of the city administration led to strong resistance by suburban satellites against being incorporated within Dublin. A conflict, parallel to that between inner and outer London, developed, with little of the economic imperative to keep inner Dublin going that existed in London.

Few railway or industrial developments took place in the nineteenth century because of the economic weakness of Dublin. There was therefore little demolition. As a result, most tenements survived into the twentieth century. A low demand for unskilled labour meant that constant emigration and a high death-rate were economically bearable, in spite of social protestations. It also meant that although there was population pressure and city expansion, it was much less than in the industrial boom areas of all other countries in this study.

Nationalist fervour was largely rooted in the land issue among the 90 per cent of Ireland's population that throughout the nineteenth century depended on the land. Therefore urban housing was not a major political issue in spite of the controversy that frequently surrounded it.

It was not until 1876, following the passing of the Artisans' and Labourers' Dwelling Act in London that any serious initiative was taken in working-class housing.

ARTISANS' DWELLING COMPANY

Less than 100 model dwellings had been built at the time of the first housing law, either through philanthropic or industrial initiatives. In 1876, the Dublin Artisans' Dwelling Company was founded, with support from major companies such as Guinness, a third of whose workforce were eventually housed by it (F.H.A. Aalen, in Bannon 1985). This company became a major builder and laid the ground for later developments by local authorities. By 1914, the Artisans' Dwelling Company had built 3,000 working-class dwellings which, in proportion to the size of Dublin, put it almost on a par with the combined efforts of all the philanthropic trusts in London at the beginning of the First World War. It thrived on the low-interest loans and below-market land sales which were allowed under the Artisans' Dwellings Acts. It paid its investors 4 per cent dividend. It charged cost rents far beyond the reach of all but the more securely employed labourers. But it tapped a need for Artisan Dwellings, fuelled by the absence of new building in the nineteenth century and the rapidly decaying conditions of tenements, the only real housing option for the low-paid.

Plate 34.1 The Dublin Artisan Dwelling Company Cottages, Rialto, Dublin

The incipient demolitions following closing orders were increasing the shortage of minimal one-room dwellings. As a result, in turn, the constant influx from the land, made tenement conditions even worse. This drove the better-off workers to search desperately for safer, more sanitary housing, fuelling demand for new, if relatively expensive, building by the Artisans' Dwelling Company.

THE GUINNESS TRUST

The Guinness Trust was founded in 1890 and built 600 dwellings by 1914. The Guinness Trust was entirely philanthropic, unlike the Dwelling Company, with a strong welfare orientation. Its historic contribution to Irish housing was on a broader social front. It built an impressive 'Recreation Hall' with a gym and leisure facilities for its tenants near St Patrick's Protestant Cathedral; it developed St Patrick's Park adjacent to its prestigious but low-cost housing schemes; and it provided teachers for adult education. It took a controlling hand in the development of central parks in Dublin, partly in an attempt to preserve the dignified atmosphere around the Cathedral, partly, according to Bannon (1985), to prevent bands and other noisy, social street-life around the centre of the city on Sundays, where gradually new working-class housing was being built.

Both the Artisans' Dwelling Company, and later the Guinness (Iveagh) Trust, were largely owned by the Protestant elite. Its ties were with the disestablished Church of Ireland and with the aristocracy of the working class. It was inevitable therefore that Dublin Corporation, with its Catholic voters and Nationalist councillors, would feel forced to do something about the problems of the very poor.

THE DECAY OF INNER DUBLIN

A combination of slow urban population growth, the continuing mass exodus of Ireland's population abroad, suburbanisation, and very limited building, led to widespread blighting of properties in city areas. Dublin Corporation began to close insanitary or derelict properties on a significant scale towards the end of the century. It had declared its first slum clearance area in 1877.

Dublin Corporation limited its scope for improving conditions unnecessarily by endorsing the principle of compensation to slum landlords whose property was deemed unfit. This was a gesture towards the small-time Irish owners who had taken over the large houses abandoned earlier by the Anglo-Irish ascendancy. Tenement ownership had often passed from Anglo-Irish, upper-class owners to publicans, shopkeepers and other small businessmen, a majority of whom were Irish Catholics and Nationalists themselves. The Nationalist cause against Britain was embedded in

Plate 34.2 The remains of the beautiful Georgian Mountjoy Square in Dublin

local Dublin politics and many Irish landlords escaped the strict enforcement of British public health laws. At one point, sixteen Dublin councillors
were tenement landlords, diluting any incentive in the city to tackle the
problem wholeheartedly and weakening the ability of the Corporation
sanitary inspectors to enforce standards (Mary Daley, in Bannon 1985). At
root, however, the problem reflected the poverty and the transience of the
rural immigrants and the largely unskilled workforce.

The policy of compensation made slum declarations and clearances,
carrying the obligation to rebuild after the 1890 Housing of the Working
Classes Act, doubly expensive. It also rewarded the worst landlords when
they were taken over.

Action by the Corporation, carrying with it expensive and unnecessary
compensations – unnecessary because acquisition was legally compulsory –
tended to increase the blighting of the worst areas, while slowing down the
rate of replacement because of the Corporation's inadequate resources.
Unlike all other British cities, the number of families in Dublin dropped in
the 1880s, reducing further any incentive to tackle housing problems
seriously.

By the end of the nineteenth century conditions in the tenements, where
over half of Dublin's entire population lived, were dire. Most tenement

dwellers had only one room – over one-third of Dublin's total population lived in a single room compared with one in ten of London's population. There were no back alleys through which refuse could be removed because of the layout of tenements within Georgian mansions around enclosed courts. The size of the previously affluent houses led to a huge density of subsequent occupation. The new occupants could rarely afford more than a room. In 1904 2.6 million pawnbrokers' tickets were issued in Dublin alone, an average of eight per household (Murphy 1991).

When, at the turn of the century, Dublin's death-rate was still extraordinarily high by British standards and there were a series of dramatic tenement collapses causing fatalities, there was a flurry of concern. But tenements continued to provide the bulk of largely unchallenged shelter for the masses right up until the Second World War.

THE EARLIEST COUNCIL HOUSING

Local authority housing was developed earlier in Ireland than in Britain because of political concerns. The limited sanitary actions of the Corporation had their main impact in displacing nearly destitute people. The Corporation was therefore pushed into building cheap housing to replace closed slum tenements. The earliest Corporation schemes built under the Artisans' and Labourers' Dwelling Act of 1876 comprised one- and two-room dwellings, built in the most infamous slum areas, with rents comparable to tenement rents and far below the Artisans' Dwelling Company rents.

In 1883 Dublin Corporation was accused – at the time of its earliest scheme – of replacing one lot of slums with another, of blighting the city with derelict, closed tenements, of conniving with and rewarding the most disreputable landlords by acting too slowly or by paying compensation, and of putting up inadequate new housing. Several early low-cost schemes, including a famous development in the red light area of Dublin known as 'the Kips' (local slang for a brothel), were unpopular and uneconomic because, even with their very low rents, the lowest-waged workers would not move there because the areas were so disreputable. Only the most marginal households moved in, creating a management nightmare and quickly recreating the old slums. Dublin Corporation lost money on several schemes through failure to let the property, to collect the rent, or to maintain the new developments due to the extreme social and economic conditions. The Corporation's housing gained from these earliest efforts a reputation for low quality and poor management that proved hard to escape.

Table 31.4 is taken from the 1911 Census survey of Dublin, showing the contrasting backgrounds of tenants in different types of housing at that time, illustrating the difference between Dublin Corporation tenants, the

Dwelling Companies, and the private tenements. The figures underline the continuing role of the tenements in housing the very poor, unskilled and insecurely employed. The Corporation, while housing many skilled and semi-skilled workers, was letting nearly half its property to unskilled and insecure workers. The Dwelling Company by contrast was housing only a minority of unskilled and a much higher proportion of secure workers and employees of the rich brewers, Watneys and Guinness.

STYLE OF DEVELOPMENT

From very early on, both the Corporation and the Dwelling Company provided mainly single-storey cottages. At the outset, they had accepted the logic of building flats in order to allow inner-city rebuilding to high density. The requirement of the housing acts was that in city sites as many dwellings had to be put back as had been removed. Thus, the earliest developments were dense tenement blocks. But flats were expensive, unpopular and difficult to manage (M. Daley in Bannon 1985), although candidates for rehousing frequently wanted to be rehoused close to the city centre, workplaces and areas of origin. Thus a further conflict was set up, lasting to this day, between the preference of tenants for inner-city locations and the impossibility of rehousing large numbers of poor families in cottage-type developments there. As a result, dense cottage estates were built with very little space at ever-increasing distance from the centre. The Corporation's first suburban cottage estate was built in 1905.

By 1908, when the Clancy Act – Housing of the Working Classes (Ireland) – was passed, both the Corporation and the Dwelling Company were constructing large cottage estates of several hundred dwellings; housing developments were from then on to be largely on the periphery of

Table 34.1 Comparison of households in Dublin by occupation and landlord, 1911

Male occupations		Tenement (%)	Corporation (%)	Artisans' Dwelling Co. (%)
Non-manual		1.9	5.3	19.3
Skilled manual		21.3	36.3	40.9
Semi-skilled manual		4.6	12.5	10.7
Unskilled manual		72.2	45.8	29.0
Brewing and distilling	(as % of	0.9	6.9	11.1
'Secure' employment	above)	3.0	11.6	22.5

Source: Census survey of Dublin 1911.

the city, with poor and expensive transport links, few amenities or open spaces, and an often monochrome aspect. Under the Clancy Act, direct subsidies to local authorities were introduced, over a decade ahead of the rest of Britain, to encourage local authority building because of the dire conditions in the tenements and the extreme poverty of the people. This attempt to increase provision signalled the growing division of Dublin into 'two hostile camps', those well housed in suburbs and those still occupying the tenement rooms. The latter formed the vast majority.

The Nationalist cause and social unrest in Ireland had finally become interlocked with urban housing conditions, and the British government decided they had to be tackled through special funding and through state building. Private landlords had no incentive to do anything because of the threat of closure due to poor conditions, the general blighting and the inability of tenants to pay rents that would allow proper repair. There was therefore no basis on which to restore Dublin through private investment.

The Clancy Act marked the end of new building by the Artisans' Dwelling Company. From then on, local authorities would be virtually the sole builders of rented housing for low-income families. The Guinness (Iveagh) Trust continued in existence, but made very small inroads. Altogether the Trust, the Dwelling Company and other small bodies provided nearly 5,100 dwellings by 1914 (Dublin Corporation 1914).

Little new private building had taken place either in towns or the countryside throughout the century, in sharp contrast to the very rapid expansion of industrial urban housing by private landlords and employers in other countries. The infamous, old, private tenements were generally in far worse condition by the end than they had been at the beginning of the century.

PERFORMANCE OF THE CORPORATION

Dublin Corporation was far behind the Dwellings' Company, though it was ahead of other British local authorities, thanks to its special funding. By 1913, it had rehoused nearly 1,400 families – 2.5 per cent of the population (Mary Daley, in Bannon 1985) – representing a higher proportion of the city population than had been rehoused by the London County Council (LCC) at the same time.

Dublin Corporation was found 'negligent' by an Inquiry in 1913 just before the war (Mary Daley, in Bannon 1985) because of conditions in the tenements. The insanitary tenement rooms, let at extremely low rents, often in a terminal state of disrepair and under life-threatening conditions, were held to offer unfair competition by decent landlords. Housing conditions in the tenements were increasingly unsafe, with only a quarter of the houses considered rescuable. But the Corporation was trapped between overcrowding, pitiful levels of poverty, dangerous conditions, legal remedies, blighting and the cost of reconstruction. Politically, Dublin had little power with the seat of government in Westminster.

ABERCROMBIE PLAN FOR DUBLIN

After more than a century of neglect, the Lord Lieutenant of Ireland launched an international competition to plan the restoration and development of Dublin in 1914. Abercrombie, famous for his 1946 Greater London Plan, won the competition. He proposed a green belt, satellite towns, and a newly planned national capital, which fitted, according to his perception, with the spirit of the Irish independence movement and with what Abercrombie had seen (and liked) of Haussman's Paris.

But both the inability of Dublin Corporation to overcome the tenement problem and Abercrombie's vision for Dublin were overshadowed by war, by the Easter Uprising and by Irish independence. So were any wider attempts to modernise Ireland's housing.

Cork, Ireland's second city, had gone through a similar process of growth, stagnation, overcrowding by emigrant peasants, transience and decline. So to a lesser extent had Limerick and Waterford. Galway became truly the gate to the West, where the Atlantic beckoned and American cities on the other side offered the promise of 'gold and silver out on the ditches and nothing to do but to gather it' (O'Sullivan 1933: 23b). In the wake of the Clancy Act, Cork Corporation, with its smaller-scale problems than Dublin, became very active in rebuilding the inner areas. Solid artisan cottages were constructed, with generous space standards for the time, across the central areas of the city. By the outbreak of the war, Cork City Council had provided several hundred high-quality dwellings, many of which are still popular today.

Irish cities in the nineteenth century changed from being administrative and trading centres of international importance to being staging posts in one of the most massive exoduses of modern history. Ten million people left Ireland's shores in the seventy years between 1845 and 1914.

Belfast, on the other hand, grew in a fashion much more integral to the British economy than the rest of Ireland. Its history and development are not included here.

Chapter 35

Irish housing in limbo – between the wars

THE EASTER UPRISING – HOUSING UNREST

The year 1914 marked a turning point in Irish housing history: Home Rule was finally agreed but not implemented, and major housing funds were earmarked by the British government for Ireland but not delivered. Tenement collapses increased and a number of residents were killed (Bannon 1985). Ireland's housing stock had remained virtually static for over 100 years. But it was the threat of army conscription in 1915, with Ireland's ambiguous loyalties, rather than poor housing conditions that created the impetus for the Easter Uprising of 1916. The poor housing conditions of Dublin's population fuelled the upheaval in the capital, where in the end the Uprising was narrowly based (Lyons 1973). The *Irish Engineer and Builder* (F. H. A. Aalen, in Bannon 1985: 181) explained the uprising in housing terms: 'The housing question today remains a burning one, as it was in 1866. Until it is remedied, there will be no real peace in this city, no real and enduring security for life and property.' In practice, conditions had deteriorated dramatically through disrepair and, in many cases, dereliction. The prosperous suburbs, largely to the south of the city, contrasted starkly with the impoverished core and north side of Dublin.

INDEPENDENCE AND THE HOUSING HIATUS

After the war, the 1919 Addison Act, passed in Westminster with the aim of building 'Homes fit for Heroes', was never fully implemented in Ireland because by then the battle for independence from Britain was well advanced. The 1919 Housing (Ireland) Act aimed specifically to compel Irish local authorities to build with direct state subsidies. This measure was targeted at curbing unrest but again, independence prevented its implementation. Only 800 houses were built by 1921. Any serious housing policy or action was long delayed, overwhelmed as the new Irish government of 1921 was with wider problems. The break with Britain prevented any further concerted housing action for several decades.

RURAL BIAS

The Nationalist government of Ireland had a strong anti-urban bias, with its historic support rooted among Irish peasants. The early spokesmen had a romantic vision of an independent Ireland that would build its wealth on successful farming, the fruits of which had previously largely gilded the homes of the occupying British aristocracy: 'since the country will rely chiefly for its wealth on agriculture and rural industry there will be no Glasgows or Pittsburghs' (P.H. Pearse 1916, quoted in Bannon 1989: 35)

Dublin, as the overdeveloped seat of an incompetent colonial administration, suffered particularly under this approach, inheriting a richly endowed but poorly maintained international capital. Not only were its housing problems seen as overwhelming, all attempts at remedies were expensive. Dreams of a well-planned and dignified national capital with spacious parks, roads, and solid, attractive housing developed in village style were to remain paper dreams, promoted by British experts like Abercrombie and the 'garden cities' architect par excellence, Unwin. The rhetoric found favour, but did not become a reality because Nationalist priorities lay in the countryside where the mass of population still lived, mainly in very poor conditions.

Between 1914 and 1928, overcrowding in the city increased, partly through increased dereliction, partly through lack of building, and there was a big rise in the number of unfit houses (Mary Daley, in Bannon 1985). Inaction was the main cause, while rural migration continued to fill the poorest homes with the poorest people. The world slump stemmed the outward flow of Ireland's people somewhat.

A NEW HOUSING PROGRAMME

In 1922, the Irish provisional government set aside £1 million for urban housing, with local authorities required to put in £1 for every £2 of grant. This generated some building. In 1924 and 1925, two Housing Acts were passed, giving grants to private individuals and bodies, encouraging owner-occupation and rural development. These two Acts led to new public and private building. In the five years from 1922–7, local authorities built 6,500 houses – adding considerably to the 9,000 they had built in the previous forty years.

COTTAGE ESTATES

Between 1927 and 1930, Dublin finally completed its first 'garden city' estate at Marino, including schools, shops, public buildings and open space. This scheme had been in gestation for fifteen years, covering over 100 acres, providing 1,000 cottage-style dwellings to generous Tudor Walters standards, these being laid out symmetrically with fewer than ten dwellings to the acre. The adoption of such low densities and such high standards meant several things – new housing was mostly built on the fringes of cities, as such generous densities would have been too expensive in central locations; the limited scale of development that was possible to these standards left a majority of poorly housed behind. Cottage developments continued, however. The City Corporation as a whole lived with the contradiction of tenants preferring cottages while preferring inner-area rehousing. It was suggested that there were five applicants for every cottage but that less than one in five wanted the suburbs (Bannon 1989). Eight thousand private homes were also built. In total, this was still a minuscule contribution in the face of terrible conditions.

The British Tudor Walters standard of twelve houses to the acre was adopted for council building. As a result 'vile jerry cottages' sprang up and local authorities firmly adopted the pattern of suburban cottage developments. Private building was also largely in the form of suburban and rural houses. The low-density building created real worries over the environment, particularly around Dublin.

DUBLIN SURVEY

An important departure in civic planning was the 1925 Dublin Survey, chaired by Horace O'Rourke, the city architect, who dramatically condemned previous approaches to the city's problems – closing orders, blighting, disrepair, overcrowding and inaction:

> ... we must resolve to shake off the unhappy desolation of the present and the past. If we continue the log-rolling methods of today, within a short period of time an address such as this will not deal with the body that is ailing, but with the corpse that is dead.
>
> (Bannon 1989: 22)

The aim of the survey was to document Dublin's needs but Dublin Corporation itself was suspended between 1924 and 1930. This meant that the capital city was poorly represented in independent Ireland and the national government frequently overlooked its problems. Figure 35.1 shows the extent of Dublin's problems.

Dublin Corporation, when it was reinstated in 1930, built several large outer estates in the period up to the Second World War, including the Crumlin estate with 4,000 cottages. These estates had similar low densities and later became popular with tenant purchasers. In all, a further 25,000 local authority dwellings were built between 1927 and 1940, most of them on outer estates.

Private initiative, encouraged by government building grants, took care of more affluent households by creating new owner-occupied housing but private-rented housing in the city was virtually untouched and inner Dublin continued to decay. With almost total lack of investment, some drastic action became inevitable. One hundred and thirty years of decline and blight had left privately owned housing in the city literally falling apart.

SLUM CLEARANCE AND INNER CITY BUILDING

In 1931, a new Housing Act introduced slum clearance measures in an attempt to rectify worsening conditions. Dublin Corporation promptly declared a large 40-acre demolition site adjacent to O'Connell Street, renamed after the destruction of Sackville Street in the Easter Uprising, in the heart of Dublin. But limited finances, deep world recession, inexperience and emigration, all continued to bedevil progress. There was a twenty-year time-lag in implementing the proposals (Bannon 1989). The City Manager aimed to build two inner flats for every outer cottage in response to the now strong desire of the national government to rescue Dublin from dereliction. But the commitment to inner-city rebuilding was undermined by the opposition to flat-building, which partly caused the delays in slum clearance and still left one-fifth of Dublin's population occupying single-room dwellings by the latter part of the 1930s. At least one-third of these were sharing with six or more people (Arnold Horner, in Bannon 1985). It was not until the 1940s and 1950s that inner-city rebuilding actually took place on any scale. None the less, slum clearance legislation did quicken the pace of local authority action.

IRELAND'S HOUSING POVERTY

The inter-war years were dominated by economic and political problems that left Ireland's housing situation very little better than before independence. From 1926 (when emigration figures were first kept) to 1945, emigration outstripped immigration by around 1,800 people a year (see Table 33.1). This made housing shortages and poor conditions seem less urgent.

About 90 per cent of houses dated from *before* the First World War. But the ground had been laid for the two most important post-war developments: the spread of suburban and rural owner-occupation, and the strong growth in local authority building. Local authorities were the 'natural' arm of the independent government rather than the industrial trusts of the nineteenth century. By 1940, 84,000 local authority dwellings had been built in Ireland, over half of which had been produced in the 1930s. This made up about 10 per cent of the total housing stock but virtually all the new rented housing.

The virtual demise of the private-rented sector had also maybe become inevitable because of the long period of decay. Table 35.1 sums up the housing situation in 1940.

Plate 35.1 Some attractive local authority housing overlooked by St Patrick's
            Cathedral, Dublin

Table 35.1 Housing situation in Ireland in 1940

Tenure	Number of units	%
Owner-occupers	440,000	54.0
Private renting	276,000	34.5
Local authorities	84,000	11.0
Trusts	4,000	0.5
Total	804,000	100.0

Sources: Estimate based on DOE (Ireland) 1986, Blackwell 1989a, Bannon 1985, 1989.

Chapter 36

After the Second World War

BUILDING OUT OF TROUBLE

In 1946, Ireland's housing conditions were a serious scandal. Only one in six dwellings had all the basic amenities. One in five families were still intensely overcrowded and the vast majority lived at more than one person per room. Table 36.1 illustrates this and also gives a comparison with 1961 regarding changes in housing conditions and tenure in Ireland.

After the war, the pace of emigration accelerated. In the 1950s, there was a net loss of 80,000 people. As the Irish population declined up to 1960, the Dublin population itself contracted. As a result, more smoke than heat was forthcoming from the scandal of the slums. Much of the indifference stemmed from the fact that Ireland's rural population lived

Table 36.1 Changes in housing conditions and tenure in Ireland between 1940 and 1961

	1946	1961
Dwellings with:		
inside piped water	36%	51%
inside sanitary fittings	23%	43%
fixed bath	15%	33%
Households with more than 2 persons per room		
as percentage of total	18%	12%
Average number of persons per room	1.01	0.90
Average number of persons per private household	4.66	3.97
Households in dwellings with 3 rooms or more	84%	91%

	1940	1961
Owner-occupiers	54%	62%
Local authority	11%	20%
Private renting	35%	18%

Sources: Blackwell 1988: 10, DOE (Ireland) 1986.

predominantly in owner-occupied housing, although much of it was traditional and lacking in any modern facilities.

None the less, some earlier slum clearance and building plans were executed and Dublin gained about 8,000 balcony block flats. The space standards were very restricted. There were no bathrooms. The blocks were built around close, congested courtyards. Many of the new flats were intensely crowded with large families. Despite this, they were popular with rehoused residents because they retained something of the close-knit character of the old tenements. At the same time, cottage dwellings were being built fairly rapidly, and between 1940 and 1960 over 60,000 council dwellings were built.

By 1960, the tenure pattern of Ireland had changed quite radically (see Table 36.1). Public renting had risen to nearly 20 per cent, while private renting had almost halved in twenty years from 32 per cent to 17 per cent. Owner-occupation was rising fast and already comprised 62 per cent of the stock. Conditions, while still poor, had begun to improve rapidly.

THE 1960s

Ireland's housing stock had shrunk from 941,000 units in 1911 to 676,000 in 1961 through clearance, rural abandonment and emigration (*Statistical Abstracts, Eire* 1966-75). The 1960s ushered in a new era of effort, with a bigger housing programme. A Consolidated Housing Act in 1966 formed the basis of the main housing developments to the late 1980s, during which time a huge amount of new housing was built.

The Act brought in three main policies. Firstly, it reaffirmed the role of local authorities in providing and managing housing; it also made them responsible for assessing housing need. Public housing was 100 per cent financed through central government grants, removing incentives to economy or efficiency and forcing the government to intervene progressively in detailed local development issues. Secondly, the Act established a system of financial assistance to individuals through loans, grants and subsidies, thus greatly accelerating progress towards modern owner-occupation, involving the almost total replacement of the rural stock. An important third policy was financial assistance to sitting tenants in local authority stock to buy their cottage dwellings. The Act attempted to establish controls over house prices, so that the incentives to buy would not cause rapid inflation. Owner-occupation, already very high, rose from 62 per cent to 71 per cent in the ten years following the Act; the number of newly built units expanded rapidly as the old stock began to be replaced.

The government set up the National Building Agency in 1963 to goad local authorities into building on an expanding scale. The National Building Agency reflected the strong centralising trend of the government at this time. One of the National Building Agency's first and most significant jobs was the construction of the only large, high-rise estate in Ireland – Ballymun, on the northern outskirts of the city adjacent to the airport. It imported the Balancey panel-building system from France, set up a factory on the site to produce the heavy concrete panels, and set about creating the twenty-three high-rise blocks of the 3,000-unit estate. The aim was to produce the maximum number of modern dwellings as quickly as possible to the highest standards possible in order to overcome the serious shortage (Blackwell 1988). Ballymun flew in the face of everything Irish politicians had believed about housing. Its strange imposition on the Irish landscape was an unrepeated experiment in modernity. With hindsight, it was widely condemned. The Dublin Area Health Authority was so concerned about 'high-rise blues' that it appointed community workers in 1973 to help the residents settle. These efforts were overwhelmed by bigger problems of segregation that we discuss later (see pp. 349–50).

NEW TOWNS

In 1967, a new-towns programme was launched, leading to three big local authority-sponsored developments around Dublin, at Tallaght (8,000 units), Clondalkin (7,000 units), and Blanchardtown (3,300 units). The aim was to clear the inner tenements. These developments harked back to Abercrombie's original 1914 plan to create satellite settlements outside a green belt around Dublin. The green belt itself had long since been over-taken by suburban sprawl. The original aim was that each new develop-ment should accommodate 100,000 people by 1991, but plans were foreshortened as low demand later underlined the isolation and economic fragility of these new communities. Tallaght, the largest, grew by 1985 to 60,000 inhabitants. But they did not gell as new towns, although owner-occupation as well as local authority renting was developed. Tallaght and Clondalkin became very low-income areas and Tallaght is considered to be one of the most depressed housing areas in the country. The poorest areas within them developed as peripheral local authority estates which came to dominate their reputation. Satellite areas such as Kilnamanagh, privately built exclusively for owner-occupation, were also developed.

LOCAL AUTHORITY LANDLORDS

Under the impact of the 1966 Act, local authorities became big builders, producing over 20,000 units a year up till 1980. But they did not develop a system of estate management to match this growth. Like British authorities, their housing operations were incorporated into their town hall adminis-trations with no estate-based services at all for their houses and a minimal caretaking service in the balcony block estates. The level of repair was minimal. Estates *per se* were expected to solve housing problems.

LOCAL AUTHORITY MANAGEMENT

There are eighty-seven local authority landlords in Ireland, most of which are small rural authorities owning very little stock. Only Dublin Corporation, owning 34,000 properties, and Cork City Corporation with 8,500, are large landlords in international terms. Between them they own one-third of all social housing.

In spite of this, only about 12 per cent of Dublin's housing was socially owned by 1989, a low proportion in a large capital city. This reflected both a high level of owner-occupation in the city and suburbs and a higher than average level of private renting. It also reflected the very high level of sales to sitting tenants. In Cork 11 per cent of the city's housing was owned by local authorities.

Town halls were the administrative arms of government. Their approach to housing was dictated by their administrative functions. Housing management was a largely clerical function and, as a service to tenants of rented housing or a requirement of landlords, was barely recognised. Functions, such as rents, lettings and welfare, were entirely centralised.

As building programmes grew, the technical departments of local authorities became much more influential and dominant. This reinforced the non-professional administration of housing management. Two other factors strongly influenced the administrative bias. Most obviously, the predominantly cottage stock built before 1970, often single-storey dwellings, was largely problem-free. It was popular with tenants and easy to maintain. Secondly, the growing sales policy meant that a very large number of tenants became owners over the years of expanding building.

Sales were widespread and far down the income ladder. The vast majority of all council dwellings built before 1970 were eventually bought by their occupants, reinforcing the view that the properties were acceptable and that intervention was unnecessary. This in turn made the sale of local authority dwellings popular with administrators, as well as with politicians and residents. Estate management only became an issue after Ireland's 'mass housing programme' of the 1970s created special problems. By then its management structures were firmly set.

Weak estate management did not mean that the amount of money spent on management and maintenance in Ireland was low. It far exceeded the rental income. In fact, by the 1980s rental income only covered about 60 per cent of the direct costs of running the stock (Blackwell 1989a). The administrative costs were so extensive as to use up half of the sum available. As a result, maintenance suffered seriously. This was a major cause of dissatisfaction and spiralling conditions. The cost of running local authority housing was around £850 per unit per year, a figure comparable to Britain and, relative to rents and Ireland's income, very high. In terms of the direct services, the spending barely showed.

ALLOCATIONS AND ACCESS

General levels of poverty and unemployment in Ireland, far higher than in other European countries, made access to public rented housing very important indeed (Combat Poverty Agency 1988).

Council housing, according to the Housing Act of 1966, had to be made available to those in need who could not afford to provide for themselves. None the less, the income limit for moving into council housing was around £12,000 (1989), above average incomes. This meant that the majority of Irish households were eligible for council housing. Residence qualifications were allowed by law and in Dublin applicants must have been resident for two years if they did not originate from Dublin or six months if they did. Very urgent priority cases took overriding precedence – for example, natural disasters (e.g. fire), demolitions, very serious health problems, compassionate circumstances, or the need to provide local housing for community reasons. Dublin's points system was weighted with the following points representing the highest level of need: overcrowding, sharing and room shortage – twenty points; time on the waiting list – thirty points; physical conditions – thirty points; medical and compassionate reasons – fifteen points.

The residence requirement, and points for time on the waiting list, might have operated in favour of more secure and more established households. However, the increase in available units led to many offers being available to whoever might take them in areas where vacancies occurred.

DIFFERENTIAL RENTS AND DISINCENTIVE TO MANAGE

From 1950, local authorities were required to set 'differential rents', depending on the tenant's income and ability to pay. Although achieving a similar outcome, it is a very different approach from the housing allowance system common in other countries in this study. Housing allowances help individuals meet rents that reflect costs, whereas differential rents cannot be related to the cost of landlord services and appear unjust to many tenants who have to pay higher rents. This system, still in place today, led to sharp discrepancies in rents for identical and sometimes neighbouring properties. It meant that an individual tenant's rent could suddenly change drastically with changed financial circumstances. It was also unfair, in that lower-income tenants paid a higher proportion of income – up to 16 per cent – than better-paid tenants – as low as 2 per cent (Blackwell 1989a). It removed any local authority incentive to cover their costs through rents, since the government met costs directly to make good the shortfall in rents caused by very low incomes. High levels of unemployment ensured that local authorities almost exclusively housed people paying very low rents, in over two-thirds of cases with no income other than state benefits. Anyone who could afford it bought their house.

Table 36.2 Changes in housing conditions in Ireland between 1961 and 1971

	1961	*1971*
Dwellings with:		
inside piped water	51%	74%
inside sanitary fittings	43%	63%
fixed bath	33%	56%
Households with more than 2 persons per room		
as percentage of total	12%	9%
Average number of persons per room	0.90	0.86
Average number of persons per private household	3.97	3.94
Households in dwellings with 3 rooms or more	91%	92%

Source: Blackwell 1988: 10.

The result was unrealistically low rents and a rent income for local authorities that bore no relation to their costs. By 1973, local authorities no longer had any input into government rent-setting at all. From 1986, rent-setting again became a local authority function, subject to government guidelines. Differential rents continued to be a controversial plank of policy, creating anomalies, reducing incentives for local authorities to perform efficiently, and giving government decisive control. In theory, it was equitable to tenants, but even that aspect was disputed because of the steep poverty trap when tenants found employment (Blackwell 1988). The cost-income imbalance became a self-fuelling problem, as local authorities were constantly constrained by a central government carrying much of the burden of cost. This gave local authorities the excuse for poor services which, in turn, gave central government ammunition with which to attack them for waste and inefficiency. It was hard for local authorities to break out of this cycle while housing management was firmly locked into the general administrative role of the town halls.

IMPROVING CONDITIONS

The 1960s saw a big change in conditions, although the total stock was stagnant due to demolitions within the cities – particularly Dublin – and abandonment and replacement of traditional croft dwellings in the country-side. This was highlighted by radically rising standards as Table 36.2 shows. By 1970, 111,000 new dwellings had been built, 87,000 of which were for owner-occupation, the rest for local authority renting.

Tenure continued to shift in favour of owner-occupation and away from private renting. The local authority share of the total stopped expanding and from then on declined steadily due to the aggressive sales policy to sitting tenants. Table 36.3 shows the changing tenure pattern.

Table 36.3 Percentage changes in tenure composition in Ireland between 1961 and 1971

	1961 (%)	1971 (%)
Public renting	20	16
Private renting	17	11
Owner-occupation	62	71
Housing association and non-profit	0.5	0.5

Sources: Blackwell 1988, DOE (Ireland) 1986.

Chapter 37

The 1970s

A PERIOD OF EXPANSION

Between 1950 and 1970, 83,600 local authority dwellings had been built – double the pre-war rate. The economy had begun to grow and emigration to slow down.

The 1970s saw a sharp change in Ireland's fortunes. For the first time since records were kept, and probably since the Famine, Irish immigration outstripped emigration. Table 37.1 shows the reversal in the trend.

The population increased on average 1.5 per cent a year throughout the 1970s, fuelling strong housing demand and generating a sustained house-building programme. A big head of steam was built up as the economy gained strength and the Irish government warmed up to entry into the European Common Market. The housing programme accelerated. Throughout the 1970s, Ireland invested a higher proportion of its national wealth in housing than Britain.

The government's role was extremely active, both on the local authority and on the private front. About one-quarter of total housing output in the 1970s was by local authorities, rising to one-third in 1975. It was the decade of greatest public building, with nearly 57,000 local authority completions. By the mid-1970s, Greater Dublin's population reached 1 million with the expansion accelerating from the mid-1960s, mainly in suburban and peripheral areas. This meant that by 1975 Dublin housed nearly one-third of Ireland's entire population.

ENTRY INTO THE EUROPEAN COMMUNITY

Ireland's entry into the EC in 1972 accentuated the centralising trend of the Irish government by turning Ireland as a whole into a region of the Community. Because of its rural economy, its poverty, its underdevelopment and its peripheral location, it enjoyed some benefits from entry into the EC.

Subsidies and grants created new possibilities, but these gains also

Table 37.1 Net migration from Ireland, 1951–80

Years	1951–60	1961–70	1971–80
Net outward/inward migration	−81,000	−27,000	+11,000

Source: DOE (Ireland) 1986.

generated some new problems. They increased the government's budget and therefore relative power, generating a public expenditure pattern that was not sustainable. The government borrowed heavily, creating a significant burden on taxpayers. Taxes rose steeply to meet a debt that absorbed over half of taxes collected. The increase in public spending encouraged rapid urban sprawl as housing subsidies were extended. The European Community farm policies generated an artificially high income for the rural economy, much of which went into newly built detached bungalows rather than into secondary industries such as food processing. The underlying state of the economy remained weak and vulnerable to international changes.

LOW-COST SCHEMES

Flats were still very unpopular. Land was not in short supply. Therefore local authorities were encouraged to use modern methods to construct low-density, low-rise houses. Breeze blocks, concrete, unconventional design, and open plan layouts became fashionable in low-cost schemes that maximised the volume of production because of unprecedented demand. The local authority stock of the 1970s was usually built on large estates, often of cheap, low-quality materials, in modern, low-rise styles. It was built to minimal standards to minimise costs and maximise units, so that those who could not manage even low-cost home-ownership could be housed and so that inner slum dwellers could at last enjoy modern conditions. An increase in slum clearance, the extremely low rents, the peripheral location of most local authority estates around cities and towns, and the strong incentives to owner-occupation all ensured that local authorities continued in their historic role of housing those on the *lowest* incomes. The growing scale of local authority developments and the sale of more popular property to existing tenants enhanced the trend towards a predominantly welfare role. The dislocation of traditional networks through clearance and the weakness of local authority management determined rapid decline in many of the new areas from the outset.

ABOLITION OF RATES

In 1973, local authority rates were abolished. All funding for housing was centrally generated from then on and local authorities were made even more dependent on central government policies. All housing capital and most housing revenue was 100 per cent government-funded. The planning role of the eighty-seven local authorities, traditionally weak and fragmented, was further limited by the loss of local income. Public administration, historically over-dependent on central government, was also weakened by this trend, providing a further justification for central government intervention. Incompetence and inaction in the inter-war years, leading to central government control and the removal of a local financial base, created greater administrative difficulties, leading to clumsy decisions, inappropriate policies and further government intervention in an attempt to resolve the problems.

SALES TO SITTING TENANTS

In 1973, the sale of local authority housing to sitting tenants was extended with higher incentives in the shape of subsidised loans, guarantees and grants. Sale prices included discounts and favourable valuations linked to the Consumer Price Index, rather than the Building Price Index (Dublin Corporation 1991).[1] Over the next fifteen years this measure was to transform much of the social housing landscape of Ireland. In all, 202,000 local authority dwellings were sold to sitting tenants as a result of this policy, nearly two-thirds of all dwellings ever built by local authorities. In spite of a very rapid public building programme throughout the 1970s, the proportion of the total stock in local authority ownership declined.

PRIVATE BUILDING

Private house production boomed – nearly a quarter of a million new dwellings were built. Today, nearly 40 per cent of all Irish housing dates from after 1970. The 1970s provided the largest increase to the total housing stock of any decade since the turn of the century. The boom in house building continued until the late 1980s, the bulk of it leading straight into owner-occupation. The level of owner-occupation soared to 76 per cent of all households by 1980.

In 1971 there were 760,000 houses. By 1980, allowing for demolitions, the stock had risen to 898,000, an increase that did not reflect the volume of building because many new houses in the countryside and in the cities replaced abandoned, older property.

IMPROVING CONDITIONS

As a result, conditions improved, overcrowding plummeted, and tenement dwellings became almost a legend rather than a reality. By 1980, tenure had evolved significantly further towards almost universal home ownership, as Table 37.2 shows, with basic amenities approaching international standards by 1981 (see Table 37.3). The number of overcrowded households had halved in ten years from 12 per cent to 6 per cent, an achievement that made the advent of the 1980s appear full of promise.

Table 37.2 Tenure composition in Ireland in 1981

Public renting	13%
Private renting	11%
Owner-occupation	76%
Housing association and non-profit	0.5%

Source: Blackwell 1988.

Table 37.3 Housing conditions in Ireland, 1981

Dwellings with:	
inside piped water	101%
inside sanitary fittings	85%
fixed bath	82%
Households with more than 2 persons per room	3%
Average number of persons per room	0.74
Average number of persons per private household	3.67
Households in dwellings with 3 rooms or more	96%

Source: Blackwell 1988: 10.

NOTE

1   By 1990, the average price of a three-bedroomed, inner-city council house was estimated at £37,000. After discounts, the sale price to a sitting tenant averaged £8,000. By comparison, to build a similar house would cost £50,000.

Chapter 38

Developments in the 1980s

Investment in housing was high throughout the decade; the stock expanded by a further 200,000 before the end of the 1980s, taking it up to over 1 million; owner-occupation reached 79 per cent. Standards rose and old housing problems were replaced by totally new ones. A number of unique initiatives were introduced in the 1980s with a number of different aims, including diversification and a mix of renovation and new build. The measures are described as closely as possible to chronological order, sector by sector.

RENOVATION

Full-scale renovation of older houses got under way. Renovation had begun on a small scale in 1958 but, for the first time in Irish history, there was a serious shift away from replacement or inaction to active inner-city renewal. Cork and Waterford were particularly imaginative in their approach, saving many attractive old buildings and integrating a traditional and vernacular style of housing with commerce, craft and tourism.

In 1982, the government set up a special Task Force for the Elderly to target improvement grants at the worst remaining slum housing in inner cities. Its aim was to carry out basic repairs and improvement on unfit and insanitary accommodation occupied by elderly or frail people. The initiative, like much housing policy since the 1960s, also involved training and job creation, employing young trainees.

URBAN RENEWAL

The Urban Renewal Act of 1986 targeted fourteen main towns and cities. It provided grants to owner-occupiers to build or renovate housing in inner areas. These grants covered 50 per cent of the cost. There was also special tax relief to owners for improvement works.

PRIVATE RENTING

Under the impact of rapid housing change, private landlords had played a small role. Incentives were virtually non-existent, since even low-income households could buy and marginal households now found a plentiful supply of low-cost, local authority housing. The abolition of rent control on most lettings and rent assistance for those receiving supplementary welfare allowances had prevented further decline but had failed to produce significant increases.

The stagnation of private renting posed problems for certain groups who were too poor to buy but who were denied access to social housing – mainly single people and childless couples, such as newly forming households, vulnerable people in need of ready furnished rooms, people leaving institutions, and a growing number of returning homeless migrants.

The increasing problems of homelessness and the virtual lack of a housing association sector as a third arm provoked a private renting initiative. The introduction in 1984 of tax incentives for new-build, private-rented accommodation was a historic break in the trend away from renting. Although small in its impact, producing under 1,500 new units in that year, it did demonstrate that incentives were necessary and beneficial if private renting was to survive alongside the financially more favoured tenures. In the event, these incentives were dropped in the following year. But in 1988 new measures to help private landlords were introduced, offering generous tax relief and helping to create nearly 2,000 new or refurbished homes. This support was extended in 1991 (DOE 1991a).

VOLUNTARY HOUSING

In 1984, direct government financial help for non-profit housing associations was introduced for the first time. Subsidised loans for 80 per cent of the cost of building with deferred repayments became available as long as the housing helped special categories of needy people such as the elderly or disabled. There was a cash limit of £20,000 per unit. High rents resulted from the repayments on the loans for 20 per cent of the total cost, which precluded the poorest households the scheme was designed for. Therefore the government later raised the proportion of housing association loans that could be subsidised to 95 per cent of the cost, as long as they helped to house homeless people. This new grant regime was introduced in 1988 in an attempt to reduce segregation of the homeless into local authority housing.

In 1980, housing associations barely provided one in 250 dwellings in Ireland, building fourteen units in 1981, but they slowly took off in the 1980s and are enjoying a bigger role as the needs of special groups grow and the role of local authorities becomes more difficult. In 1990, they built

500 dwellings. The total stock had reached 1,600 and was bound to increase rapidly.

Co-operative housing initiatives also began to gather momentum under this new funding regime.

THE £5,000 SURRENDER GRANT

Probably the most innovative and certainly the most controversial initiative of the 1980s was the local authority surrender grant, introduced in 1984. The aim of this grant was to induce tenants who wanted to buy, but were not prepared to buy their local authority dwelling, to move, thereby freeing up a low-cost unit for more needy households on the waiting list. They were given a £5,000 grant to help them do this. It aimed to kill two birds with one stone; extend owner-occupation and help homelessness; however, its impact was almost entirely negative.

Many tenants on the worst estates took up the £5,000 surrender grant, leaving local authorities with a massive problem of empty property, impoverishment, difficulties in letting, economic decline and greatly increased stigma. After nearly 9,000 tenants had received the grant between 1984 and 1987 – one in twelve of all council tenants – the scheme was abandoned in 1987 in a flurry of denunciations by virtually every housing lobby group in Ireland, including groups representing the homeless who should have benefited from it (see Chapter 39 for detailed discussion).

TENANT INCENTIVES

A new tenant incentive scheme was also introduced in 1984, making the terms for sale to sitting tenants even more favourable.

Throughout the 1980s, the sale of council property continued, but very few dwellings built after 1970 were purchased. Thus the large, low-cost estates were a special target for the surrender grant because their unpopularity prevented tenant purchase while encouraging flight. The tenant purchase and surrender grant schemes combined to intensify the polarisation within social rented housing because tenant purchase was only successful in more stable and popular areas, while the take-up of the surrender grant was concentrated in areas where people with ambitions to become owners did not want to buy.

THE 1988 PURCHASE SCHEME

After the surrender grant was abolished, the incentives for tenant purchase were increased even further to keep up the pace of expanding owner-occupation, while reducing the expensive and problematic local authority role. Discounts were introduced regardless of length of tenancy;

repayments were indexed so that repayments in the first five years became minimal (DOE 1988b, 1988c); and market values were introduced to replace sale prices based on the outstanding debt on the property, enhanced in line with the Consumer Price Index. This was aimed at helping sales of newer properties which were among the most expensive and least popular. Market values for some of these properties were significantly lower than either the historic cost of construction or outstanding loans. Previously tenant purchase had to cover the cost of the outstanding debt.

Chapter 39

The impact of tenant purchase schemes on local authority estates

Sales had both positive and negative effects on council housing. They appear to have helped many estates, particularly some large inter-war cottage estates. Tenant purchase strengthened the sense of 'belonging' (Corcoran 1987) and enabled many more low-income households to become owner-occupiers than would otherwise have been possible. Government and local authorities alike claimed that it improved the maintenance of those houses and also the general appearance of the estates (Corcoran 1987). Some council estates that were previously unpopular, such as Ballybrack in outer Dublin, became more settled as a result of some tenants being able to buy. However, the most unpopular and poorest housing areas had very few sales to existing tenants. Low levels of economic activity, poor community stability and high turnover resulted, as those who wanted to purchase invariably moved elsewhere (SUSS Centre 1987).

Both the Combat Poverty Agency, funded through the EC, and organisations working with the homeless and rootless, such as Threshold, argued that the very strong tenant purchase policies had a seriously dislocating effect on public rented housing. The surrender grant scheme was uniquely unpopular with local authorities, while sales to sitting tenants were invariably supported. In its first year of operation, about 75 per cent of the take-up of surrender grants was concentrated within the three most stigmatised and disadvantaged areas in Dublin – Ballymun, Darndale and Tallaght. The vacancy rate on these estates multiplied five times over five years. Ballymun was the huge high-rise estate near Dublin airport; Darndale was a low-cost peripheral scheme, built of cheap and ugly breeze blocks; and Tallaght was the largest new town development, with 40 per cent of its houses owner-occupied but the rest a vast municipal holding. All three were built in Ireland's short mass housing boom. In Tallaght, one in six tenants – virtually everyone with a job – moved away as a result of the surrender grants (Threshold 1987). Ballymun lost one-third of its population in three years.

One of the motives behind the surrender grant scheme was to encourage low-income but ambitious tenants to find houses they were prepared to buy in more promising areas. They were given a unique chance to exercise choice and realise aspirations towards home-ownership. But the impact of such a policy on the remaining community was devastating.

A further aim was to help previously excluded, seriously disadvantaged households to gain access to local authority housing. But the accelerated rehousing of the homeless often failed to meet their deeper needs. Lack of support, community divisions, weak management, sheer poverty, and social instability made settling in very difficult.

The report by Threshold (1987) on the impact of the surrender grant underlined the glaring problems facing local authority tenants and managers on the worst estates. In the study, the effects of the surrender grant were found to be as follows:

- Communities where unemployment was already high suffered the loss of many of those people who had jobs.
- Income levels dropped and services in the areas deteriorated.
- Many of the vacant houses were vandalised.
- The community was substantially deprived of its leaders as they were the ones to move out with the grant.
- The large number of vacancies enabled the waiting list for Dublin Corporation housing to be reduced.
- But an already fragmented community now had to encompass the addition of other marginalised groups, such as single homeless people, single-parent families – many of whom were also unemployed – which the Corporation now found it possible to house.
- The community spirit, essential if such problems are to be combated, had taken demoralising punishment.
- Stigma, particularly in relation to job opportunities, provided a strong motive to change address.
- The overall effect gave further impetus to those who could do so to move.
- As the level of unemployed in the community consequently rose further, the circle was completed.

(Threshold 1987: IX)

The impact on management problems was not anticipated but was devastating, catapulting both the government and local authorities into finally launching a management initiative, beginning with Dublin Corporation's first local office in 1985 (IPA Conference 1988).

DROP IN DEMAND

A combination of factors – the general matching of supply with demand through a high rate of building, the high level of owner-occupation, the very favourable purchase schemes offered to local authority tenants, the high rate of sales of more desirable council property, the continued high rate of house building for private ownership till recently, and renewed high levels of emigration meant that demand within the local authority sector dropped significantly in the mid-1980s, with growing social need among those who were left behind in the race for owner-occupation.

One consequence of low demand was that up to one-quarter of waiting list applicants failed to take up their offers of accommodation if they were to the least popular estates (Kelleher *et al.* 1988). As a result, there was an intensification in the hierarchy of estates. Allocations were increasingly geared to matching the poorest and most needy groups with the worst housing areas. These were the only people likely to accept. At the same time, the characteristics of those applying changed. The proportion of applicants in work fell to less than 30 per cent. The proportion of one-parent families rose steeply.

One-parent families were explicitly denied access to better housing for a long time. In Dublin, until recently, they were only eligible for flats. More recently, they were rehoused either in flats or in other hard-to-let areas. The proportion of all lettings in Dublin going to one-parent families rose to nearly half of the total in the late 1980s.

Homeless people, many of whom were single, vulnerable men with a history of alcohol problems, mental illness and institutional living, were also generally only offered unpopular properties on large, hard-to-let estates (Kelleher *et al.* 1988).

The result of these allocations policies was a rapid rise in social and management problems in the least popular estates. Ballymun illustrated the segregating tendencies of these policies most vividly. It represented 8 per cent of Dublin's local authority housing stock. Yet in 1990, of the 2,800 flats, 800 were occupied by one-parent families. In 1985, 45 per cent of all rehoused single-parent families in Dublin were allocated to Ballymun. Sixty per cent of all homeless single people went there. Between 1980 and 1986 the proportion of new lettings that went to two-parent families dropped from 78 per cent to 22 per cent. Lettings to single people rose from 20 per cent to 68 per cent. Half of these were to single women with children (SUSS Centre 1987). The impact of lettings problems on the worst estates was so severe that Dublin Corporation adopted a radical new approach to lettings, allowing tenants to have an input into screening applicants on Ballymun where complete social breakdown threatened.

Patricia Kelleher put it baldly in *Settling in the City*, a report on home-lessness funded by the Combat Poverty Agency:

The crude forces of housing supply and demand have helped to create segregated and disadvantaged public housing estates in many parts of Dublin. Housing alone may not be able to redress the human consequences of poverty and unemployment ... 88% of homeless applicants were totally dependent on social welfare payments. The great majority were housed in Ballymun, Tallaght and inner city (81%). 60% of single men were housed in Ballymun – a year later one quarter of [homeless] people housed had vacated their tenancies.

(Kelleher *et al.* 1988: 9)

DUBLIN'S WAITING LIST

The size of Dublin Corporation's waiting list fluctuated between 5,000 and 5,500 from 1978 to 1983. After that it dropped steadily and was down to 3,000 by 1986. The number of completions throughout the period was between 1,000 and 1,500 a year. The overall drop in the size of the waiting list might show that new applications did not keep pace with completions. However, the number of new lettings far outstripped the level of completions and was made possible by the increasing vacancy rate in the public stock rather than by a fall in the number of applicants. The Threshold study showed that many of these vacancies arose through the take-up of the surrender grant and were concentrated in the areas of lowest demand (Threshold 1987). Table 39.1 gives exact figures from the Department of the Environment; most parts of Ireland experienced a fall in demand (see Table 39.2 for overall figures).

The combination of sales, reduced demand, increased poverty, changing family structure, the nature and location of the council stock, and weak management and maintenance, created a large number of empty units and high turnover in the most unpopular estates. Figure 39.1 illustrates the increase in empty dwellings up to 1987 and the continued high levels to 1990.

Table 39.1 Impact of surrender grant – changes in size of Dublin Corporation waiting list, housing supply, and lettings from waiting list, 1983–9

| | New lettings | Local authority houses completed | Vacancies: | |
			Surrender Grant	Casual
1983	8,943	6,190	–	2,753
1984	10,164	7,002	–	3,162
1985	11,791	6,523	2,226	3,042
1986	11,921	5,517	3,100	3,304
1987	9,600	3,074	2,400	4,126
1988	7,252	1,450	5,802	
1989	5,775	768	5,003	

Source: DOE (Ireland) 1991b.

Table 39.2 Drop in demand for local authority renting in Ireland, 1981–8

Year	Waiting list applicants	Year	Waiting list applicants
1981	29,000	1985	22,500
1982	30,000	1986	20,500
1983	29,000	1987	18,500
1984	26,500	1988	17,700

Sources: Blackwell 1989a, DOE (Ireland) 1991a.

CUTS IN TAX BENEFIT TO OWNER-OCCUPIERS

All these problems, and particularly the evidence of adequate supply, led to a change in the support for owner-occupation. Up to 1985, extending owner-occupation down the income ladder was an overriding objective. From 1985, there was a gradual reduction in the government incentives to

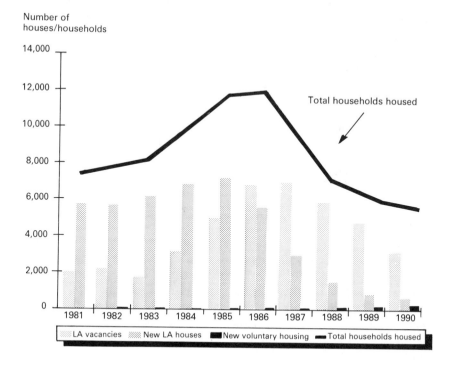

Figure 39.1 Social housing programme in Ireland, 1981–90

Source: DOE (Ireland) 1991a: 7

build or buy housing for owner-occupation. This erosion culminated in 1987 with fairly sharp curtailment of privileges. Tax relief on mortgage interest was reduced from a maximum value of £4,000 to £3,600 a year; it could now apply to 90 per cent rather than 100 per cent of the interest payable. A ceiling on the amount of relief was also introduced (DOE 1986). One of the purposes of these changes was to slow down the rate of new house-building in order to maximise the use of the existing stock, to cut demand, and hopefully to reduce some of the back-up problems.

BALANCE BETWEEN NUMBER OF HOUSEHOLDS AND NUMBER OF DWELLINGS

The big housing programme had led to a growing equilibrium between the number of households and the number of housing units. As a result, since 1985 there was a drop in production which was particularly marked in the public sector but also applied to the private sector (see Table 39.3). Local authority construction dropped from one-third of all completions in 1975 to only 10 per cent in 1988. The public building programme had virtually disappeared by 1990, but the emerging problems of homelessness led to renewed initiatives in 1991.

Table 39.3: House completions in Ireland in 1981, 1984, 1986, 1987 and 1988

	1981	*1984*	*1986*	*1987*	*1988*
Local authority houses	5,500	6,500	5,516	3,074	1,450
Other houses	22,500	18,500	17,164	15,376	14,200

Source: DOE (Ireland) 1988a.

HOUSING ACT 1988 – HOMELESSNESS

In 1988, in an attempt to tackle problems among vulnerable households in gaining access to local authority housing, a new Housing Act was passed requiring local authorities to give priority to the needs of the homeless. However, the Act fell short of giving local authorities the direct duty to house homeless people. In practice, local authorities rehoused increasing numbers of such households, particularly in unpopular areas, because of the drop in demand from better-off or more stable households for social housing.

THE ABANDONMENT OF THE SURRENDER GRANT

The surrender grant scheme had achieved the desired effect in opening up lettings to the homeless, but the policy was abandoned in 1987 under a battery of protest over its segregating, polarising and denuding impact. Irish housing policy debates became dominated by the seemingly overwhelming problems of large publicly-rented, peripheral estates and inner-city tenement flats where little tenant purchase took place but where the surrender grant had had a particularly devastating impact.

When once the surrender grant no longer existed and council building was brought to a virtual standstill – Dublin Corporation completed only fourteen houses in 1988, none in 1989 and twenty-five in 1990 – demand for low-income social housing rose again, waiting lists were growing and vacancies were dropping. Emigration still removed many potential applicants but this too would change if the economy picked up or, as proved to be the case, if there were fewer opportunities abroad.

POLARISATION

The 1980s had the effect of intensifying problems in local authority housing. The profile of tenants shows how deep the social problems had become.

Large families dominated the local authority sector. Although household size had fallen generally, with an average for Ireland of just over three, there were 4.4 people per household in the local authority sector (Blackwell 1988). *Over half* of the residents in council housing were under 16, with all that this implied in wear and tear, noise, energy, and need for special facilities and activities. This compared with 30 per cent for Ireland as a whole. The proportion of unemployed and of welfare recipients was also much higher – around 70 per cent of households had no breadwinner, compared with one-third in the population as a whole. About four times the level of one-parent families were housed in the local authority sector compared with the whole population. Poverty made the division between owners and renters stark and clear (EC 1990b).

UNPOPULAR ESTATES

One of the most important developments in the 1980s was the recognition that government action was needed to reverse the declining fortunes of local authority housing. The initiative was targeted at the unpopular estates where very little tenant purchase had taken place, principally the modern estates, but also the early balcony block estates. Table 39.4 shows how tenant purchase affected properties built in different periods, reflecting partly their historic cost but mainly their style, popularity and the income

Table 39.4 Dwellings built, sold, and owned by local authorities in Ireland, pre-1940 to 1989

Period built	Number built	Number sold to tenants	Retained stock
Before 1940	83,824	73,325	10,499
1940–49	21,314	17,988	3,326
1950–59	38,716	30,136	8,580
1960–69	23,621	14,504	9,117
1970–79	56,572	19,748	36,824
1980–85	37,066	1,431	35,635
1986–89	10,700*	2,542*	8,158*
Total	271,813	159,674	112,139

Source: Blackwell 1989a.
Note: *Estimates only.

of occupants. This in turn was reflected in social and environmental conditions. Only one-fifth of the stock built since 1970 had been sold (Blackwell 1988), whereas four-fifths of the pre-1970 stock had been sold – almost everything except the inner-city flats. This was partly because of the higher cost until recently of purchasing newer houses, partly because the stock itself was less attractive. There is now a growth in applications for the purchase of flats with 1,200 in the pipeline in Dublin.

In all, between 1970 and 1987, over 106,000 new council units were built, mostly on large peripheral estates, often using poorly insulated concrete which was highly unsuited to the damp Irish climate, and which were not attractive to potential buyers or even, in extreme cases, to potential tenants.

Outside the major cities, local authority housing tended to be built in small clusters. None the less, even small council estates were usually built quite separately from private developments. Local authority housing was generally considered stigmatised and separated out by the income of its occupants. Social isolation and decline were encouraged by the physical construction and layout of the estates. While tenant purchase had helped to stabilise some estates, much of the remaining stock was difficult to sell and would largely remain for low-cost renting.

REMEDIAL WORKS SCHEME

In 1985, the Irish government announced a long-sought-after initiative to tackle problem-prone local authority estates; the Remedial Works Scheme. There were about eighty run-down estates needing remedial works in Ireland and large sums were made available as grants to local authorities in order to tackle problems of disrepair and structural defects.

Twenty thousand council properties formed part of the programme to restore the most difficult estates – 20 per cent of the unsold council stock. In these areas tenant purchase was virtually non-existent. There were other areas not included in the programme which were also socially very deprived or suffering from the problems on a less intense scale.

The eighty estates in the Remedial Works Programme fell into three main categories: inner-city pre-war or early post-war developments; large modern outer estates; small clusters of unpopular modern developments in small local authority areas. Half of all the problem housing was in Dublin; half of the rest was in Cork. The average size of estates in the programme was 250 units, much smaller than in other countries. However, this belies the fact that the most serious problems were in large estates of a thousand or more dwellings.

The programme offered £1.6 million in its first year (1986); it had risen to £16 million by 1990. Between 1986 and 1991, £63 million had been given to local authorities. A further £131 million was still required. Two thousand units were renovated at an average cost of £15,000 per unit. This represented a very heavy reinvestment, certainly far beyond the estimated sale price under tenant purchase.

PHYSICAL CONDITIONS

Overall, very few flats were built by local authorities in Ireland – about 6 per cent of the total local authority-built stock or about 15,000 flats (Blackwell 1989a: 83). Nearly 3,000 of these were on the high-rise Ballymun estate on the northern edge of Dublin. The inner-city blocks of flats were often not modernised and lacked bathrooms and adequate heating. There were only a few thousand of these dwellings in the country.

Very few of the flats had been sold. As a result, about two-thirds of pre-1950 council dwellings were flats, many of them extremely run-down. Most were in the new scheme.

The biggest physical problems related to the modern, non-traditionally-built stock, often on large outer estates. The physical problems on these estates related to security, the environment, the estate layout, common parts, as well as problems of heating, insulation, defective windows and cheap, ugly finishes to basic building elements. Estates where these problems were concentrated were often built on a cut-price basis in an attempt to catch up on the 'mass housing' needs of Ireland's period of rapid growth. They were also targeted.

THE GOVERNMENT INTERVENES

The government rescue programme was based on several factors:

- There were certain areas of council housing where properties could neither be sold nor were being managed effectively. These estates were clearly in a spiralling cycle of decline.
- The surrender grant had revealed a much deeper and wider alienation than anticipated.
- Its impact accelerated management problems in the worst areas beyond anything previously experienced.
- The condition of the 'low-cost' housing schemes of the early 1970s was so serious as to make some of them appear unviable.
- Meanwhile, demand for housing from people in marginal or greatly disadvantaged groups was growing, particularly one-parent families and transient single men.
- The social and management problems of the worst estates were visibly accelerating; pressure was mounting across Ireland for something to be done. Housing lobby groups singled out these areas as much the most difficult and disadvantaged.

The government was directly responsible for any attempt to address these problems because of its dominant funding and promotional role in relation to local authority housing.

The European Community's Poverty Programme had helped focus attention on these areas by funding an ambitious initiative in Ireland. Much of its effort went into the poorest local authority estates, including Ballymun and Darndale, two of the most problematic areas. This programme emphasised the housing and social problems of these areas, stressing the extreme polarisation that was going on.

A NEW EMPHASIS: TENANT INVOLVEMENT IN MANAGEMENT

Alongside the physical renovation grants, the government took a strong lead in encouraging approaches to involving tenants and improving management. Their guidelines stressed that past practices had created many of the problems and should now be changed:

> Imposing solutions on the tenants will not succeed. They must be allowed to participate fully if the problems, many of which are of a social nature, are to be tackled and ultimately resolved. Applications [for Remedial Works funding] must include proposed initiatives in this regard.

(Corcoran 1987)

But the DOE remained sceptical of the local authorities' ability or willingness to innovate: 'little progress appears to have been made in many areas in improving management and maintenance arrangements, or increasing the genuine involvement and co-operation of residents in a structured way' (Corcoran 1987).

Historically, tenants had played a minimal role in council housing. The assumption was that they would simply be glad to be housed and the problems would end there. However, the rapid deterioration of estates, the increasing poverty of communities, as well as the increased difficulties in letting, led to initiatives to engage tenants more. It was now becoming accepted that physical improvements and management initiatives would not work without the backing of the tenants.

The Minister, Padraig Flynn, on 31 January 1989, announced a new allocation with the following conditions:

> I wish to emphasize that in conjunction with the physical improvements to the dwellings, it will be necessary for the local authorities concerned to put in place improved management and maintenance arrangements with particular emphasis on tenant involvement as a means of achieving the long-term improvements which these estates so badly need.
>
> (DOE 1989)

THE IMPROVEMENTS

When the Remedial Works Programme was launched by the government in 1985 it offered 100 per cent funding to cover the cost of restoring the eighty estates that were targeted. The government was aware that a 100 per cent grant to cover the capital costs of renovation gave local authorities little incentive to economise and maximum incentive to include every possible improvement.

By contrast, local authorities were saddled with housing problems they felt unable to tackle themselves, and the Remedial Works Scheme offered real promise of 'sorting out the houses once and for all'. The money covered four key elements:

- upgrading of internal conditions, usually heating and, in the case of older flats, kitchens, bathrooms and room layout;
- external works including windows, cladding, roofs and sometimes doors and porches;
- the environment and security measures, usually the access areas and land immediately adjacent to buildings;
- lifts, refuse disposal, staircases, entrances, balconies and other communal areas.

PHYSICAL RENOVATION

The buildings and their immediate surroundings had such serious problems that most of the money was used on them. Very little was left over for the general environment of the estates. None was allocated for the provision of social facilities.

Tenants were consulted in a most detailed way about priorities, which inevitably focused on the dwellings. Representatives came regularly to liaison meetings with architects. The retention or introduction of coal fires was of central importance to tenants and was widely adopted. Architects were willing to listen and produce costly plans which included environmental planting and enclosure, as well as the introduction of colours and modern standards of insulation. Local authorities asked for special funds. Because the estates had such serious physical problems in the dwellings, it was hard to give priority to the environment. The physical works were very popular but a question remained over the overall impact, even though the need for environmental upgrading was recognised, particularly on the very large estates, and the scheme met the full costs of any work carried out.

Local authorities achieved some progress in restoring dwellings on estates that appeared unrescuable, but had not always focused their efforts on managing the improved areas. The Irish government was pressing for this change as a follow-through to its investment in the buildings, but several large estates had no caretaking or cleaning, no local office and no plan to upgrade the repair or management service.

MANAGEMENT CHANGE

The most ambitious and innovative experiment in Irish housing management was on the Ballymun estate. In desperation as empty units soared to over 400, the Corporation collaborated with the Combat Poverty Agency in breaching every convention of housing management in the city. They set up a local office for the area of the estate where tenants were creating the most pressures over spiralling conditions. A local manager and a community worker were based in the office, funded by the Corporation. Tenants were allowed to have an input into vetting 75 per cent of lettings. A local repairs team was set up. Flats were made available for community purposes. It was enough for the Corporation to see the cost of repairs and the number of empty flats dropping to extend the experiment to two other areas of the estate. The number of empty dwellings at Ballymun was halved by 1991 and there was a waiting list for the areas with local offices. Relations between the tenants and the Corporation were improving after years of conflict. However, the offices had not been made permanent and there was as yet no major decentralisation of management throughout the city.

Cork City, by contrast, continued to run its housing from the town hall with a very detached approach to management problems. The Department of the Environment's message had not yet percolated into action on management, although tenants' views on improvements were certainly being heard. To this extent, the new emphasis seemed to be working. The test would be whether longer-term repairs, caretaking and tenant liaison could be installed and maintained on the estates themselves, which were still very remote from the main administrative centres of the local authorities.

The aim of the Remedial Works Programme in Ireland was to create a revolution in tenant involvement and experiments in local management. But the Remedial Works Programme alone could not reverse the acute social problems faced by extremely marginal households in the worst estates. Nor could management change. There remained a serious question over whether these estates were at all suitable for housing in large numbers the most marginal households to the virtual exclusion of economically active households.

Plate 39.1 The building of Ireland's only high-rise estate at Ballymun, North Dublin

Chapter 40

An overview and conclusions

A NEW SOCIAL HOUSING PROGRAMME FOR IRELAND

In April 1991 the Irish government issued a new 'Plan for Social Housing' to tackle housing problems (DOE 1991a). It included several important recommendations:

- New council housing should be bought from the private sector in scattered units throughout city areas with mixed social groups. Estates of local authority housing should no longer be built.
- Voluntary housing associations would receive higher grants and rent support to provide special housing for vulnerable, needy groups.
- Private landlords should continue to receive special help to stay in business.
- Low-income owner-occupation should still receive special help.
- Tenants should be more involved in the running and improvement of their estates.

These measures firmly reversed the main post-war trends. It will no longer be recommended to concentrate homeless single people in large, unpopular estates. Nor will single tenure estates be built. The new programme also committed the government to extending the Remedial Works Scheme.

However, the new emphasis may not resolve problems on those estates already housing very vulnerable people in large numbers. It rests very much with the local authority landlords to devise and sustain new experiments in breaking up large estates, in creating mixed tenures there, in providing concentrated management services, in maintaining improvements, and in continuing to work with tenants' representatives. Turning back the tide might only be possible if the wider economic situation improves, although that of itself would create new strains, demands and pressures.

All forms of rented housing had begun to expand again by 1991, whereas owner-occupation had reached a plateau. Local authorities were no longer selling so much of their stock, as most of the best properties had gone and most tenants had insufficient income. Therefore they were confronted with having to improve their management and win the confidence of their tenants. Their share of the total stock had dropped to 10 per cent by 1991.

The voluntary housing movement was building at fifty times the rate of the previous eighty years – still only 500 units a year! Private landlords were offered extended tax relief, as incentives to invest in private renting were improved (DOE 1991a).

OWNER-OCCUPATION

In Ireland, support for owner-occupation is long-standing and uncontroversial. Each government statement of objectives clearly underlines the support for owner-occupation as a universal goal. The obligation is placed on individuals, with the help of the state, to meet their own housing needs, so long as they are able. Governments have supported owner-occupation in a consistent and single-minded way, not just with mortgage interest tax relief but also with special cash grants for first-time buyers; cash grants to help individuals build replacement and new housing in rural areas; the underwriting of private loans by local authorities for low-income households; mortgage support for unemployed owner-occupiers; the direct provision of mortgages by local authorities for low-income households; the long-standing practice of selling council stock to sitting tenants; and other strong incentives. In 1992, new measures were proposed to make the purchase of council flats easier (DOE 1992).

The two main parties in Ireland both support the aim of universal home ownership. The Irish Labour Party, while more cautious and more critical of the distortions created by over-favouring owner-occupation, is also committed to the principle. The two main assumptions are that virtually everyone wants to own their own home and that housing will be better cared for and therefore cheaper in the long run if individuals feel deeply committed to it (Corcoran 1987).

LOCAL AUTHORITY OWNER-OCCUPATION

The government has used local authority rented housing as a stepping-stone into owner-occupation. One-third of Ireland's stock was built by local authorities, but over two-thirds of it has been sold to occupants. Local politicians and administrators express virtually unanimous support for the sale of council housing. Dublin Corporation has sold 38,000 council dwellings, nearly 60 per cent of all the houses it has ever built. Cork City Council has sold 5,000, 40 per cent of what it built. Sales are expected to cut into local authority stocks even further.

The government makes it as easy as possible for the maximum number of sitting tenants to become owners, through a number of special features.

Tenants are eligible to buy their home from the first day of occupancy. There are no legal fees or down-payments. The local authority handles the transfer. The cost of purchase is on average less than half the value of the property.

In Dublin, the average sale price of council housing was £8,000, whereas the average price of an inner-city terraced property of similar size was £50,000 (1989). None the less, repayments were usually much higher than rents – maybe £17 per week.

Local authorities themselves assess whether a tenant can afford the repayments. Tenant-owners pay their mortgages to the rent collectors in the same way as they pay their rent. Purchasers hitting an economic crisis or defaulters can opt to revert to being tenants. In Cork, about one household per week does this. One major gain from the approach, integrating tenant purchase with renting through the local authority system, is that owning itself is less distinct from renting as a form of tenure.

The Irish housing ownership pattern is unique in Europe in the extent to which people who are poor have been helped to buy good housing. It demonstrates how far owner-occupation can be pushed and how popular and attractive it can be. But home ownership in Ireland is strongly linked with the national sense of identity, based on historic struggles over land ownership and, as such, may be unique.

NEW PRESSURES

By the end of the 1980s a number of new housing problems were piling up in Ireland. Low-income owner-occupation was producing some serious strains in terms of meeting repayments and repair bills. Older inner-city areas were improving but private-rented accommodation was under pressure, worsening the homelessness situation. The unpopular, peripheral local authority estates were ever poorer, housing increasingly transient people as the economic situation deteriorated. On the other hand, the Remedial Works Programme offered new hope and some prospect of better conditions in the long run. Also, the steep drop in building created new demand from low-income households. Waiting lists were growing again and the number of empty dwellings was dropping. As a result, local authority building was expanding again – only very slightly with 150 new units in 1991!

OVERVIEW OF HOUSING DEVELOPMENTS

In spite of still serious problems, remarkable progress in Irish housing conditions had been achieved. As the 1980s progressed, the traditional problems of obsolescence, overcrowding and lack of basic amenities were reduced to very low levels. Meanwhile the pattern of ownership continued to evolve. Table 40.1 summarises the main developments.

Much of Ireland's housing stock is modern and has greatly expanded in the last twenty years. Over 400,000 units have been built since 1970 (see Table 40.2).

Table 40.1 Change in types of house tenure in Ireland, 1951–90

Tenure	1951[a] (%)	1961 (%)	1971 (%)	1981 (%)	1990[b] (%)
Public renting	11	18	16	12	13
Private renting	32	17	11	10	9
Owner-occupation	54	60	71	74	78
Housing association and non-profit	0.5	0.5	0.5	0.5	0.5
Total stock ('000)	–	676	760	898	1,057[c]

Sources: Blackwell 1988, DOE (Ireland) 1991b.
Notes: [a]Estimate based on DOE sources.
   [b]We have given DOE figures for 1990.
   [c]Total population in 1988 (estimate DOE), 3,500,000; total housing stock in 1988 (estimate DOE), 1,057,000; new housing units 1980–90, 190,000; net additions to stock 1980–90, 100,000.

Table 40.2 Housing stock in Ireland by period, pre-1919 to 1988

	Number of dwellings	Percentage of total stock built
Pre-1919	246,591	24
1919–40	141,354	13
1941–60	142,643	13
1961–70	110,985	11
1971–80	227,763	21
1981–86	147,400	14
1987–88	40,000	4
Total	1,056,736	100

Source: DOE (Ireland) 1986.

CONCLUSIONS

The economy

The economic boom of the 1970s and 1980s had tailed off and Ireland's foreign debt situation had seriously worsened, leaving Ireland with a higher ratio of debt to income than most other EC countries. It also had a higher level of taxation in order to maintain its public services. Both these factors led Irish governments to seek economic retrenchments by cutting public spending. Unemployment rose to nearly 20 per cent, and emigration by the late 1980s was extraordinarily high – 30,000 or more were leaving each year, compared with net immigration of 13,600 throughout the 1970s (Blackwell 1989b). This was the highest post-war level.

European links

However, 1992 and Community developments are bringing further funds to Ireland. One hope is that Ireland's low economic development, its good housing stock, and its relatively clean environment should provide attractive possibilities for investment and growth, though these assets are hampered by her small internal market, her distance from, and weak transport links with, Europe, and the historic problems of unresolved sectarian conflict in the North.

Ireland, a single region within the European Community

The division of Ireland makes progress on many fronts more difficult. Most importantly, it diverts political leadership away from national economic and social issues.

The divisions between North and South often seem as bitter today as ever. It is hard to know whether Northern Ireland is any nearer to self-government now than it was when the Stormont government was suspended in the province in 1970. It is clear that the Dublin and Westminster governments favour some kind of compromise outcome. The fact that Britain and Ireland are partners within the European Community could begin to blur historic barriers across Ireland. Ireland, North and South, could move closer together as a result of regional developments. The wider political and economic framework of the European Community provides a new kind of umbrella that could be more neutral and more sensitive to the historic problems of Ireland than Britain has ever been able to be.

The Danish model

The Irish government is currently examining how the small Scandinavian countries like Denmark and Norway revolutionised their economies after the Second World War from a peasant, agricultural background into successful, modern industrial states. One important ingredient was their 'market oriented' system of welfare, delivering housing and other social services through businesslike units of organisation. The other was their emphasis on production, modernisation and export. It may be that Irish local authorities will keep their general administrative and political role but relinquish their management services to more autonomous bodies, based around large estates or housing areas. This process has partially begun at Ballymun in Dublin and will possibly be extended to other areas (DOE 1991a, 1992).

Emigration

The almost continual outward flow of the Irish population has generated a depressed and debilitated feeling among those left behind at all levels of Irish society but particularly in the poorest estates, where the desire to leave may be strongest, the exodus most dramatic, and those remaining the most powerless.

Ireland as a whole faces similar problems to the declining regions of other European countries. The social housing sector often takes care, not of the newcomers, but of those who are left behind, those who failed to move on and often those who did not emigrate but had poor opportunities and few resources at home.

One of the most striking features of poor housing areas in Ireland is the homogeneity of the population, the Celtic features of virtually all the children, and the almost total absence of anyone who is not Irish. In that sense, the Irish housing estates are reminiscent of the South Wales areas which have a population 'left behind' after coal-mining disappeared. But because Ireland has been losing people steadily and almost unremittingly for seven generations, the effects are more deepset and more difficult to reverse.

Social change

The powerful role of the Church and its teaching on family and marriage are increasingly contradicted by the concrete reality that Irish communities face – namely, rising rates of marital breakdown; rising numbers of one-parent families; and unprecedented levels of emigration with its associated 'non-Catholic' influences. The Catholic Church, an institution that has dominated much of Irish life – holding it back or holding it together, depending on one's perspective – now looks remarkably vulnerable. This in turn enhances the fragility of many social and cultural institutions. Almost all schools and many medical and welfare institutions depend largely on the Church. Large church buildings dominated the landscape of every major estate. Social activities on estates often revolved around Church-sponsored facilities. Dublin Corporation, as well as tenants' organisations, relied heavily on the churches for support. A process of secularisation, common across the Western world, is maybe more destabilising in its impact in Ireland because of the extreme contrasts between the social and religious objectives and the reality of social problems, and also because of religious, cultural and political cross-currents in the struggle for independence.

Politics and management

In Ireland, politics are a very local affair, with many TDs (MPs) also being elected local authority councillors. They intervene in the minutiae of allocations and are often accused of operating 'the parish pump' instead of running the country.

Irish housing policy and housing tenure are far from being the burning issues they have been in Britain. In that sense, Ireland is much more akin to Europe, with a stable political framework for housing development. As this section has illustrated, that framework is heavily biased in favour of individual ownership, though not to the exclusion of social housing. Local authorities are assumed to be long-term landlords, if in a very welfare-oriented role. There is no talk of major stock transfers to housing associations or of rapid growth in either private renting or voluntary housing, although new rented housing will increasingly be provided by them.

Local authorities remain relaxed in their approach to management. There is little sense of urgency over the acute management difficulties, and few management initiatives. There are as yet few financial incentives for local authorities to change their approach. Central government remains very much the main actor and it is hard to envisage major changes. But the government is pressing for tenant-based initiatives on marginal estates and, if the estates become more stable, this approach may grow.

Major government housing developments started late in Ireland. Not until the 1970s did Irish local authorities build on a large scale. Local authorities have been happy to sell a majority of their stock over a long period. They may now be happy to parcel out management into more direct, more tenant-oriented service organisations along the lines of Britain's Priority Estates Projects (Ballymun Task Force 1991). Non-profit organisations which have a strong tradition in Ireland may well seize the initiative, rather than simply castigate local authorities for housing the very poor so poorly.

Part VI

Summary and conclusions

Chapter 41

Summary of five country studies

State-sponsored housing went through common cycles of development in all countries:

- private and philanthropic initiative led to state-funded and regulated housing provision;
- the high-rise boom after the Second World War provoked a shift in emphasis to renovation;
- the strong development of owner-occupation cut demand for 'mass' estates and encouraged access by vulnerable groups (often immigrants) to the most unpopular housing areas;
- attempts to shore up private renting and curb the costs of owner-occupation through changing incentives were coupled with renewed shortages and homelessness problems, leading to an increasing reliance on social housing for marginal groups.

Each country, within these common trends, had unique features. The following summaries are drawn from the five country studies.

FRANCE

Developments in France were state-driven, with powerful and often top-heavy institutions orchestrating events (Dauge 1991). Housing was neglected until well after the Second World War as a result of late urbanisation, and the effort to catch up produced massive-scale, high-rise, peripheral urban developments to overcome terrible conditions. Housing providers by law had to be autonomous because of early fears of political favouritism, corruption and conflict (Quilliot and Guerrand 1989). But within individual *communes*, state-sponsored but legally independent landlords sometimes enjoyed a near-monopoly position. All tenures grew rapidly in the post-war period under big government incentives, near universal savings, and investment (Ministère de l'Equipement et du Logement 1989).

Immigration since 1960, mainly from the African continent and the

Caribbean, greatly affected urban conditions. Vast peripheral estates became segregated and unpopular. France experienced serious disorders in several peripheral social housing estates in the early 1980s and 1990/91 (*Le Monde* 1991).

The French took a strong lead in tackling unpopular estates and initiated European Community-wide efforts to combat social segregation in cities (Jacquier 1991).

GERMANY

Germany has a federal state with widely dispersed powers and responsibilities. Huge population upheavals, war and border problems made Germany's housing situation extremely difficult. Tenure and ownership were not dominant issues but building on an unparalleled scale was. Subsidies were strong and fairly evenly distributed between private and non-profit organisations. All kinds of landlords played a social role in relation to local authority responsibility for access and overall developments. Mass housing became important because of huge shortages. But the diversity of landlords and the decentralised system of government prevented large-scale peripheral estates from becoming over-dominant. Owner-occupation played a relatively small role in Germany, while private renting remained the main tenure.

German efficiency generated swift responses to seemingly overwhelming problems. But a devastating corruption scandal in non-profit housing in 1982, following the bankruptcy of the largest, German-based, social landlord in Europe, led to the abolition of non-profit laws and a greater emphasis on private initiative (Fuhrich *et al.* 1986).

Housing shortage and poor conditions became a major political issue again in 1990 in the former Eastern and Western regions of Germany after a short period of balance in supply and demand. Re-unification of the two Germanies and changes in Eastern Europe affected Germany far more than elsewhere, leading to rapid immigration, renewed house building, and intense conflict in the late 1980s and early 1990s.

BRITAIN

The early decline of private landlords through tight rent controls in intensely urbanised Britain led to an emphasis on owner-occupation and strong state intervention (Thompson 1990). The fact that cities were old and urban housing obsolete led to massive state-run slum clearance, dense, inner-area rebuilding, a strong welfare role for social housing, and early polarisation. The scale of local authority landlords made social housing politically contentious but also increasingly valuable to urban authorities. Rapid immigration and racial tensions were interwoven with slum clearance and renewal programmes, provoking periodic, intense conflict in major cities throughout most of the post-war period (Scarman 1986).

The political control and ownership of social housing, and the link with government-run slum clearance programmes, led to short-term objectives – numbers over quality, low rents over repair – and weak management, as well as extreme swings of policy, depending on elections (OECD 1988). Housing associations did not play a major role until the 1970s, leading to a near monopoly role for local authority landlords, not just within neighbourhoods as in France, but within whole cities. Residents played an unusually powerful role, attacking local authority standards but voting down government alternatives. By 1991, social and private renting were enjoying something of a come-back, low-income owner-occupation was facing a financial crisis, and property values were in absolute decline. State-run housing was being increasingly diversified and privatised.

DENMARK

Multi-party, long-term political stability, a strong welfare state, market orientation and an emphasis on quality, made Danish social institutions unusually responsive and impressively effective. The commitment to participation and the co-operative tradition encouraged a strong focus on social integration, generous social and community facilities, and a much greater say for tenants than in other countries. Danish tenants must buy a stake in their social housing and elect a majority of representatives to the controlling boards of all social housing companies. The belief in 'tenants' democracy' coloured the Danish approach to housing problems.

Denmark built its urban housing rapidly after the Second World War, specialising in prefabricated industrial systems and creating many 'mass' housing areas which became problematic more recently. Remedial programmes for these areas were developed in line with the Danish commitment to efficiency, integration, and quality provision.

Danish social housing organisations had an ethos of independence, based on partial self-financing, local budgetary control and accountability that made them more like businesses than welfare bodies in other countries. Their urban social problems have been contained, if not eliminated, through enviable standards of service (Scott 1975, Jones 1986).

IRELAND

Ireland's still predominantly rural base, its reliance on Britain, and the border troubles with the north, influenced housing greatly. Owner-occupation and self-build housing overwhelmingly dominated developments. The vast majority of the Irish population lived in comparatively high-quality, recently built, single-family, privately owned houses (DOE (Ireland) 1991a;

Blackwell 1989a). The reversal of a hundred years of almost continuous emigration in the 1970s led to major developments of low-cost social housing estates through local authorities (Blackwell 1989a). Local authorities sold their stock to sitting tenants on a massive scale – over two-thirds of all they built.

The rural dominance, long-term emigration and politically uncontroversial sale of better quality social housing all helped create intensely segregated and run-down social rented estates. Irish social housing was more marginal than in any other country.

The decline of these areas was exacerbated by large-scale unemployment and poverty, the dominance of central over local government, and the return to high levels of emigration in the late 1980s. The government and the European Community funded special programmes to help the poorest areas, with residents playing an active role.

Ireland's economic problems dominated her national life and her ties with the European Community. Ireland, as an outer island of the Community, was more like Greece and Portugal than Denmark in many respects, and understanding her difficulties was important if European harmonisation was to become a reality.

There is a strong local colour to housing developments: state-inspired, monumental developments in France; multi-landlord and privately oriented, mass building for rent in Germany; large local authority landlords, political controversy and mass owner-occupation in Britain; co-operatively based, tenant-oriented, socially advanced housing in Denmark; almost universal, rural and suburban owner-occupation and impoverished social housing in Ireland.

But housing experience also converges: architectural aberrations, outer estates, racial tensions, inner-city decline, polarisation, segregation, decay and neglect, private market orientation, and government re-engagement in the face of intense problems. The summary of main findings which follows explores the unanticipated convergence of experience.

Chapter 42

Main findings

FORMS OF HOUSING PROVISION

Pre-Second World War, private landlords in the five countries provided the bulk of homes in urban areas. However, rapid urbanisation had created such insanitary and overcrowded conditions for many of the poorest households that government intervention to speed up building programmes and to regulate urban conditions accelerated everywhere.

The Second World War was a housing catastrophe for Germany, France and, to a lesser extent, Britain, creating large tracts of devastated or badly damaged housing in big cities. In all five countries building virtually halted for five years and after the war there was a chronic housing shortage everywhere. Over time, this led to a massive scale of state-aided building and, between 1950 and 1960, a majority of units were directly sponsored by the state. This was the only time in this century when such significant state intervention occurred. The relentless drive since the war to produce more housing units continued virtually unabated until fairly recently. But since 1960, about three-quarters of all dwellings in the five countries have been built by private owners, although directly government-sponsored housebuilding continued to be significant till the late 1970s, producing between 25 per cent and 45 per cent of new units in the five countries.

Thus throughout modern times, in spite of huge pressures on governments to intervene and in spite of very large government programmes in times of crisis, housing has been built and owned predominantly privately, though not without many indirect forms of government help.

Private renting

The level of private renting has declined in all five countries, though much more slowly on the Continent than in Britain or Ireland. France, Germany and Denmark have sought to preserve private renting as a valuable source of cheap, flexible and popular urban housing. They have provided tax incentives and even direct subsidies to encourage private landlords to stay in business. This has slowed rather than arrested the steady decline. But it has preserved private renting as a very large sector, generally larger than the social sector. In Germany private renting dominates.

By contrast, in Britain and Ireland private renting has shrunk so fast that it is in danger of disappearing and major problems have emerged in the social rented sector as it attempts to house within its more rigid and uniform structures growing numbers of households no longer catered for in the private-rented sector.

An element in the survival of a large private-rented sector on the Continent – 20 per cent in Denmark, 31 per cent in France, 45 per cent in Germany – is the performance of private landlords. Government protection and financial incentives led to a much later shift to owner-occupation and ensured that a wide income range went on being housed in private-rented flats. The role of landlords in maintaining standards therefore appears to have remained more active, including reliance on the 'concierge' system. At the same time, very cheap and dilapidated private renting has been valued for the reasons of low cost, flexibility and openness to marginal groups. The idea of wholesale slum clearance or urban renewal was not countenanced in any of the three Continental countries.

The experience of Britain was almost the reverse, where well down the income ladder people shifted over to owner-occupation from the late nineteenth century; where private landlords were starved of incentives to maintain their property through long-enduring and tight rent controls without other compensating incentives; and where the older urban housing stock hit obsolescence faster. Massive slum clearance appeared the obvious answer, both socially and politically, thereby wiping out the majority of private renting that was not converted to owner-occupation.

Ireland is a case of its own because of its near universal owner-occupation and its almost exclusively 'welfare' social housing, with a residual private-rented sector.

All countries accept the important role of private renting with its varied, often small, and local operations; with its frequently low standards and consequent low rents. It is often needed by newly formed households; by incoming job-seekers and migrants, by young people, including growing numbers of students; by those who do not fit into any conventional category and can be absorbed in a many-sided urban network offering different styles of ownership but who are difficult to absorb short-term in single-tenure, communal and uniform estates of social housing.

Cities attract and create a large variety of households who do not fit into clear categories or who are on the move. Cities have for centuries provided messy and sometimes shady, quick-in, quick-out rented housing. Private landlords have been the providers and their decline squeezes precisely those groups that have most easily been accommodated by them – people at a transitional stage in life or in difficulty, who need cheap and immediate shelter, who cannot simply wait in a queue for something better but require an interim home (Donnison 1987).

Because of this need, Denmark has restricted the sale of private-rented property into owner-occupation; France and Germany have protected and extended the incentives for private landlords; and Britain and Ireland have for the first time provided short-term tax relief for new private landlords. Private renting may not grow, but it will not be allowed to disappear.

Owner-occupation

In all countries the pattern of private housing provision changed radically after 1960. Owner-occupation increasingly replaced private renting, and then social renting, in the provision of *new* dwellings. This trend to owner-occupation became so strong that the older private-rented sector began to fall prey to owner-occupiers too; improvement grants for inner-city renewal made the conversion of dilapidated rented property more attractive. The sale of some social housing into owner-occupation became a common feature too, though the right to buy only exists in Britain and Ireland.

The desire to own a home appears to be both strong and deep-set in most people in a range of countries. It represents the traditional form of ownership in most peasant societies. In particular, families with children prefer houses and in many cases prefer to own. Owner-occupation more often takes the form of single-family houses than is the case with renting in all the countries visited. By contrast, renting is often in the form of flats because landlords can concentrate more people on scarce urban land and install more modern amenities at higher space standards in multi-storey blocks.

There is a strong debate in all countries over:

1 the desirability of ever higher levels of owner-occupation;
2 the cost of supporting owner-occupation;
3 the strength of the desire for owner-occupation.

The reality is that all countries support it, electorally and financially; in all countries it is the most sought-after form of housing because it provides the largest number of single-family houses, the highest space standards and greatest individual control.

Although owner-occupation is more extensive in Britain and Ireland, it has grown rapidly in the three Continental countries. Even Germany, with only 40 per cent owner-occupation, has experienced heavy demand and rapid growth. While it is true that flats have been more popular and less problematic generally on the Continent than in Britain or Ireland, it is also the case that the majority of new units in France and Denmark are now single-family houses. The drift is possibly less strong in Germany because of huge housing pressures through immigration. Even there, there is a strong shift towards single- or two-family houses for owner-occupation. Experience from the five countries underlines the growing wealth and therefore the rising standards and increased individual spending on better quality, more individualised housing.

An important continuing growth area everywhere will be the second homes market, already a significant sector as urban congestion and environmental degradation continue and wealth increases.

Another growth area will involve forms of shared ownership and 'semi-ownership', where owner-occupiers enjoy some of the features of tenants, thereby making a form of owner-occupation more accessible to lower-income groups. The case of new leaseholders of flats under British Right to Buy legislation, where structural responsibilities and communal maintenance remain with the landlord, illustrates this. A similar system applies to private leaseholders and condominium ownership in Europe. Danish private co-operatives, where member-tenants own a stake in their home but can only take that stake up to 20 per cent of the total value of the property, offer a different model. The individual property remains as part of the co-operative, available to future members, while still representing a capital asset and a high level of individual control to the shareholding tenant/member.

Owner-occupation is likely to continue to grow in Europe, though its rate of expansion has already slowed because of its high cost and therefore problems of affordability. Government-led financial incentives have been limited or cut recently in all countries.

State-owned housing

Britain and Ireland are almost unique in Western Europe in providing the bulk of their social housing through direct public ownership. In France, Germany and Denmark, the amount of public housing is negligible, a small residue from the post-war crisis.

Local authorities in Britain dominate rented housing. They got directly involved in large-scale redevelopment programmes coupled with house building, which made them grow rapidly and forced them into rehousing *en masse* households from the worst areas of old housing, where many marginal households had previously lived. They increasingly took over from private landlords in 'twilight' areas. There are few pressures for good performance where there is a monopoly, particularly a monopoly service to low-income households.

In Britain and Ireland, public housing appears to have certain characteristics that make it difficult to manage and have led to a decline in public support:

- until 1990 it had not had discrete budgets, being part of the local authority operations, and there was therefore no clear allocation of funds for management and maintenance;
- it has had low rents because electoral support is closely tied to 'political ownership' of social housing;
- it lacks a single-purpose landlord structure and therefore has complex management arrangements;
- its access and allocation systems are highly bureaucratic and insensitive to varied individual needs;
- its staff are often untrained in housing management and represent town hall administrations rather than tenant-oriented services;
- its level of disrepair is high because it is provided as a welfare service to minimal standards rather than a business service to satisfy tenants' aspirations.

These characteristics make the visible dilapidation of council housing in Britain and Ireland acute, with levels of disrepair much higher than are evident in France, Germany or Denmark. Currently there are moves to reform the public sector in Britain, shifting it more towards a European model of single-purpose housing associations, tenant-based organisations, and 'approved' landlords.

Social non-profit landlords and co-operatives

In all five countries, the process of urbanisation threw up self-help and philanthropic housing organisations to provide decent housing for working people late in the nineteenth century. By the beginning of this century, when state intervention became inevitable, non-profit housing companies, associations, and co-operatives were flourishing in all five countries.

In Britain, through a series of historical accidents and quirks of political and industrial history, local authorities were chosen by central government to build the 'homes fit for heroes' and to clear slums. Ireland followed suit. Elsewhere the existing housing organisations were used and new ones were created, modelled on the non-profit or co-operative companies, largely to fill the same role as local authorities in Britain and Ireland. These associations were sometimes directly publicly or semi-publicly sponsored, but they often originated in private initiative and sometimes were co-operatively owned by members. They grew very rapidly after the war, being used directly by European governments to build 'mass housing', until they provided up to about 20 per cent of all housing. They offered numerous organisational advantages which increased their strength and level of public support over time:

- they were unitary single-purpose housing organisations whose sole rationale for existence was the building and management of housing;

- they had to balance their income and expenditure through discrete budgets – otherwise they became insolvent;
- they therefore had to manage their stock efficiently in terms of rent arrears, lettings, empty property and repairs – otherwise they would be bankrupted;
- they charged rents that related in some measure to their costs and to the requirement for repairs, resulting in rents roughly *double* the equivalent in the late 1980s for British council housing;
- they were removed from direct political interference in most cases, though this was *not* universally the case, as has been seen in the country studies;
- they were regulated by their own statutes and they had an autonomous legal framework rather than being directly controlled by elected politicians – the local authority could only play a strictly limited role in the affairs of the housing company;
- they controlled their process of allocations and were therefore able to respond more closely to individual and local pressures (though they did not define the categories of people who could be nominated to their units by public bodies);
- they put the management needs of the housing service first, before wider social or political commitments because this was their *raison d'être*;
- they were dependent on winning the goodwill and support of the government and therefore had an incentive to perform well;
- they were in competition with each other for 'good tenants' – they therefore 'sold themselves' in ways that local authorities never had to do;
- they were, with few exceptions, small, discrete organisations with a local base, generally controlling a few thousand properties;
- their rents rose to reflect the higher quality of service, without electoral considerations having a direct bearing;
- their spending on repairs was higher than in Britain (OECD 1988).

These factors made European social housing organisations more efficient, better managed, more prosperous, less welfare-oriented, more 'up-market', and more able to maintain a wide income band among their tenants than council housing in Britain and Ireland. It also meant that they played a smaller role in housing the most needy groups than British and Irish local authorities. On the whole, the poorest groups were left in the private-rented sector. This has been a source of criticism and debate in the countries concerned, as well as in Britain, where councils face acute polarisation through the rehousing policies. None the less, the strengths of non-profit housing were outlined in the OECD report on urban housing finance in member countries in the following clear terms:

Small-scale, non-profit organisations may offer an attractive alternative tenure in the rented sector. They possess a number of features such as relative autonomy, using a combination of public and private finance, often decentralised management structures, and frequently incorporating tenant participation, which make them particularly suited to the changing housing situation.

(OECD 1988: 101)

VALUE OF SOCIAL HOUSING

In spite of the cost of providing social housing, the problems it poses for long-term management and its socially segregating tendency, it has proved to be both resilient and useful in housing those very groups that tend to become marginalised or a heavy cost on the society as a whole. It appears more useful in periods of rapid transition than its private alternatives. It is also possibly cheaper (OECD 1988).

Big demographic changes in the five countries have led to a radical revision of views about housing need, owner-occupation and social housing. Needs are expanding and there are growing shortages in all countries. The boom in owner-occupation has been slowed and governments are increasingly targeting resources on social housing.

There are a number of causes. Firstly, the very large increase in the elderly dependent population represents a major challenge. It is clear that much more small, semi-sheltered or supported accommodation will greatly relieve some of the other problems associated with the frailty, dependence, handicap, poor health and isolation of growing numbers of elderly people. Freeing up larger properties for needy younger families would be a by-product.

Secondly, the rapid growth in numbers of one-parent families in all countries represents a very different kind of need. These families tend to be poor, vulnerable and insecure, often heavily dependent on state support and on rented housing. Social housing in all countries increasingly caters for separated families. Any alternatives would be far more costly and difficult to provide on the scale required for the families in greatest need. This creates special demands, both for access to social housing and on the management structures. But there are few alternatives to social housing for many in this group in any of the five countries and, since large numbers of children are involved, it is inescapable for governments to respond – at least at the level of basic need for shelter. It is widely recognised that many of these families simply cannot house themselves.

The third major growing need is for households from racial minorities to be housed better and on more equal terms with the majority population, although this barely applies in Ireland, where emigration, not immigration, is the issue. Unless this happens, inter-communal tensions in urban areas may escalate seriously.

Demand for social housing from minorities has increased dramatically in the last ten years, mainly as a result of their becoming more permanently established. Large numbers of second-generation immigrants are beginning to form new households, further expanding demand. Their weaker economic position makes owner-occupation less accessible and the pressures of inner-city renovation and gentrification push increasing numbers of minorities to apply for social housing or actually makes them homeless in the areas where they first settled on arrival.

There are other changes too, such as the rapid growth in single-person and young adult childless households. It is possible that these will be more easily accommodated in the private-rented sector, including within the owner-occupied households that often let out a single room or two. Short-term, single-room lettings may actually increase significantly.

Therefore social housing, after ten years of declining demand and growing management problems, is becoming an important and seriously over-subscribed provider for growing sectors of the community. Governments increasingly recognise this.

Social house-building will therefore continue to take an upward turn after hitting an unprecedented low point in the 1980s. Denmark and Germany have already expanded their programmes. France is maintaining a steady 50,000 new units a year. Even in Britain, with its strong emphasis on private initiative and its anti-public housing stand throughout the 1980s, there is a planned expansion in government-sponsored housing association building. The same is true of Ireland.

THE ROLE OF LOCAL AUTHORITY ENABLERS

The role of local authorities on the Continent is very different from Britain and Ireland, but they are crucial to social housing organisations in all countries:

1   They *channel money* from central government and fight for resources to build social housing in their areas (Wollmann 1985, Salicath 1987, Quilliot and Guerrand 1989).
2   They either *nominate families* to housing companies or regulate access overall, as well as nominating emergency cases; they help to determine rehousing priorities through representation on the allocation committees of housing associations.
3   They exercise *planning powers*, though possibly less strongly than in Britain.
4   They provide directly most of the social and educational *services and facilities* that are linked to estates.
5   They support *community initiatives* and are vital partners in government programmes to tackle problems such as crime, unemployment,

youth problems (Dubedout 1983).

6   They are the major vehicles for special housing programmes, such as the emergency housing programme in Germany, the Cities and Suburbs initiatives in France, elderly and youth housing in Denmark (Ville et Banlieue de France September 1990; Boligministeriet 1988d).

As the new problems and new demands emerge – the elderly, the minorities, the unemployed and untrained, the one-parent families – the role of local authorities grows rather than shrinks. In Germany it was the local authorities that stepped in and rescued much of Neue Heimat's 300,000 properties through the establishment of new social housing companies. German local authorities are pivotal organisations in the new emergency housing programmes, announced at the end of 1989 in response to the disintegration of the 'Iron Curtain'. In France, it is local authorities that the government and the centrally appointed *préfects* have to lean on to implement rescue programmes for distressed estates. In Denmark, local authorities handle urban renewal and ensure that the homeless get housed. They also provide political leadership for social housing construction. They provide the link between need and provision, even though they do not provide housing directly.

Civic pride and urban integration may be the most important roles local authorities on the Continent see themselves as playing. The huge mass housing estates, with all their problems, enjoy a prominence that attracts resources. The talk everywhere is of making them viable assets. Money goes into art and sculpture, youth activities and festivals, face-lifting and external paint, planting and redesign. Progressive and dynamic local authority mayors actively support this work in the hope that it will enhance their civic status and bring the large social housing estates into the mainstream. Reintegration and higher standards are the talk everywhere. The level of investment is often no higher than in Britain, but the profile of the local authorities is.

In Britain, the enabling role of local authorities is just emerging. Cities like Glasgow, Birmingham and Cardiff are already active in new roles. Local authorities on the Continent are being leant on by central governments to play this role more actively. In Britain, however, the process of denuding local authorities of some of their direct landlord role and transferring it to other bodies often stands in the way of constructive thinking about the enabling role they might play. No country seriously considers dispensing with the local authority role, even though it may change radically.

THE ROLE OF GOVERNMENT

In all countries, governments continue to have a major role in housing:

1   Governments regulate the *housing environment* in urbanised, industrialised societies to cope with basic problems such as sewage, water, fire, access, transport and economic needs. Private organisations need government and strongly support the provision of an ordered and regulated infrastructure (Gay, March 1990).

2   The *inequalities of pure market systems* are nowhere more stark than in housing. Homelessness, racial tensions and urban decay are huge problems in all countries. It is hard for government to escape responsibility for these issues. The British and French government reactions to urban disorders in the 1980s underline the stresses that inequality creates and the pressure on governments to intervene, regardless of political or economic policy (Scarman 1986, Dubedout 1983). The disorders and the subsequent interventions were closely tied to social housing in both countries.

3   *Affordable, reasonable housing* for all has so far eluded all five countries. Just as the goal appears reachable, new demands and problems emerge. All five countries are experiencing acute housing problems of one kind or another. In each country, the move away from direct provision and building subsidies has been matched by increases in income support for housing and growing programmes of urban renovation, co-operative building and special rescue programmes in social housing. The role of government seems inescapable. Even in the USA it was concluded that affordability was *the* crucial problem where provision was concentrated in the private sector (Rossi 1989).

4   The special problems of *managing modern cities* with their complex networks, economic interaction and social make-up require governments to play a dynamic role. On the Continent, governments see local authorities as enhancing the ability of government to address urban problems (Ville et Banlieue de France Conference September 1990).

Not only have governments played a crucial role, but they will find it ever harder to extricate themselves from their role in ensuring workable conditions in urbanised areas and in providing support to very disadvantaged households and groups. There are no signs that private individuals or bodies will fulfil either of these roles, alone.

Problems with government's role

However, government intervention, much as it may be needed, creates or leads to many problems. Firstly, government programmes have produced *unwieldy and inefficient housing structures.* This can happen indirectly, as in the German experience of Neue Heimat (Fuhrich *et al.* 1986), or more directly, as in the French experience of ZUPs (*zones à urbanisation prioritaire*) (HLM 1989a) or the British inner-city local authorities (Audit

Commission 1986). Because intervention has often been costly and clumsy it has also become less popular.

Secondly, *social and environmental problems* that became concentrated in mass social housing areas have also continued to fall to government to be addressed. The private sector has little incentive to involve itself directly in urban problems for purely commercial reasons; Michael Heseltine discovered this in his Merseyside initiatives. The British government ended up footing most of the bill for urban regeneration and creating another range of government-sponsored quangos in the shape of urban development corporations.

Thirdly, the role of government in housing strategies and provision has underlined the difficulty of *targeting help* effectively at those who most need it. Governments have not always helped the very poor, but have often concentrated on the middle ground. To the extent that governments aim to help the very poor, as British and Irish homelessness legislation requires, their services have often been insensitively provided, uneven, and so minimal that integration does not result. Targeted help often means poor quality, stigmatisation and rejection of the welfare role. Therefore governments are often between the devil and the deep blue sea.

The move away from direct provision

The combination of factors created a climate for change within government:

- Why provide social housing on a broad front when better-off households wanted and were able to provide housing independently?
- Why run large bureaucracies that slowed up progress, were probably inefficient and were unresponsive to customers?
- Why build on a large scale when small-scale developments were clearly more popular?
- Why retain tight government control when people clearly wanted to exercise more choice and more control themselves?
- Why provide and regulate publicly sponsored housing when the majority were sufficiently prosperous to pay more for better and more varied housing?
- Why not help the very poor who were still often badly housed by both targeting funding and providing incentives for the better-off to move out of existing social housing?

These questions, raised by governments everywhere, were often framed as offering the prospect of government disengagement, falsely, since owner-occupation was receiving higher and higher levels of support at ever-increasing cost to the government. The more indirect form of help, involving individual households in greater expense to meet rising standards

Table 42.1 Changes in goverment housing policies, 1976–91

	Targeted housing allowances and reforms of finance for housing	Stronger incentives to owner-occupation	Unpopularity of mass housing estates	Government rescue programmes	Serious decline in private renting	Access by immigrant groups to social housing	Special features	Renewed interest in social housing
France	1977	Late 1970s	Mid-1970s	1976	From 1970s	Late 1970s	Barre reforms 1977. Prime Minister's distressed areas 1982. Disorders 1981, 1990.	1982 1990
Germany	1980	1980s	Early 1980s	1984	–	Early 1980s	'Neue Heimat' goes into crisis, 1982. Abolition of non-profit status 1989. Collapse of iron curtain and re-unification 1990.	1990
UK	1972 1982 1989	1960s	Mid-1970s	1979	From 1945	Mid-1970s	Difficult-to-let investigation 1976. Right to buy 1980. Tenants' choice 1988. Urban disorders 1981, 1985, 1987, 1991.	1991

Denmark	Late 1970s 1984	1970s	Late 1970s	1985	1980s	Early 1980s	Tenant boards 1970. Cuts in support to owner-occupation 1984. Steady support for social housing. Throughout 1980s
Ireland	1977 1990	1970s	Late 1970s	1987	From 1950s	–	Tenant surrender grants – 1984. Tenant purchase. Remedial works scheme – 1987. 1991

and aspirations, allowed for much greater choice and escaped some of the organisational problems of direct forms of support and provision. The economic and social marginalisation of significant areas of social housing went hand in hand with the preferred treatment and growing economic and social importance of owner-occupation.

These policies were pursued most vigorously in Britain and Ireland, but continental countries, while later and slower, followed suit with growing enthusiasm.

Thus the role of governments in all five countries moved rapidly from the late 1970s onwards towards:

- encouraging independent private housing through cash incentives and relying on independent housing companies and associations rather than on direct state provision;
- helping the most needy to get better housing through regulating access to social housing and targeting housing allowances;
- improving the management and maintenance of social housing through forcing rent rises and through special rescue initiatives;
- enhancing the efficiency of social landlords through creating a climate of competition;
- supporting tenants' initiatives and community-based organisations as a form of self-help and more individualised control;
- establishing special programmes to upgrade physical conditions on mass housing estates and to redress the social and management problems of these areas;
- cutting direct building programmes until new shortages forced a change in emphasis back to social provision.

The core idea of governments was to narrow the role of social housing; to improve existing social housing; to continue to provide it on a limited basis to more narrowly defined groups; to address wider housing needs *indirectly* through continued and sometimes new incentives to owner-occupiers, part-owners, private landlords and semi-private organisations like co-operatives and independent housing companies. Britain, Ireland, France, Germany and Denmark all went through these transitions in various ways with remarkably similar drifts of policy and parallel, if varied, outcomes. There are major transitions still under way in all countries, with governments still trying to move away from direct provision. But for all the problems that result from their interventions, governments are still the main protagonists in the direction of housing policy, the main facilitators for low-cost housing, the main vehicle for ensuring access to the very poor, and still major contributors directly or indirectly, at least in part, to the cost of new housing. Table 42.1 summarises the changes.

Chapter 43

General themes

URBANISATION AND SUBURBANISATION

In Europe, two countervailing trends dominated twentieth-century urban housing patterns – the move into cities and the move outwards. Urbanisation led to dense congestion and slum development. This created the conditions for state intervention to protect the urban workforce and prevent the collapse of social order under the impact of rapid change. State-sponsored social housing was one of the results.

Urban pressures in turn led to rapid suburban growth, both of social housing estates and increasingly of single-family, owner-occupied housing. As state-sponsored renting and owner-occupation grew, the relative position of private landlords declined, although to variable degrees in different countries.

'MASS' ESTATES

Both social housing and owner-occupation threw up difficult problems. In the face of chronic shortages and deteriorating urban conditions, post-war governments everywhere encouraged industrialised building methods, leading to 'mass' housing estates that unified and replicated basic modern designs, producing cheap, prefabricated, and easily reproduced units on a vast scale. Industrialised, flatted estates were built almost exclusively for social renting, though a few attempts at diversification were made in most countries. This housing was initially popular but quickly declined through over-concentration of people, poor quality materials, and bald external appearance. The physical aspect of many estates created a stark and alien atmosphere that was often hard to humanise. The number of units prevailed over the quality of the environment, and social, physical and management problems rapidly accumulated.

OVER-EMPHASIS ON OWNER-OCCUPATION

Suburban owner-occupation with government incentives was the most popular individualised response to 'mass' housing in the 1960s and 1970s. Each unit invariably had its own direct access and controlled space. Individuals could add, decorate and adapt the basic housing form, both externally and internally. 'Grey suburbia' and 'little boxes' were condemned by social protesters but had an intrinsic and lasting appeal in all countries because of the sense of control and individual potential they offered. But they led to urban sprawl, growing financial problems among low-income purchasers, greater immobility, and access problems. Also, the more heavily subsidised owner-occupation became, the more rapid the flight of economically stable households from private and social renting. Thus, solutions bred new problems in all tenures and were closely interconnected.

INNER-CITY RENEWAL AND CONFLICT

Two other related changes intensified these problems. Firstly, competition for inner-city space grew again after decades of decline. From the 1960s onwards, governments gave incentives to renovation in the face of mass housing decay, shortage of suburban land, and deteriorating inner-city conditions. As richer households bought up potentially attractive, older property, they displaced the poorest groups. Large minority populations that had moved into declining inner areas in the years of economic boom after the war were often concentrated in neighbourhoods ripe for gentrification. Increasingly, they were pushed out into the most unpopular estates.

Many inner areas became popular again, housing diverse social groups and a complex mixture of black and white, rich and poor, successful and unsuccessful, new and old, recreating traditional urban juxtapositions in a new social frame. At the same time, private renting declined as properties were converted and upgraded. Popularity invariably created pressure.

The rise and decline of inner areas, their preservation or demolition, their contrasts of wealth and poverty, created some of the greatest social tensions and occasionally actual disorders, often focusing on housing issues. Sometimes, slum clearance or urban renewal schemes were the focus or trigger; sometimes, poverty and policing; sometimes, gentrification. Inner cities became a byword for social – and increasingly racial – conflict.

INCREASES IN MARGINAL GROUPS

At the same time, there were growing pressures among the poorest social groups – family break-up (EC 1989a); racial discrimination (*The Guardian* 1991); rises in levels of unemployment; decline in private renting in inner areas (Pfeiffer 1989). From the mid-1970s, unpopular social housing estates – as they emptied of the more skilled, more able-bodied, more independent – offered a refuge for the more vulnerable, less skilled, more dependent. Governments largely encouraged these trends, providing income support and rent allowances on a growing scale to ensure that the most needy would be accepted, often reluctantly, by social and private landlords, while also providing cash support or other incentives for low-income owner-occupation. As the mass housing estates changed from being a visionary solution into a necessary liability, worries grew about the development of ghettos (EC 1991a). Finding a balance between keeping mass housing estates for needy residents and retaining more secure tenants in order to provide balanced and mixed communities became a new objective of governments in all five countries.

A LOCAL EMPHASIS

There emerged an emphasis on local initiatives to help re-integrate the estates. Integration, it was hoped, would prevent the breakdown of order, would reaffirm a common purpose among residents, and would turn dependent communities back into a workforce. Very local effort would involve residents, restoring a sense of confidence and generating new skills (*Ensembles* 1991). There were no longer clear, generalised or 'mass' solutions to the myriad individual needs that made up each local community. Small scale became increasingly attractive over large scale; local control was increasingly favoured over complex central control; and mechanised solutions no longer inspired confidence (DOE 1987). How housing was used, run, maintained and decorated had come to matter as much as how it was designed and built (Vestergaard 1991).

COMMON URBAN PATTERNS

Between 1850 and 1990, urban housing patterns and the role of the state in their development had followed a clear trend. Different factors reinforced each other and sometimes ran in parallel rather than in chronological sequence. The dates at which different stages were reached differed in different countries, as did the intensity with which different problems were experienced. None the less, Figure 43.1 broadly reflects developments in the five countries.

GOVERNMENT STRATEGIES

Government strategies were complex and varied within this common framework. The subsidies that underpinned policies were firmly bedded in national housing finance systems (McGuire 1981, OECD 1988, Hills *et al.*

Figure 43.1 Pattern of twentieth-century urban housing developments and state intervention

Table 43.1 Government housing policies adopted in five countries

Strategy	France	Germany	Britain	Denmark	Ireland
Support for autonomous social landlords	✔	✔	✔ Recently	✔	✔ Recently
Local authority landlords	✔ A few	–	✔	✔ A few	✔
Slum clearance	–	–	✔	–	✔
Mass industrialised building	✔	✔	✔	✔	✔ Limited
Sponsorship of peripheral estates	✔	✔	✔	✔	✔
Subsidies to private landlords	✔	✔	✔ Recently	✔ Recently	✔ Recently
Inner Area Renewal	✔	✔	✔	✔	✔
Access measures for minorities	✔	✔	✔	✔	–
Raising rents towards market levels	✔	✔	✔	✔	*
Housing allowances related to income	✔	✔	✔	✔	*
Improvement programmes for social housing	✔	✔	✔	✔	✔
Emphasis on tenant involvement	✔	Embryonic	✔	✔	✔
Subsidies to owner-occupation	✔	✔	✔	✔	✔
Indirect encouragement of single-family houses	✔	–	✔	✔	✔
Special support for low-income owner-occupiers	✔	✔	✔	Until 1980s ✔	✔
Reduced incentives to owner-occupy by 1991	✔	–	✔	✔	✔
Support for social housing to meet new needs	✔	✔	✔	✔	✔

Note: *Differential rents according to income in Ireland.

1990). Their impact invariably carried disadvantages as well as advantages. There was a strong similarity in programmes between the five countries. Table 43.1 attempts to summarise the main housing policies uncovered in the study, showing the countries where they were adopted as a major strand. The impact varied according to local conditions and the other policies they were combined with. There was, however, a clear pattern of advantages and disadvantages emerging in relation to the different strategies. Some advantages proved over time to offer disadvantages. For example, mass building solved shortages but created management problems. The pattern of advantages and disadvantages occurred with varying degrees of intensity in the countries where the policy had been adopted. Table 43.2 attempts to summarise the impact of the policies.

ORGANISATIONAL PROBLEMS

While policy initiatives by governments helped to shape housing developments (Boelhouwer and van der Heijden 1992), organisational problems arose from the growing complexity of government involvement in urban systems. The organisational problems appeared most starkly in social housing.

All five countries experienced the following problems to a greater or lesser extent in the wake of state-sponsored housing:

- bureaucratisation in social housing organisations;
- over-concentration of ownership in key urban areas;
- paternalistic and remote relations between landlords and low-income tenants;
- growing concentrations of social need in particular housing areas;
- racial and community tensions in some of the most unpopular areas;

Each country was at a different place on the spectrum of problems, ranging from mild to intense. But all countries in the study had a significant number of estates with these problems on a considerable scale, even though each country was starting from a different level of wealth, development and need (*The Economist Pocket Europe* 1992).

OUTCOMES

The outcomes of these patterns were not clear. But certain developments in state-sponsored housing seemed likely in all countries:

- *Renewal of estates* Modern, unpopular housing estates would be preserved and renewed in spite of their obsolete style and conditions, their social tensions and problematic physical and management requirements. Governments did not have the resources or power to

Table 43.2 Advantages and disadvantages of the main approaches to state intervention

Strategy	Advantages	Disadvantages
Support for autonomous social landlords*	– Balanced budgets – Business principles – Emphasis on social mix – Custodial services – Enforcement of rent payment – Repair and cleaning to maintain viability – Private finance – Customer orientation	– Access problems and barriers – Government support generates bureaucracy – Numbers game – Mass estates – Separation from local political decisions
Local authority landlords*	– Strong welfare role – Strong local support – Access for homeless	– Political change and pressure – Poor management incentives – Part of wider town hall bureaucracy – Mass estates – Remote services
Slum clearance*	– Destruction of outmoded housing – Impetus for better conditions – Rebuilding in inner cities	– Destruction of some popular areas – Time-lag and other costs – Socially disruptive – Hard to manage – Inner mass estates – Unpopular with older residents – Destroyed jobs and facilities
Mass industrialised building*	– Volume of units – Quick short-term solution – Higher internal standards	– Unpopular – Hard to manage – Building problems – Hidden costs – Socially disruptive
Sponsorship of peripheral estates*	– Green field sites – Garden city ideal – Better internal conditions – Space	– Segregated from city – Far from jobs and amenities – Rapid decline – Often unpopular
Subsidies to private landlords*	– Supply increases – Diversity – Competition with social landlords, therefore higher standards	– Poor conditions in old housing – Insufficient return – Can lead to profiteering on shortages

Table 43.2 (continued)

Strategy	Advantages	Disadvantages
	– Houses marginal groups – Sustains older, poorer areas	– Expensive subsidies – Less control
Inner Area Renewal*	– Halts decay – Saves attractive housing – Regenerates economy – Mixed residential patterns	– Creates pressure on vulnerable and on minorities – Aggrevates social tensions
Access measures for minorities	– Equitable – Improves conditions – Helps integration – Contains tension	– Doesn't remove discrimination – Still concentrated in worst areas – Can lead to ghettos because of style and location of estates
Raising rents towards market levels	– More resources – More market approach – More spending on maintenance – More landlords	– Poverty trap for low-paid – Deterrent to better-off – Higher cost of allowances – Pushes employed out
Income-related housing allowances*	– Helps access by poor – Increases landlord income – More spending on maintenance	– Costly – Hard to limit rents – Increases government role – Open to abuse – Poverty trap
Improvement programmes for social housing*	– Direct impact on conditions – Popular with residents – More resources – Generates management and social initiatives	– Costly – Limited (e.g. by structure) – Ignores deeper problems – Hard to manage – Requires detailed consultation
Emphasis on tenant involvement	– Reduces alienation – Helps local staff – Makes landlords more responsive – Generates care over spending	– Sometimes unrepresentative – Difficult to take responsibility for biggest problems – Lack of training
Subsidies to owner-occupation*	– Expands investment – Popular – Indirect state role – Allows choice and control	– Inflation – High cost – Flight from renting – Barriers for poor – Default on payments

Table 43.2 (continued)

Strategy	Advantages	Disadvantages
Indirect encouragement of single-family houses	– Encourages greater individual investment and responsibility – Reduces areas of dispute – Popular	– Suburban sprawl – High energy and land costs – Distance from jobs, transport
Special support for low-income owner-occupiers*	– Redistributes wealth – Diversifies ownership – Popular	– Marginal buyers – Foreclosures and mortgage arrears – Flight from renting
Reduced incentives to owner-occupiers	– Cuts prices and inflation – Helps landlords – Reduces government cost	– Financial pressures on lower-income owners – Unpopular – Adjustment problems
Support for social housing to meet new needs*	– Helps very needy – Reduces homelessness – Helps minorities – Reduces urban tensions	– Costly – Politically controversial – New problems of engagement

Note: *All strategies marked with asterisk involved direct government subsidies.

tackle these problems on the required scale through new-building. They had to conserve.

- *Expansion of social housing programmes.* Social housing would continue to play an important role at the bottom of the market because social needs were multiplying, in spite of generally higher standards and growing affluence.
- *Cuts in welfare benefits.* The fact that the dominant majority enjoyed rising standards would lead to continued restrictions on welfare that were damaging to marginal households and created such social scandals as 'cardboard city' in central London and violent street riots in suburban France (Harris 1990). In turn, the limits to welfare, which hit the most marginal households most harshly, would increase demand for social housing and create new pressures for state action, to which the state would at least partially respond (*The Guardian* 1990, DOE 23 May 1991).
- *A role for government.* Governments would remain involved in social housing because housing built with state funds required government financial and organisational backing if 'mass' estates were to be saved from terminal decline. Private sources of funds were not available on the scale required.

- *Reintegration.* Fears of unrest and alienation would lead to a growing emphasis on community involvement and increasing attempts to re-integrate marginal communities. Training and experimental initiatives would flourish. It remained to be seen whether such locally responsive initiatives would survive as a basis for wider and longer-term policy because such local experiments were often vulnerable to political changes.

Therefore certain trends – the decline of mass estates and cuts in welfare – would be in seeming conflict with other policies: a growth in estate improvement programmes and an expansion in social housing programmes after a decade of cut-backs. But the upward trend in support for social housing and the rescue of peripheral housing areas would be similar in all countries.

Chapter 44

Why state-sponsored housing will survive

There are two possible explanations for the convergence of experience and policy direction in the five countries. Firstly, state-sponsored housing is indispensable to modern urban society, just as government itself is, because of the growing economic complexities, greater population movements, and greater instability. European governments, unlike the USA, have not stood back from these problems. Even in the United States, with its emphasis on market sovereignty, there have been virtually continuous government programmes to provide or support low-cost housing of one kind or another. Private investors cannot gain from housing families dependent on state benefits unless the state itself pays the return to the investors. Therefore, the state cannot escape responsibility very easily.

Secondly, if governments do sponsor and support housing for poor urban households, there will be a constant struggle to renew and upgrade its social and physical conditions as they continually fall behind the rising standards of the majority. Governments invest such effort in social housing in an attempt to integrate their diverse societies, reduce urban conflict in a situation of sharp and often growing polarisation, and make best use of their increasingly scarce resources of land, buildings, and young people. They also want to escape responsibility for slum conditions, where they have played a direct role in building, by helping restore conditions. Pressures to equalise conditions will often prove intense (CECODHAS 1990).

GOVERNMENT ACTION IN EUROPE'S CITIES

European urban experience diverges sharply from that of the United States in the active role of governments. Population density and urban containment determine the interrelatedness and interdependence of groups and areas. While sharp divisions remain and even grow, there are constant pressures and attempts to reintegrate, upgrade and harmonise conditions. This, in turn, is based on a combination of fear, self-interest, and democratic traditions.

DEMOCRACY OR EXCLUSION?

Democracy is commonly interpreted as the right to participate and thereby influence decisions. The failure to exercise that right, or to influence conditions among disadvantaged groups, to some extent at least reflects on the wider society, its values, its attitudes, and its openness to all residents. The interdependence of its members and the need for harmony make the majority uneasy within a democracy when minorities are visibly excluded. Exclusion is a limited and short-term solution, advocated by few, although the pressures towards exclusion are always present in the name of a narrow kind of harmony. Therefore European governments continue to worry about the threat to urban cohesion posed by socially segregated housing areas. And they continually invest in social programmes to help integrate marginal groups (EC 1991a).

The cost of welfare may be electorally unpopular but the density of European cities determines the constant pressure to tackle urban problems since urban unrest is possibly even more electorally unpopular (*The Independent Magazine* 1990).

HELPING THE BOTTOM LAYER

As long as attributes and achievements are differentially rewarded, there will inevitably be a bottom layer and the tendency will be for that bottom

Table 44.1 The role of social housing, the scale of its problems, and the level of support for its tenants

Country	Per cent stock owned by social housing organisations[a]	Estimated % of socially owned stock difficult to let or manage[b]	Social housing tenants in receipt of housing allowances (%)
France	17	33	50
Germany	25	15[c]	25
Britain	27	20	65
Denmark	21	15	50
Ireland	13	20	70[d]

Source: Country studies.

Notes: [a] Including housing associations.

[b] These estimates are based on information provided by national governments, by research bodies evaluating the government rescue programmes, and by landlord organisations.[1]

[c] Includes all outer estates (*Grossiedlungen*), some of which have far less severe problems than the worst, but which none the less pose financial and management problems (Bundesministers für Raumordnung, Bauwesen und Städtebau 1988).

[d] In Ireland the system of differential rents is in place of housing allowances (see country study pp. 336–7).

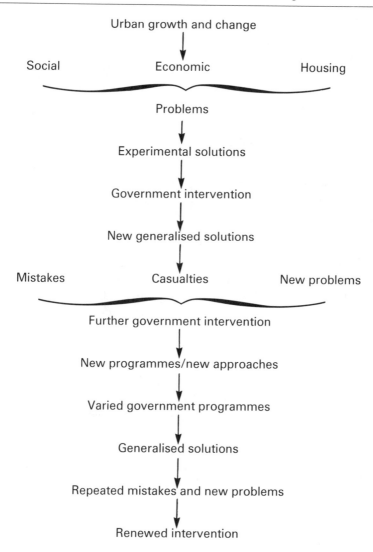

Figure 44.1 The cycle of government intervention

layer to participate less in the wider society. As the number of casualties grows, through accelerating change and movement, the need for a safety net and for 'reintegration' becomes stronger. Efforts at dismantling the government's role have failed, even if its character has changed. Within the state benefit system, social housing plays an increasingly important role. The proportion of households dependent on benefit in social housing has risen significantly in all countries. Therefore state-sponsored housing is likely to continue, with significant state support for its effective management and maintenance, at least for the foreseeable future.

Table 44.1 illustrates the importance of the social housing stock, the majority of which does not pose particular problems. It also illustrates the high levels of dependence on state support.

OWNER-OCCUPATION ECLIPSED?

Owner-occupation was no longer *the* dominant solution to modern housing needs, in spite of its near universal popularity. Private renting was increasingly protected and promoted as an essential part of the housing market and a valuable adjunct to social housing, but was under increasing financial pressure everywhere – a problem which all governments were being drawn to address. Its less regulated nature and fast response made it a useful vehicle for the range of household needs but attempts to halt its decline were far from certain of success. Therefore, forms of government support were changing, but withdrawal was the least likely scenario. Figure 44.1 illustrates the process of renewed state intervention.

NOTE

1   Estimates for the proportion of the social housing stock that is classed as difficult to let or manage are based on the following sources:
     France – French government programme announced in 1991.
     Germany – Report of the Ministry of Housing into the problems of large outer estates, 1988.
     Britain – Estimate based on local authority reports, Priority Estates Project information, Estate Action 1991.
     Denmark – Information from the Danish Building Research Institute 1991.
     Ireland – Information from the government remedial works programme.

Chapter 45

Conclusions

Several important conclusions emerge from this history of social housing in five European countries.

SUCCESSFUL SOCIAL HOUSING

Firstly, *the majority of social housing* works well, although it often houses low-income households (see Table 44.1). This fact is probably more influential than any other in its continuing role in spite of serious difficulties. Despite the effects of poverty, a majority of social housing is not posing special problems.

OWNERSHIP AND MANAGEMENT

Secondly, *the form of ownership and the style of management* do not make sufficient difference to overcome the bigger social problems facing the wide variety of state-sponsored landlords in all countries. On the other hand, the more autonomous continental system appears to have limited the decay and disrepair, and created pressures to perform that seemed far weaker under direct state ownership in Britain and Ireland (OECD 1988).

AGAINST MASS HOUSING

Thirdly, there is a *reaction against mass* and in favour of individual solutions. This showed up most clearly in the popularity of owner-occupation in single-family houses.

MOVING ON

Fourthly, while cities represent the most extreme form of human herding, they also represent the most *individualistic effort.* 'Psychological reserves' are usually focused on moving on and upwards, rather than on preserving communities, leading to constant change. Those who stay put tend to be the older generation or those who are left behind. This of itself has created polarised and unstable areas over many decades (White 1986).

IMMIGRATION

Fifthly, cities constantly attract *new waves of immigrants* who move into the vacated areas (which they share with the elderly, the failed, the under-achievers) in the least popular housing. Thus poor areas constantly recreate and replenish themselves unless they are actually wiped out. They require continuous support, rather than one-off initiatives. This issue is likely to become more dominant following European integration.

RACIAL DIVERSITY

The issue of *race and cultural diversity* is becoming more pressing because of the scale of immigration, the maturing of second-generation households from racial minorities, and the opening of frontiers. Without new tolerance and understanding, tensions will continue to mount. The children and young people, who are increasingly of minority origin in many poor estates, require access to education and training that will help them join main-stream society and compete on equal terms in an increasingly wealthy Europe. Confidence would help build the foundations of equality. It would be a mistake to wait for further violence and break-down to force the pace of change, as the United States Report of the National Advisory Commission on Civil Disorders stressed in 1968:

> Our nation is moving towards two societies ... To pursue our present course will involve the continuing polarisation of the American Community and, ultimately, the destruction of basic democratic values ... It is time now to turn with all the purpose at our command to the major unfinished business of this nation ... to make good the promises of American democracy to all citizens.
>
> (National Advisory Commission on Civil Disorders 1968: 1, 2)

SEGREGATED OR MIXED COMMUNITIES

State-sponsored social housing is gradually replacing privately owned urban slums as the least-favoured housing in the five countries of this study.

The 12.5 million post-war social housing units of the five countries are likely to remain in use to meet new needs; it is therefore important to reverse the intense problems social landlords face on the segregated and uniformly poor 'mass' estates, which threaten stability in all countries. Governments are developing strategies to make the most unpopular estates part of mixed neighbourhoods, with mixed uses, mixed populations and modern services (DOE 1988). More social housing is needed to make such diversification possible.

POSTSCRIPT

Hovels to high rise has shown the central role of social housing in meeting widespread needs. This role is growing again – after a decade of decline – and estates that were threatened with obsolescence have been rehabilitated to house the very poor and the most marginal households. These estates, across Europe, have become the focus of urban tension and unrest. They have also become the target of special programmes devised and supported by national governments and by the European Commission to reverse their decay and their isolation, to enhance their management and to re-establish social stability. A separate study will describe these Europe-wide programmes in more detail.

Social housing has acquired a permanent and unforeseen role, responding to pressures of immigration, segregation and marginalisation. Owner-occupation is no longer a clear and simple housing solution for all. Protests over unstable and decayed conditions in state-sponsored housing have evoked a quick response. Governments have found that they could not readily extricate themselves from this responsibility, having become so deeply involved.

Bibliography

Abel-Smith, B. (1991) *Final report on the 2nd Poverty Programme 1985–89*, Com (91), 29 Final, Brussels.
—— and Townsend, P. (1965) *The Poor and the Poorest*, London: G. Bell.
Adsboel K (1988) *Social Life in Housing Areas*, Presentation ICA Seminar, Stockholm, 4 July, The Federation of Non-Profit Housing Companies, Denmark.
Alestalo, M. and Kuhnle, S. (1984) *The Scandinavian Route – Economic, Social, and Political Developments in Denmark, Finland, Norway and Sweden*, Research Report No. 31, Research Group for Comparative Sociology, University of Helsinki.
Andersen, H. S. (1985) 'Danish Low-Rise Housing Co-operatives (bofaellesshaber) as an Example of a Local Community Organisation' in *Scandinavian Housing and Planning Research*, No. 2.
Andrews, C. L. (1985) *Notes on a Visit to Vridsloeselille Andelsboligforening (VA) Housing Company, Denmark, April 1985.*
Anson, Lady (1991) Institute of Housing, Midlands Branch Annual Conference, Birmingham, November.
Arbejdernes Kooperative Byggeforening (1989) *Aarsstatistik 1988*, Copenhagen: AKB.
—— (1991) Information received from AKB, Copenhagen.
Assemblée Nationale (1990) *Projet de Loi Visant à la Mise en Oeuvre du Droit au Logement*, Seconde Session Ordinaire de 1989–90, 3 May, Paris.
Association of Metropolitan Authorities (1984) *Defects in Housing Part 2: Industrialised and System Built Dwellings of the 1960s and 1970s*, London: AMA.
Audit Commission (1986) *Managing the Crisis on Council Housing*, London: HMSO.
—— (1987) *The Management of London's Authorities: Preventing the Breakdown of Services*, Occasional Papers No. 2, London: HMSO.
Aughton, H. with Malpass, P. (1990) *Housing Finance – a basic guide*, London: Shelter.
Avery, D. C. (1987) *Civilisation de Corneuve – Images Brisées d'une Cité*, Paris: L'Harmatten.
Ball, M., Harloe, M., and Martens, M. (1988) *Housing and Social Change in Europe and the USA*, London: Routledge.
Ballymun Task Force (1988) *A Programme of Renewal for Ballymun – An Integrated Housing Policy*, Dublin: Ballymun Task Force.
—— (1991) *Report*, Dublin: Ballymun Task Force.
Bannon, M. J. (ed.) (1985) *A Hundred Years of Irish Planning Vol. 1, The Emergence of Irish Planning 1880–1920*, Dublin: Turoe Press Ltd.

—— (ed.) (1989) *Planning – The Irish Experience 1920–1988*, Dublin: Wolfhound Press.

Barraclough, G. and Stone, N. (eds) (1989) *The Times Atlas of World History*, 3rd edn, London: Times Books.

Bautätigkeit & Wohnungen, Fachserie 5, Reihe 2 (1975, 1980, 1985, 1987) 'Bewilligungen im sozialen Wohnungsbau', Wiesbaden: Statistisches Bundesamt.

Bautätigkeit & Wohnungen, Fachserie 5, Reihe 4 (1955/6, 1960, 1965, 1970), Wiesbaden: Statistisches Bundesamt.

Behar, D. (Acadie) (1987) *Les Régies de Quartier, Suivi – Évaluation*, Paris: Commission Nationale de Développement Social des Quartiers.

Best, R. (1992) NFHA Conference speech reported in *Housing Associations Weekly*, 27 March.

Bevölkerungsstruktur und Wirtschaftskraft der Bundesländer 1976–1987, Wiesbaden: Statistisches Bundesamt.

Blackwell, J. (1984) 'Housing finance in Ireland in the 1980s', *The Irish Banking Review*, December.

—— (1988) *A Review of Housing Policy – No. 87*, Dublin: National Economic and Social Council.

—— (ed.) (1989a) *Towards an efficient and equitable housing policy*, Special Issue of *Administration*, 36/4, 1988, Dublin: Institute of Public Administration.

—— (1989b) *Housing Finance and Housing Subsidies in Ireland*, Department of Environmental Studies, University College, Dublin – paper for Rowntree Memorial Trust Workshop 28–30 June.

—— and Kennedy, S. (eds) (1988) *Focus on homelessness – a new look at housing policy*, Dublin: The Columba Press.

Blüthmann, H. and Kahlen, R. (1986) 'Hypothek der frühen Jahre. Der Niedergang des grössten Wohnungsbaukonzerns in Europa', *Die Zeit*, No. 44, November.

Bochum Conference (1991) on 'Wohnen und Leben', Ministerium für Bauen und Wohnen des Landes Nordrhein-Westfalen.

Boelhouwer, P. J. and van der Heijden, H. M. H., University of Delft, Holland.

Boligministeriet (1984) *Financing of Housing Denmark*, Copenhagen.

—— (1985) *Housing Finance in Denmark with Special Reference to the Non-Profit Sector*, Contribution to the ICA Seminar in Portugal, 10 December, Copenhagen.

—— (1986) *Cirkulaere om organisation og udlejning af almennyttigt byggeri*, Boligstyrelsens Cirkulaere af 14 Marts, Copenhagen.

—— (1987a) *Co-operation between Landlords and Tenants*, Act on Urban Renewal and Improvement of Housing, Copenhagen.

—— (1987b) *Den almennyttige boligsektors rolle paa boligmarkedet*, Copenhagen.

—— (1987c) *Financing of Housing in Denmark*, Copenhagen.

—— (1988a) *Boligmarkedet og boligpolitikken – et debatoplaeg*, Copenhagen.

—— (1988b) *Bygge – og boligpolitisk oversigt*, Copenhagen.

—— (1988c) *Co-operative Housing in Denmark*, Copenhagen.

—— (1988d) *Housing for the Elderly in Denmark*, Act on Urban Renewal and Improvement of Housing, Copenhagen.

—— (1988e) *Non-Profit Housing in Denmark*, Copenhagen.

—— (1988f) *The Human Settlements Situation and Related Trends and Policies*, Copenhagen.

—— (1989) Statistics and information received.

—— (undated) *Housing Finance in Denmark – especially by mortgage credit*, Copenhagen.

Boligministieret – National Housing Agency (1987), *Rehousing Schemes and*

Financial Assistance for Higher Rents, Act on Urban Renewal and Improvement of Housing.

Boligselskabernes Landsforening (1989a) *Budgetoverslag – Aarsregnskab For 1988*, Copenhagen.

——— (1989b) *Statistik 88.*

Bonnevay, L. (1912) Report No. 50, 16/19 July, Chambre Annexe no. 1847: 806.

Briggs, A. (1983) *A Social History of England*, Harmondsworth: Penguin Books.

Bundesanstalt für Arbeit Monatsbericht (February 1992).

Bundesbaublatt, December 1981 and April 1988.

Bundesministers für Raumordnung, Bauwesen und Städtebau (1986) *Der Wohnungsbestand in Grossiedlungen in der Bundesrepublik Deutschland*, Heft Nr. 01/076, Bonn.

——— (1988) *Städtebaulicher Bericht, Neubausiedlungen der 60er und 70er Jahre. Probleme und Lösungswege*, Bonn.

——— (1989a) *Sozialwohnungsbestand und Belegungsrechte*, Ref: W II 4 – 20 22 11 (29 April), Bonn.

——— (1989b) *Sicherung der Wohnungsversorgung durch kommunalen Erwerb von Belegungsrechten im Wohnungsbestand*, Ref: W I 5 – 30 09 03 – 2 (24 May), Bonn.

——— (1989c) *Dringlichkeitsfalle und Belegung von Sozialwohnungen*, Ref: W II 4 – 20 11 30 – 2 (24 August),   Bonn.

——— (1989d) *Sachstandsbericht Projektgruppe 'Belegungsrechte'*,   Ref: W II 4 – 20 11 – 33 – 1 (12 September), Bonn.

——— (1989e) *Ergebnisse der Gespräche mit Städtevertretern am 24. August und 31. August 1989 zum Thema 'kommunale Belegungsrechte'*, Ref: W II 4 – 20 11 33 – 1 (22 September), Bonn.

——— (1989f) *Wohngeld - und Mietenbericht 1989*, Ref: W I 5 – 30 09 03 – 29/7 (1 December), Bonn.

——— (1989g) *Überblick über die derzeit geltende Förderung einschliesslich der aktuell beschlossenen Massnahmen*, Ref: W II 1 – 21 04 04 – 83 (8 December), Bonn.

——— (1989h) *Statistics 1989.*

——— (1991) *Vitalisierung von Grosssiedlungen*, Bonn-Bad Godesberg, June.

——— (undated) *Zur künftigen Rolle der Gemeinnützigen Wohnungsunternehmen*, Bonn.

Burbidge, M. (1985) *Scandinavia 1985. Notes of a Visit to Sweden and Denmark, May/June 1985*, London: Department of the Environment.

Burbidge M., Wilson, S., Kirby, K. and Curtis, A. (1981) *An Investigation of Difficult to Let Housing*, Vol. 1: *General Findings*; Vol. 2: *Case Studies of Post-war Estates*; Vol. 3: *Case Studies of Pre-War Estates*, London: DOE.

Burgeat, Y. (1989) Report of visit to Britain by the National Commission for the Social Development of Neighbourhoods, July.

——— (1991) Speech at LSE European Workshop, Cumberland Lodge, Windsor.

Burnett, J. (1980) *A Social History of Housing 1815–1970*, London: Methuen.

Byfornyelsesselskabet Denmark (1987) *Urban Renewal and Redevelopment*, Copenhagen.

Cancellieri, A. (1989) *Rapport Reflexion sur l'Habitat et le Logement Social en Ile de France dans les Perspectives Nationales et Européennes*, Comité Economique et Social de la Région d'Ile de France, Commission de l'Habitation.

———, Foscoso, J., Lemoine, J., Mahaut, M., and Paoli, R. (1986) *40 Ans de Maitrise d'Ouvrage du Logement Social en France*, Evolution du Savoir-Faire des Organismes HLM, Paris: Economica.

Cassell, M. (1984) *Inside Nationwide – One hundred years of co-operation 1884–*

1984, produced by Namemaker Ltd., co-ordinated by M. H. Seeley, London: Nationwide Building Society.

CECODHAS (1990) *Les Organismes au Logement Social et la lutte contre l'exclusion dans les Pays de la CEE: Principaux Extraits des documents édités à l'occasion de la rencontre des Ministries du Logement de la CEE – Lille 18–19.10.89*, Brussels.

Census Survey of Dublin (1911).

CHAC (1939) *The Management of Municipal Housing Estates*, London: CHAC.

Chamberlayne, P. (1990) Unpublished essay on former East German co-operatives, N.E. London Polytechnic.

Chapman Hendy Associates (1992) *Waltham Forest Feasibility Study 1991*, London. London.

City of Cologne (1989) Report on the problems of Kölnberg.

Christiansen, S. P. (1985) *Housing and Improvement – a comparative study, Britain–Denmark*, Vols. I and II, University of York Advanced Architectural Studies.

Christiansen, U., Kristensen, H., Prag, S., and Vestergaard, H. (1991) *Blev bebyggelserne bedre? Foreloebige ergaringer fra forbedringen af 5 nyere etageboligbebyggelser*, Hoersholm: SBI.

Clark, P. (1991) Unpublished report on practical work experience in Dresden, London: LSE.

Cole, I., Arnold, P., Windle, K. (1988a) *Decentralisation – The Views of Elected Members*, Research Working Paper 4, Housing Decentralisation Research Project, Sheffield: Sheffield City Polytechnic, Department of Urban and Regional Studies.

—— (1988b) *Decentralisation in Mansfield: The Response of Housing Staff*, Research Working Paper 6, Housing Decentralisation Research Project, Sheffield: Sheffield City Polytechnic, Department of Urban and Regional Studies.

—— (1989) *Decentralisation in Newcastle*, Research Working Paper 7, Housing Decentralisation Research Project, Sheffield: Sheffield City Polytechnic, Department of Urban and Regional Studies, April.

Combat Poverty Agency (1988) *Poverty and the Social Welfare System in Ireland*, No. 1, September, Dublin: Combat Poverty Agency.

Commission des Communautés Européennes (1987) *La Communauté Européenne, Population*, Luxembourg: L'Office des Publications Officielles des Communautés.

Connolly, M. and Knox, C. *Policy differences within the United Kingdom: the case of housing policy in Northern Ireland 1979–1989*, University of Ulster.

Cope, H. (1990) *Housing Associations – Policy and Practice*, London: Macmillan.

Copenhagen Statistical Office (1983–7) *Statistical 10-Year Review, Municipality of Copenhagen.*

Corcoran, T. (1987) *An overview of the provision and subsidisation of local authority housing*, Background paper for seminar 'Paying for Housing – is there a better way?', Dublin, 30 November.

Couch, C. (1985) *Housing Conditions in Britain and Germany*, London: Anglo-German Foundation for the Study of Industrial Studies.

Cronberg, T. (1985) *Tenants Involvement in the Management of Social Housing in the Nordic Countries*, OECD/Sweden Seminar on Community Involvement in Urban Service Provision, 28–30.5.85, OECD. See also in *Scandinavian Housing and Planning Research*, No. 3, 1986.

Cullen, B. (1989) *Poverty, Community and Development – A report on the issues of social policy that have arisen in the work of the nine projects of the Second European Programme to Combat Poverty, (1985–1989*, Research Report Series

Dublin: Combat Poverty Agency.

Cullingworth, B. (1969) *Council Housing Purposes, Procedures and Priorities*, Report of the Government's Advisory Committee on Housing Management, London.

Curci, G. (1988) 'Les HLM: une vocation sociale qui s'accentue', in *Revue Economie et Statistiques*, January, INSEE.

Czasny, K. (1987) *Vergleich der Wohnungspolitik in 6 Europaischen Staaten*, Wien: Institut für Stadtforschung.

Dahrendorf, R. (1982) *On Britain*, London: BBC.

—— (1985) *On Freedom*, London: BBC.

Danish National Urban Renewal Company (1989) Information received.

Danish Statistical Office (1982) Information received, April.

Darley, G. (1990) *Octavia Hill – A Life*, London: Constable & Co. Ltd.

Dauge, Y. (1990) Speech at Ville et Banlieue Conference at CNIC, Paris, September.

—— (1991) 'Riots and Rising Expectations in Urban Europe', Speech at London School of Economics, 6 March, (trans. Anne Power), London: LSE Housing.

Délégation Interministérielle à la Ville (1991) 'Halte à la segrégation', *Ensembles*, July, No. 32.

Délégation Interministeriélle à la Ville et au Développement Social Urbain (1989) *Justice et Quartiers*, Paris: DIV.

Délégation Interministérielle à la Ville et Délégation à l'Aménagement du Territoire et à l'Action Régionale (1990) *148 Quartiers, Bilan des contrats de développement social des quartiers du IXe plan 1984/88*, Paris.

Department of the Environment (DOE) (1968) *Housing and Construction Statistics*, London: HMSO.

—— (1974) 'Difficult to Let', Unpublished report of postal survey, London: DOE.

—— (1977a) *Housing Policy – A Consultative Document, Part 1*, Cmnd 6851, Green Paper, London: HMSO.

—— (1977b) *Inner Area Study – Unequal City* on Birmingham; *Inner Area Study – Change and Decay* on Liverpool; *Inner Area Study – Inner London: Policies for Dispersal and Balance* on Lambeth; London: DOE.

—— (1987) *Tenants in the Lead – The Housing Co-operatives Review*, London: HMSO.

—— (1988) *Tenants' Choice*, London: HMSO.

—— (1990) *Housing and Construction Statistics 1979–89*, London: HMSO.

—— (23 May 1991) *City Challenge*, Press Release from Action for Cities, London: DOE.

—— (1991a) *Estate Action Annual Report 1990–1991*, London: HMSO.

—— (1991b) *Estate Action – New Life for Local Authority Estates – Guideline for local authorities on Estate Action and Housing Action Trusts and links with related programmes*, London: DOE.

—— (1991c) *Housing and Construction Statistics*, London: HMSO.

—— (1991d) *Housing Policy Technical Volume*, London: DOE.

—— (1991e) Press Release, 11 February, London: DOE.

—— (1992a) Press Releases on City Challenge, London: DOE.

—— (1992b) *Housing and Construction Statistics*, London: HMSO.

—— (undated) *Estate Action: The Way Ahead*, Briefing Note – The Government's Proposals for the Development of the Estate Action Programme, London: DOE.

Department of the Environment (Ireland) (1986) *The Human Settlements Situation and Related Trends and Policies – Ireland 1987, in accordance with the decision of the Economic Commission for Europe Committee on Housing, Building and Planning at its 47th Session in September 1986,* DOE monograph with additional material from the Department of Tourism, Fisheries and Forestry and the National Institute of Physical Planning and Construction Research, Dublin: DOE.

—— (1987) *Annual Bulletin of Housing Statistics 1987, incorporating the Quarterly Bulletin for Quarter ended 31st December, 1987,* Dublin: Government Publications Sales Office.

—— (1988a) *Annual Bulletin of Housing Statistics – 1988,* Dublin: Government Publications Sales Office.

—— (1988b) *1988 Tenant Purchase Scheme* leaflet, Dublin: DOE.

—— (1988c) *Tenant Purchase Scheme for Local Authority Dwellings,* DOE Circular HRT 2/88, Dublin: DOE.

—— (1989) 'Flynn Announces Allocations for Local Authority Remedial Works Scheme', Government Information Services for the Department of the Environment, Dublin: DOE (Press Release, 31 January).

—— (1991a) *A Plan for Social Housing,* Dublin: DOE.

—— (1991b) Information and estimates received, Dublin: DOE.

—— (1992) Press Release 'Major New Housing Bill', 27 March, Dublin: DOE.

Der Spiegel (6/1982) 'Gut getarnt im Dickicht der Firmen.   Neue Heimat: Die dunklen Geschäfte von Vietor und Genossen'.

—— (7/1982) 'Da mussten langst die Staatsanwalte hin'.

—— (20/1982) 'Das Geld auf dem Acker'.

—— (41/1985) 'So schnell wie möglich alles verkloppen. Die Wohnungsverkäufe der Neuen Heimat drücken   Sozialdemokraten wie Gewerkschaften politisch in die Defensive'.

—— (4/1986) 'Neue Heimat. Sehr geschockt'.

—— (7/1986) 'Neue Heimat. Noch Knackpunkte'.

—— (17/1986) 'Verlust verschleiert, Gewinn nach Bedarf. Hamburger Untersuchungsausschuss belegt die Tricks und Manipulationen der Neuen Heimat'.

—— (22/1986) 'Der DGB und die Katakomben Firma'.

—— (23/1986) 'In einem halben Jahr ist alles aus. Bei der Auflösung der Neuen Heimat wird weiter getrickst und geschoben'.

—— (39/1986) 'Herr, sie wissen nicht was sie tun'.

—— (40/1986) 'Neue Heimat: Das wird mächtig reingehauen'.

—— (44/1986) 'Meinungsumfrage'.

—— (50/1988) 'Wohnungsnot im Wohlstands-Deutschland'.

—— (9/1989) 'Wohnungsnot – Goldener Osten'.

Deutscher Bundestag (1987) Beschlussempfehlung und Bericht des 3. Untersuchungsausschusses Neue Heimat nach Art. 44, Drucksache 10/5575, 'Einsetzung eines Untersuchungs-ausschusses' Drucksache 10/6779, 7 January.

—— (1988a) Antwort der Bundesregierung auf die Kleine Anfrage der Abgeordneten Frau Osterle-Schwering und Fraktion Die Grünen, Drucksache 11/1764, 'Diskrepanz zwischen der Zahl der sozialen Mietwohnungen und der Zahl der Wohnberechtigten', Drucksache 11/1862, 23 February.

—— (1988b) Gesetzentwurf der Abgeordneten Frau Osterle-Schwerin, Frau Teubner und der Fraktion Die Grünen 'Entwurf eines Gesetzes zur Förderung gemeinschaftlicher Wohnungsunternehmen (FGW)',   Drucksache 11/2199, 25 April.

—— (1989a) Antrag der Fraktion der SPD 'Sofortprogramm für eine aktive Wohnungspolitik', Drucksache 11/4083, 24 February.

—— (1989b) Antrag der Abgeordneten Frau Osterle-Schwerin, Frau Teubner under der Fraktion Die Grünen 'Okologische und soziale Offensive gegen

Wohnungsnot', Drucksache 11/4181, 13 March.

Dick, E. (1989) 'BM für Raumordnung, Bauwesen & Städtebau', Interview on 18 December, Bonn.

Dickens, P., Duncan, S. Goodwin, M., and Gray, F. (1985) *Housing, States and Localities*, London: Methuen.

Die Zeit (44/1986) 'Millionen für eine Mark. Auf Druck der Banken müssen die Gewerkschaften die Neue Heimat zurückkaufen'.

Donnison, D. (1987) Workshop on the Future of Social Housing, Broadway, Worcs., April.

―――― (1991), *Democracy, Bureaucracy & Public Choice – Economic Explanations in Political Science*, Hemel Hempstead: Harvester Wheatsheaf.

―――― and Middleton, A. (1987) *Regenerating the Inner City*, London: Routledge & Kegan Paul.

Dourlens, C. and Vidal-Naquet, P. A. (1987) *Ayants Droits et Territoire*, L'Attribution de Logements Sociaux dans le Champ de l'Expérimentation, Aix-en-Provence: CERPE.

Dubedout, H. (1983) *Ensemble Refaire la Ville – Rapport au Premier Ministre du Président de la Commission nationale pour le développement social des quartiers*, Paris: La Documentation Française.

Dublin Corporation (1914) *Report of the Departmental Committee appointed to Inquire into the Housing Conditions of the Working Classes of the City of Dublin 1914*, Cd 7273, Dublin: Dublin Corporation.

―――― (undated) *Darndale – Residents' Brief and Darndale Proposals*, Architects' drawings, Housing Architects Department, Dublin: Dublin Corporation.

―――― (1984) *Renting a Home – Dublin Corporation – Scheme of Priorities for Letting Housing Accommodation*, Dublin: Dublin Corporation, November.

―――― (1986) *Local Authority Revenue Accounts Summary Review Account*, Dublin: Dublin Corporation.

―――― (1987a) *Oliver Bond House – Survey and Refurbishment Proposals July 1987*, Housing Architects Department, Dublin: Dublin Corporation, July.

―――― (1987b) *St. Mary's Mansions Refurbishment Proposals November 1987*, Housing Architects Department, Dublin: Dublin Corporation, November.

―――― (1988) *Cost Plan for refurbishing of 96 No. Flats (Phase 1) – Blocks C D E F at Oliver Bond House – Cost Plan Drawings October 1988*, Dublin: Dublin Corporation.

―――― (1991) Information received from Dublin Corporation.

Dubral, C. (1987) 'Sozialer Wohnungsbau 1986' *Wirtschaft und Statistik*, September.

Dunleavy, P. (1981) *The Politics of Mass Housing in Britain 1945–1975*, Oxford: Clarendon Press.

Duport, J.P. (ed.) (1989) *Réhabilitation de l'Habitat en France*, Paris: Economica.

Duvigneau, H. J. and Schönefeldt, L. (1989) 'Wohnungspolitik und Wohnungswirtschaft in der Bundesrepublik Deutschland', Köln: Gesamtverband Gemeinnütziger Wohnungsunternehmen.

Dybbroe, O. (1989) Interview, Roskilde University, Denmark.

Dyer, A. *A Report – Darndale Redesign Pilot Project 1987/88*, Dublin: Darndale Community Philosophy.

EC (1989a) *Lone-parent families in the European Community*, London: Family Policy Studies Centre, January.

―――― (1989b) *Conditions de Vie dans les Villes d'Europe*, Dublin: The European Foundation for the Improvement of Living and Working Conditions.

―――― (1989c) *Eurostat Review (1977-1986)*, Luxembourg Office for Official Publications of the European Community.

—— (1990a) *Family Budget Tables.*
—— (1990b) *Ville Insertion Emploi*, Brussels: European Service Network.
—— (1991a) *Towards a Europe of Solidarity: Urban Social Development, Part-nerships and the Struggle against Social Exclusion*, European Commission of Social Affairs Conference at Lille, May.
—— (1991b) *Labour Force Survey.*
—— (1991c) *New Social Programme 10.*
Emms, P. F. (1990) *Social Housing – a European dilemma*, Bristol University: SAUS.
Emsley, I. (1986) *The Development of Housing Associations with special reference to London*, New York and London: Garland Publishing Inc.
Ensembles (1991) Paris, May.
Eurostat (1970–86).
—— (1989) Demographic Statistics.
—— (1986) *Social indicators for the EC 1977–86.*
—— (1992) *General Statistics.*
Ferris, J. (1972) *Participation in Urban Planning – the Barnsbury case*, Occasional Papers on Social Administration No. 46, London: G. Bell & Sons.
Fishman, W. J. (1988) *East End 1988 – A year in a London borough among the labouring poor*, London: Duckworth.
Fleischer, H. (1987) 'Ausländer im Bundesgebiet 1986', *Wirtschaft und Statistik*, March.
Foot, M. (1975) *Aneurin Bevan 1945–1960*, London: Paladin.
Forrest, R. and Murie, A. (1991), *Selling the Welfare State – The Privatisation of Public Housing*, London: Routledge.
Foster, R. F. (1989) *Modern Ireland 1600–1972*, London: Penguin Books.
Franke, H. (1990) GDR – 'Chances for the German Housing Market', *Wirtschafts-dienst* 1990/VI, June, No. 6.
Fuerst, J. S. (ed.) (1974) *Public Housing in Europe and America*, London: Croom Helm.
Fuhrich, M, and Meuter, H. (1987) 'The Legacy of Community Housebuilding in the Federal Republic of Germany', in M. Bulos and S. Walker (eds), *The Legacy and Opportunity for High Rise Housing in Europe: The Management of Inno-vation*, London: South Bank Polytechnic, Housing Studies Group.
—— Neusüss, C., and Petzinger, R. (1986) *Neue Heimat*, Hamburg: VSA Verlag.
Gardiner, K., Hills, J. and Kleinman, M. (1991) *Putting a price on council housing – valuing voluntary transfers*, Welfare Estate Programme Discussion Paper WSP/62, London: LSE.
Gay, A., Director of the London Regeneration Consortium (1990) Speaking at Science/Po Urba 20th Anniversary Conference, March, Paris.
Geindre, F. (1989) *L'Attribution des Logements Sociaux*, Rapport au Ministre de l'Equipement, du Logement, de Transports et de la Mer et au Ministre Délégué au Logement, St. Clair.
Geodaetisk Institut (1988) *Building Guide for the Copenhagen Area – Danish Housing and Construction Year 1988*, Copenhagen.
German Census (W. Germany) (1987).
Gesamtverband der Wohnungswirtschaft (German National Federation of Housing Associations) (1989, 1991) Information collected.
—— (1990) *Berliner Memorandum*, July.
Gesamtverband Gemeinnütziger Wohnungsunternehmen e.V. (1988) *Das Neue Recht für die Gemeinnützige Wohnungswirtschaft*, Vorschriften und Erläuter-ungen zum Steuerreformgesetz 1990, Schriftenreihe 29.
Ghekière, L. (1988) *Housing policies and subsidies in France* (English trans.), Paris: UNFOHLM.

Gibbins, O. (1988) *Grossiedlungen, Bestandspflege und Weiterentwicklung*, Munchen: Callwey.

Gifford, Lord (Chairman) (1986) *The Broadwater Farm Inquiry Report: Report of the Independent Inquiry into disturbances of October 1985 at the Broadwater Farm Estate, Tottenham, chaired by Lord Gifford QC*, London: London Borough of Haringey.

Glasgow District Council (1986) *Inquiry into Housing in Glasgow*, Glasgow District Council.

—— (1992) Information collected.

GLC (1974) Minutes of the Greater London Council.

—— (1976) Minutes of the Great London Council.

Gould, B. (1991) Speech at Institute of Housing Conference, Harrogate, June.

Gray, P. F (1989) Paper presented to National Conference on Estate Management for Senior Officers of Housing Authorities, Dublin: Institute of Public Administration, 14 April.

Greiff, R. and Ulbrich, R. (1992) *Country Profile – Germany: Housing supply and housing policy in the Federal Republic of Germany and the future of large housing estates in the East and West German Länder*, Darmstadt: Institut Wohnen und Umwelt, Project Group on Housing, Social Integration and Livable Environments in Cities.

Greve, J. with Currie, E. (1990a) *Homelessness in Britain*, York: Joseph Rowntree Memorial Trust.

—— and Currie, E. (1990b) *Homelessness and Young People*, A Report to The Prince's Trust, Institute for Research in the Social Sciences, University of York.

Gruber, W. (1981) *Sozialer Wohnungsbau in der Bundesrepublik*, Köln: Pahl-Rugenstein.

Guinchat, P., Chaulet, M. P., and Gaillardot, L. (1981) *Il Etait Une Fois l'Habitat*, Chronique de Logement Social en France, Collection Urbanisme, Paris: Editions du Moniteur.

Hachmann, C. J. (1989) 'Gesellschaft für Wohnungswirtschaft', Interview on 21 December, Köln.

Hall, P. (1988) *Cities of Tomorrow – An Intellectual History of Urban Planning and Design in the Twentieth Century*, Oxford: Basil Blackwell.

—— (1989) *London 2001*, London: Unwin Hyman.

Hallett, G. (1977) *Housing and Land Policies in West Germany and Britain*, London: Macmillan.

—— (1988) *Land Housing Policies in Europe and the USA: A Comparative Analysis*, London: Routledge.

Halsey, A. H. (ed.) (1988) *British Social Trends Since 1900 – A Guide to the Changing Social Structure of Britain*, Basingstoke: Macmillan.

Hambleton, R. and Hoggett, P. (eds) (1987) *The Politics of Decentralisation: theory and practice of a radical local government initiative*, Working Paper 46, Bristol University: SAUS.

Hamm, H. (1988) 'Der soziale Wohnungsbau 1987/88', *Bundesbaublatt*, October 1988, No. 10.

Hamnett, C. and Randolph, B. (1988) *Cities, Housing and Profits: flat break-up and the decline of private renting*, London: Hutchinson.

Hardwick, N. (1991) Talk at LSE Seminar, February, based on Centrepoint survey, London: LSE.

Harloe, M. (1988) 'The Changing Role of Social Rented Housing', in M. Ball, M. Harloe, and M. Martens, *Housing and Social Change in Europe and USA*, London: Routledge.

Harris, J. (1990) 'Society and the State in Twentieth Century Britain' in *Cambridge*

Social History 1750–1950, Vol. 3, Social Agencies and Institutions (ed. F. M. L. Thompson), Cambridge: Cambridge University Press.

Hasselfeldt, G. (Housing Minister) (1989) 'Entschlossene Antworten auf wachsenden Nachfragedruck', *Wirtschaftsdienst*, Nr. 11, November.

Heckscher (1984) *The Welfare State and Beyond: Success and Problems in Scandinavia*, Minneapolis: University of Minnesota Press.

Henderson, J. and Karn, V. (1987) *Race, Class and State Housing*, Aldershot: Gower.

Hervo, M. and Charras, M. A. (1971) *Bidonvilles l'Enlisement*, Paris: Maspero.

Hillier, B., Kühne-Büning, L., Rahs, R., Tsoskounoglou, H., and Marshall, C. (1987) *Problem Housing of the 60s and 70s in Britain and Germany*, Bochum: GFW, and London: UAS.

Hills, J. (1990) *Housing Finance: Problems and Solutions*, Report for the Joseph Rowntree Foundation, York.

—— (1991a) *From Right-to-Buy to Rent-to-Mortgage: Privatisation of Council Housing Since 1979*, Discussion Paper WSP/61, STICERD, London: LSE.

—— (1991b) *Unravelling Housing Finance – Subsidies, Benefits and Taxation*, Oxford: Clarendon Press.

—— and Mullings, B. (1990) *The State of Welfare – The Welfare State since 1974*, Oxford: Clarendon Press.

—— Hubert, F., Tomann, H., and Whitehead, C. (1990) 'Shifting subsidy from bricks and mortar to people – experiences in Britain and West Germany', *Housing Studies*, Vol. 15, No. 3.

Hirche, K. (1984) *Der Koloss wankt?*, Düsseldorf: Econ Verlag.

HLM (Habitations à Loyer Modéré) (1989a) 'Un Siècle de l'Habitat Social – 100 Ans de Progrés', *HLM Aujourd'hui*, Numero hors-Serie, Supplement au no. 13, Paris: HLM.

—— (1989b) 'Le Progrés par l'Habitat', *HLM Aujourd'hui*, No. 15, Paris: HLM.

—— (1989c) 'Les Orientations de la Politique Sociale de l'Habitat', *Actualités HLM*, Cahier No. 4, 28 September, Paris: HLM.

—— (1989d) 'La Ville, Mode d'Emploi', *Actualités HLM*, Numéro Special, Supplement No. 426, 30 September, Paris: HLM.

—— (1990a) *Rapport du Groupe Interféderal – Les Organismes d'HLM et la Qualité du Service Rendu*, Marseille: Congrés National HLM.

—— (1990b) 'Danemark. Le Pays de Bien Loges', *HLM Aujhourd'hui*, No. 17, Paris: HLM.

Holland, Sir Milner (Chairman) (1965) *Report of the Committee on Housing in Greater London*, Cmnd 2605, London: HMSO.

Holmans, A. E. (1987) *Housing Policy in Britain*, London: Croom Helm.

—— (1992) Unpublished information, London: DOE.

Hope, T. and Foster, J. (1991) *Conflicting forces: changing the dynamics of crime and community on a problem estate*, London: Home Office.

Howard, E. (1898) *Tomorrow: A Peaceful Path to Real Reform*, London: Swan Sonnenschein.

Hvidtfeldt, H., Rohde, P., and Skov, E. (1986) 'New Goals and Methods in Connection with the Reshaping of Multi-Storey Housing Areas – the Gellerup Project', *Scandinavian Housing and Planning Research*, No. 3.

Informationen zur politischen Bildung (201/1984) *Ausländer*, München: Franzis Verlag.

INSEE (1988) Housing Survey.

—— (1989) *Données Sociales 1989*.

Institut Français d'Architecture (1985) 'Logement Social 1950–1980', *Bulletin d'Informationes Architecturales*, Supplement au No. 95, May, Paris.

Institute of Public Administration (IPA) (1988) Conference on Estate Management, Dublin: IPA.

Irish National Census (1986).

Islington Borough Council (1974) Minutes, London Borough of Islington.

Jacquier, C. (1991) *Voyage dans dix quartiers Européens en Crise*, Paris: l'Harmattan.

Jantzen, E. B. and Kaaris, H. (1984) 'Danish Low-Rise Housing', *Scandinavian Housing and Planning Research*, No. 1.

Jencks, C. and Peterson, P. E. (1991) *The Urban Underclass*, Washington: Brookings Institution.

Jones, W. G. (1986) *Denmark, a modern history*, London: Croom Helm.

Joyce, L. and McCashin, A. (1982) *Poverty and Social Policy*, Institute of Public Administration – The Irish National Report presented to the Commission of the European Communities, Dublin: IPA.

Kelleher, P. with the assistance of Deehan, A., McCarthy, P., Farrelly, J., and Maher, K. (researchers) (1988) *Settling in the City – the development of an integrated strategy for housing and settling homeless people in Dublin*, Dublin: A Focus Point Report.

Kemp, P. (ed.) (1986) *The Future of Housing Benefits*, Studies in Housing 1, Centre for Housing Research.

Kennedy, D. (1984) 'West Germany', in M.Wynn (ed.), *Housing in Europe*, Kent: Croom Helm.

Kjeldsen, M. K. (1976) *Industrialised Housing in Denmark 1965–76*, Copenhagen: Danish Building Centre.

Knop, W. (1989) 'Bestand an Gebäuden und Wohnungen 1987. Ergebnis der Gebäude- und Wohnungszahlung', *Wirtschaft und Statistik*, August.

Kreibich, V. (1986) *The End of Social Housing in the Federal Republic of Germany? The Case of Hanover*, Dortmund: Institut für Raumplanung, Universität Dortmund, September.

Kristensen, H. (1989a) *Danish Post-War Public Housing in Trouble; Trends and Strategies. Main Trends and Discussion of some Strategies*, Hoersholm: The Danish Building Research Institute.

—— (1989b) *Improvement of Post-War Multi-Storey Housing Estates*, Hoersholm: The Danish Building Research Institute.

Kroes, H., Yunkers, F., and Mulder, A. (1988) *Between Owner-Occupation and Rented Sector Housing in 10 European Countries*, The Netherlands Christian Institute for Social Housing, De Bilt.

Lambert, J., Paris, C., and Blackaby, B. (1978) *Housing Policy and the State*, London: Macmillan.

Lee, J. J. (1989) *Ireland 1912–1985 – Politics and Society*, Cambridge: Cambridge University Press.

Lemann, N. (1991) *The Promised Land*, New York: Alfred Knopf.

Lemoine, J. (1985) *Le Maitre d'Ouvrage du Logement Social*, Paris: HLM.

Le Monde (1991) 'Un Entretien avec le Maire (PS) de Mantes-la-Jolie', Paris, 12 June.

Lewis, J. (1991) *Women and Social Action in Edwardian England*, London: Edward Elgar.

Lloyds Bank (1987) *Denmark – Economic Report*, London.

Local Authority News (Autumn 1989).

London Weekend TV (1991) *A Summer on the Estate*, Documentary on Kingshold Estate, LB Hackney, Wild and Fresh Productions, September.

Lundquist, L. J. (1986) *Housing policy and equality: a study of tenure conversions and their effects*, London: Croom Helm.

Lyons, F. S. L. (1973) *Ireland since the famine*, London: Fontana Paperbacks.

Macey, J. (1982) *Housing Management* (4th ed), London: The Estates Gazette Ltd.

McGuire, C. C. (1981) *International Housing Policies: a comparative analysis*, Toronto: Lexington Books.

Maclennan, D., Gibb, K., and More, A. (1990) *Paying for Britain's housing*, York: Joseph Rowntree Foundation Housing Finance Series, in association with National Federation of Housing Associations, London.

Magnussen, J. (1979) *Urban Policy and Change in Denmark*, Curb, Rapport 11, Department of Geography, University of Copenhagen.

Mainwaring, R. (1988) *The Walsall Experience*, London: DOE.

Malpass, P. and Murie, A. (1987) *Housing Policy and Practice* (2nd ed), Basing-stoke: Macmillan Education.

Marklund, S. (1988) *Paradise Lost? The Nordic Welfare States and the Recession 1975–1985*, Lund: Arkiv Poerlag.

Mayer, M. (1978) *The Builders*, New York: W. W. Norton & Co Inc.

Mead, W. R. (1981) *An Historical Geography of Scandinavia*, London: Academic Press.

Mearns, A. C. (1970) *The Bitter Cry of Outcast London*, Leicester University Press.

Merrett, S. and Cranston, R. (1992) *A National Housing Bank*, Fabian pamphlet 552, London: Fabian Society.

Mingioni, E. (1989) *Fragmented Societies*, London: Blackwell.

Ministère de l'Aménagement du Territoire, de l'Equipement, du Logement et du Tourisme (1991/2) Estimates based on figures from these Ministries, Paris.

Ministère de l'Equipement et du Logement – Direction de la Construction (1989) *Le Role du Parc HLM et l'Habitat des Populations Défavorisées ou Modèstes*, Nancy: Laboratoire Logement.

Ministry of Foreign Affairs, Press and Cultural Relations Department (1988) *Facts about Denmark*, Copenhagen.

Mitchell, P. (1990) *Memento Mori – The Flats at Quarry Hill, Leeds*, Otley: Smith Settle Ltd.

Moller, Helge (1989) 'Facts om Engelsk boliglovgivning', *Boligen*, September, 56 Argang, Kobenhavn.

Muehlich, E. (1988) *Differentiation of the Housing Provision System according to the ongoing Internationalisation and Individualisation of the Welfare State*, Prepared for the international conference 'Housing Policy between State and Market', 15–19 September, Dubrovnik.

Müller, W. (1974) 'Städtebau', S.95, Seminar *Entsorgungskonzepte für Grosswohn-siedlung*.

Murphy, L. (1991) Unpublished information on Dublin's housing and social con-ditions, London: LSE, Dept. of Geography.

NABCO Education and Advisory Service (undated) *The Co-operative Housing Option – Information about co-ownership housing co-operatives – mutual-ownership housing co-operatives – creating an ownership stage in apartments/ flats – facilitating management of common property – sharing maintenance costs – promotion of a good housing environment*, Dublin: National Association of Building Co-operatives.

—— (undated) *The Co-operative Housing Option – Information about co-ownership housing – self-help housing groups – local authorities – tenant owner-ship groups – private developers*, Dublin: NABCO.

—— (undated) *The Co-operative Housing Option – Information about tenant managment co-operatives – landlord/local authority/tenant partnerships for estate management*, Dublin: NABCO.

National Advisory Commission on Civil Disorders (Otto Kerner, Chairman) (1968)

Report of the National Advisory Commission on Civil Disorders – US Riot Commission Report, New York: Bantam Books.

National Federation of Housing Associations (1986) *Housing legislation – a guide to the consolidated Acts*, Compiled by Adrian Moran, edited by NFHA (Beverley Markham), London: NFHA.

—— (1989) *The Future of Housing Associations: The Housing Bill/Impact of higher-rents on new housing association tenants*, London: NFHA.

—— (1990) *The Housing Act 1988 and summary of Local Government and Housing Act 1989 – a guide for Housing Associations* (2nd edn), London: NFHA.

—— (undated) *NFHA Jubilee 1935–1985 Album*, Peter Jones, Produced by Liverpool Housing Trust Information Services, Liverpool, London: NFHA.

National Federation of Non-Profit Housing Companies, Denmark (1989 and 1991) Information received.

Nationwide Building Society (1987) *Background Bulletin, House Prices in Europe*, London: Nationwide Building Society, January.

Nielsen, G. (1984) *Housing for Young Adults and Single Parents*, Contribution to the ICA Housing Conference in Hamburg on 12 October, Boligselskabernes Landsforening.

—— (1985) Information received from the National Housing Federation, Copenhagen.

Niner, P. (1991) 'Housing Management' in A. Norton and K. Novy, *Low cost social housing in Britain and Germany*, London: Anglo-German Foundation.

North Islington Housing Rights Project (1976) *Street by Street – Improvement and Tenant Control in Islington*, London: Shelter.

Nygaard, J. and Thomassen, S. (1989) 'Tenant Democracy in the State of Denmark', speech at Priory Estates Project Annual Conference, October, Liverpool.

O'Cuinn, S. (1991) Information from Sean O'Cuinn on Remedial Works Schemes, Dublin: DOE.

O'Sullivan, M. (1933) *Twenty Years A-Growing*, London: Oxford University Press.

OECD (1986) *Living conditions in OECD Countries*, Social Policy Studies No. 3, Paris: OECD.

—— (1988) *Urban Housing Finance*, Paris, OECD.

—— (1989) *National Accounts Statistics 1970–1989*, Vol. 1: 145, Paris: OECD.

Office of Population Censuses and Surveys (1979, 1988) *General Household Survey*, London: HMSO.

—— (1992) *Population Trends 1992*, No. 67, Spring.

Osenberg, H. (1988) *Aims and Measures for Social Housing on the Three Levels of Policy and Administration in West Germany*, Bonn.

—— (undated), *Tenancy Legislation Since 1945: From Housing Rationing to Social Market Economy*, Bonn.

Owen, R. (1970) *Report to the County of Lanark – A new view of society*, edited by V. A. C. Gatrell, Harmondsworth: Pelican Books.

Parker, J. and Dugmore, K. (1976) *Colour and the Allocation of GLC Housing*, GLC Research Report No. 21, London: GLC.

Parker, T. (1983) *The People of Providence*, London: Hutchinson.

Petzinger, R. and Riege, M. (1981) *Die Neue Wohnungsnot*, Hamburg: VSA Verlag.

Pfeiffer, U. (1989) 'Auf Sand gebaut. Ein Programm gegen den Wohnungsmangel', *Die Zeit*, No. 12, 17 March.

Pinto, R. (1992) 'Estate Action', Ph.D. submitted to the University of London.

Poverty Today (1989) Vol. 3, No. 1 (July).

—— (1989) No. 7 (September).

—— (1989) Vol. 3, No. 8 (December).

Power, A. (1972) *I woke up this morning*, London: British Council of Churches.

—— (1973) *David and Goliath – Barnsbury 1973*, London: Holloway Neighbourhood Law Centre.

—— (1977a) *Racial Minorities and Council Housing in Islington*, London: North Islington Housing Rights Project.

—— (1977b) *Holloway Tenant Co-operative – Five Years On*, London: North Islington Housing Rights Project.

—— (1982) *Priority Estates Project 1982: Improving Problem Council Estates: A Summary of Aims and Progress*, London:DOE.

—— (1984) *Local Housing Management*, London: DOE.

—— (April 1987) *The PEP Guide to Local Housing Management*, Vols 1, 2 and 3, London: Priority Estates Project Ltd.

—— (1987a), *Property before People – The Management of Twentieth Century Council Housing*, London: Allen & Unwin.

—— (1987b) 'The Crisis in Council Housing – is public housing manageable?', *The Political Quarterly* July/September, Vol. 58, No. 3.

—— (1988) *Under New Management – The experience of thirteen Islington co-operatives*, London: Priority Estates Project Ltd.

—— (1990) *Mass housing estates and spirals of social breakdown in European cities*, London: LSE Housing/CECODHAS.

—— (1991) *Running to Stand Still – Progress in local management on twenty unpopular housing estates*, London: Priority Estates Project Ltd.

Provan, B. (1991) Unpublished research on French social housing estates, London, LSE.

Quartiers (1986) *Actes de Colloques – Les Cités en Question*, Paris: Commission Nationale pour le Développement Social des Quartiers.

Quilliot, R. and Guerrand, R.H. (1989) *Cent Ans d'Habitat: Une Utopie Réaliste*, Paris: Albin Michel.

Rahs, R. (1982) *Die Gemeinnützige Wohnungswirtschaft in Dortmund – 1hr Beitrag zur Beliebung der lokalen Wohnungsnot*: 9, 16-17, 26, Dortmund.

—— (1986) *The Privatisation of the Social Housing Stock in the FRG – The Example of the 'Neue Heimat'*, Dortmund.

—— (1989, 1991, 1992) Unpublished information.

—— Osenberg, H., and Muehlich, E. (undated) 'Housing Policy between State and Market', Briefing Paper on Housing Policy in The Federal Republic of Germany, prepared for the international conference, 15–19 September, Dubrovnik.

Raynsford, N. (1992) Seminar 'Arm's Length Housing Companies', 12 March, London.

Renaudin, G. (1991) Information on economic initiatives, Paris: National Union of HLMs.

Roof (1991), London: Shelter Publications, September.

Rose, E.J.B. in association with Deakin, N., Abrams, M., Jackson, V., Peston, M., Vanags, A.H., Cohen, B., Gaiskell, J. and Ward, P. (1969) *Colour and Citizenship*, London: Oxford University Press.

Rossi, P. H. (1989) *Down and Out in America – The Origins of Homelessness*, Chicago and London: University of Chicago Press.

Royal Commission on the Housing of the Working Classes (1885) First Report, BPP 1884–5, Vol. XXX, London: Stationery Office.

Salicath, N. (1987) *Danish Social Housing Corporations*, Vols I and II, Co-operative Building Industries Ltd. with the support of the Danish Boligministieret, Copenhagen.

Saunders, P. (1990) *A Nation of Home Owners*, London: Unwin Hyman.
Scarman, Lord (1986) *The Scarman Report – The Brixton Disorders*, Harmondsworth: Pelican Books.
Schaefer, J. P. (1989) *Housing Finance and Subsidy System in France*, York: Joseph Rowntree Memorial Trust.
—— (1992) Unpublished research.
Schroeder, K. (1985) *Zwischen Aufenthalt und Integration*, Sozialreport SR 9-85, Bonn: Inter Nationes.
Scott, F. (1975) *Scandinavia*, Cambridge: Harvard University Press.
Seewald, H. (1987) 'Wohngeld 1986', *Wirtschaft und Statistik*, October.
Smith, M. E. H. (1971, 1989 edns) *Guide to Housing*, London: Housing Centre Trust.
Social Trends (1992), London: HMSO.
Stahl, K. and Struyk, R. J. (1985) *US and West German Housing Markets*, New York: Springer-Verlag.
Statens Byggeforskningsinstitut undated), *Sammenfatning – Lokalomradeundersogelse af Boligforhold og Beboersammensaetning*, Hoersholm: SBI.
—— (undated) Documentation on worst estates: *Esbjerg – Bilag; Esbjerg – Lokalomraadeundersogelse af boligforhold og beboersammensaetning; Aalborg – Bilag; Aalborg – Lokalomraadeundersogelse af boligforhold og beboersammensaetning; Karlebo – Bilag; Karlebo – Lokalomraadeundersogelse af boligforhold og beboersammensaetning.*
—— (1986) *Boligomraader i Krise*, Notat fra Statens Byggeforskningsinstitut til udvalget vedroerende den almennyttige sektors fremtidige rolle paa boligmarkedet, Hoersholm: SBI.
—— (1988) *Sakdan forbedres etageboligomraader*, Hoersholm: SBI.
Statistical Abstracts, Eire (1966–75).
Statistisches Jahrbuch (1929–89).
Statistisches Bundesamt (1983) *Strukturdaten über die Ausländer in der BRD*, Wiesbaden.
—— (1987) *Bevölkerungsstruktur & Wirtschaftskraft der Bundesländer 1976–1987*, Wiesbaden.
Stewart, J. (1988) *A new management for housing departments?*, Luton: The Local Government Training Board.
Superior Council for Integration (Chairman, Marceau Long) (1991), French government.
SUSS Centre (1987) *A Block of Facts – Ballymun twenty-one years on*, Dublin: SUSS Centre.
Svensson, O. (1988) *Planning of low-rise urban housing areas*, Hoersholm: The Danish Building Research Institute.
Swenarton, M (1981) *Homes fit for Heroes*, London: Heinemann.
Tableaux de l'Economie Française 1991–2, Paris.
The Economist (16 January 1988) 'Poorest of the rich', in *A Survey of the Republic of Ireland.*
—— (24 June 1989) 'Packed and gone', p.62.
—— (2 September 1989) 'If two Germanies become one', pp. 45–6.
—— (9 September 1989) 'Haugheyzelski', p. 51.
—— (16 September 1989) 'Goodbye to Berlin', pp. 67–68.
—— (28 October 1989) 'When the wall comes down', pp. 5–28.
—— (16 March 1991) 'Waiting for the next wave'.
The Economist Intelligence Unit (1989) *West German Country Profile 1989–90*, London.
—— (1989–90) *Denmark, Iceland, Country Profile*, London.

The Economist Pocket Europe (1992) London: The Economist Books/Random Century.

The Economist Supplement (1988) 'Denmark – The smug debtor', 3 September.

The Guardian (1990) 'In Search of a Miracle', Europe section, 28 December.

—— (1991) 'Exclusion Zone', Jean-Yves Lhomeau, 21 June.

The Independent Magazine (1990) 'The Rough Edge of Paris', 15 December.

The Independent on Sunday (1991) Godfrey Hodgson, 'Death Throes of the old order', 13 October.

The Irish Times (1989a), F. McDonald, 'Sheriff Street flats to be demolished', 18 May.

—— (1989b) 'Sheriff Street residents "to be rehoused" in area', 19 May.

—— (1989c) R. Foster, 'Yuppies blamed for demolition', 19 May.

—— (1989d) Leader: 'Inner City Priorities', 19 May.

The Peabody Donation Fund Report for the twelve months ended 31 March 1984, London.

The Weekend Guardian, (1989), V. Horwell, 'Life goes on in Northern Ireland?', 25–6 March.

—— (1989) 'The new East Enders', 8–9 April.

Thompson, F. M. L. (1990) *Cambridge Social History of Britain 1750–1950*, 3 Vols, Cambridge: Cambridge University Press.

Thomson, D. (1950) *England in the nineteenth century (1815–1914)*, The Pelican History of England 8, Harmondsworth: Penguin Books.

Threshold (1987) *Policy Consequences – A Study of the £5,000 Surrender Grant in the Dublin Housing Area*, Dublin, June.

Threshold News (1989) No. 1, May, Dublin.

Tomann, H. (1988) *Recent developments in German housing policy*, Welfare State Programme Seminar paper, London: LSE.

—— (1991) *German Housing Policy in View of Transition*, Prepared for Third ECE Workshop on housing policies in the context of socio-economic transition, Berlin: Institute of Economic Policy and Economic History, Freie Universität, October.

Tuppen, J. N. and Mingret, P. (1986) *Suburban Malaise in French Cities*, TPR 57 (2).

Ude, C. (1990) *Wege aus der Wohnungsnot*, München: Piper.

UK Census (1911), London: HMSO.

UK Census (1986), London: HMSO.

United Nations (1957, 1960, 1965, 1970, 1975, 1980–6) *Annual Bulletins of Housing & Building Statistics for Europe.*

—— (1992) *Monthly Bulletins of Statistics.*

UNIOPSS, DIV., Fondation de France (1990) *Accueil et Insertion dans la Ville par le Logement*, Paris: FNC Pact-Arim.

Van der Cammen, H. (ed.) (1988) *Four Metropolises in Western Europe*, Assen/ Maastricht: Van Gorcum.

Van Vliet, W. (ed.) (1990) *International Handbook of Housing Policies and Practices*, New York, Westport County, London: Greenwood Press.

Vestergaard, H. (1982) *Social Context of Housing Policy – Main Trends in Housing Policy – Size of Households in a Historic Perspective*, Hoersholm: SBI.

—— (1985) *Organisation og oekonomi i den almennyttige boligsektor*, Hoersholm: SBI.

—— (1988) 'Revitalization of Danish Post-War Public Housing in Crisis – Background, Programme, Scope and Evaluation of Effects', Lecture given at the International Research Conference on Housing Policy and Urban Innovation, Amstersam, 27 June to 1 July, Hoersholm: SBI.

—— (1989) Information collected from Danish Building Research Institute, Hoersholm: SBI.

—— (1991) *Bevolkningens Bevagelser*, June.

Ville et Banlieue de France (1990) Conference. Paris, September.

Von Beyme, K. and Schmidt, M. (1985) *Policy and Politics in the FRG*, London: Gower.

Wagner, G. (1987) *The Chocolate Conscience*, London: Chatto & Windus.

Walsh D. (1986) *The Party Inside Fianna Fail*, quoted in J. J. Lee, (1989) *Ireland 1912 – 1985 – Politics and Society*, Cambridge: Cambridge University Press, p. XIV.

Wedel, E. (1989) 'Wohnraumversorgung der Haushalte 1987.   Ergebnis der Gebäude- und Wohnungszahlung' *Wirtschaft und Statistik*, August.

Weigall, D. and Murphy, M. (1982) *Modern History (1815 to present day)*, London: Letts.

Werner, J. W. (1974) 'Trade Union Housing in Western Germany', in J. S. Fuerst, (ed.), *Public Housing in Europe and America*, London: Croom Helm.

White, J. (1986) *The Worst Street in North London: Campbell   Bank, Islington, between the Wars*, London: Routledge & Kegan Paul.

Wibb, S. and Wilcox, S. (1991) *Time for Mortgage Benefits*, York: Joseph Rowntree Foundation.

Willmott, P. and Murie, A. (1988) *Polarisation and Social Housing. The British and French Experience*, Report 676, London: Policy Studies Institute.

Wilson, D. M. (ed.) (1980) *The Northern World – the history and heritage of northern Europe*, London: Thames & Hudson.

Winter, J. M. (1986) *The Great War and the British People*, London: Macmillan.

Wirtschaft und Statistik (1966, 1976, 1983, 1984, 1987).

Wolfe, T. (1991) *From Bauhaus to Our House*, New York: Farrar Straus Giroux.

Wollmann, H. (1985) 'Housing Policy: between state   intervention and the market' in K. Von Beyme, and M. Schmidt, *Policy and Politics in the FRG*, London: Gower.

Women's Group on Public Welfare (1939–42) *Our Towns: a close-up study made during 1939–1942*, London: Oxford University Press.

World Bank Development Report (1991).

Wullkopf, U. (1992) *Wohnungsprobleme in den Neuen Bundesländern*, Darmstadt: Institut Wohnen Umwelt.

Wynn, M. (ed.) (1983) *Housing in Europe*, London: Croom Helm.

Zeldin, T. (1991) 'France's Great Fantasy', *The Guardian*, 3 May.

Zipfel, T. (1989) *Estate Management Boards – An Introduction*, London: Priority Estates Project Ltd.

Zulauf, M. (1990) Unpublished information collected by M. Zulauf and reports of visits to Berlin, Dresden, Leipzig, Cologne, Bonn and Bremen.

Index